Process Development Division

AA003132

Presentations at the 2010 AIChE Annual Meeting

Salt Lake City, Utah, USA
7-12 November 2010

ISBN: 978-1-61782-158-5

Printed from e-media with permission by:

Curran Associates, Inc.
57 Morehouse Lane
Red Hook, NY 12571

Printed by Curran Associates, Inc. (2011)

AIChE
3 Park Avenue
New York, NY 10016-5991

Phone: (203) 702-7660
Fax: (203) 775-5177

www.aiche.org

Additional copies of this publication are available from:

Curran Associates, Inc.
57 Morehouse Lane
Red Hook, NY 12571 USA
Phone: 845-758-0400
Fax: 845-758-2634
Email: curran@proceedings.com
Web: www.proceedings.com

TABLE OF CONTENTS

(33a) Succeeding through Alliances: A Business Strategy

Anthony Nickens, Ceramatec, Inc., Salt Lake City, UT

In today's world of complex innovation challenges, alliances are indispensible. Sooner an entity can realize this, faster it can progress. Formation of technology commercialization alliances has been important at Ceramatec, a high technology company engaged in solving tough R&D challenges. Ceramatec has succeeded at spinning off companies, licensing technologies out and in-house commercialization of several technologies due to having alliances at the very core of business strategy. The presentation will discuss case studies and share lessons learned from several years experience.

See more of this Session: Managing Innovation
See more of this Group/Topical: Management Division

The Soft Side of Innovation: What Engineers Need to Know and Do for Corporate Success

Jeff Lindsay, Innovation Edge, Neenah, WI
jlindsay@innovation.com

Abstract

Chemical engineers often find that excellent problem solving skills and creativity do not translate into success inside a corporation. Some grow frustrated when technically sound solutions and clever innovations fail to have the impact they expected. Some grow disenchanted, concluding that it's all about "politics" instead of logic. In reality, innovation success involves technical expertise plus a network of relationships in order to drive adoption of the innovation. Failure to understand the "soft side of innovation" can contribute to innovation fatigue and unnecessary frustration. This presentation will consider the intangible elements that need to be understood for innovation success, including five key steps for success.

The Problem

Engineers are technically trained and skilled in solving problems–problems that involve mass, energy, chemical reactions, liquid flow, and other non-humanoid materials and products. Engineers often come into corporations, find exciting opportunities for innovation, and then become disappointed and frustrated when unexpected forces seem to quash the concept. The journey of engineers as innovators can be a lonely and aggravating one.

Learning from a Cosmic Analogy: Dark Energy and Dark Matter

In the past decade or so, scientists have been astounded to discover that the universe that we can see represents only a tiny fraction of the matter and energy that governs the cosmos. Based on the motion of stars and galaxies, strange "dark matter" must be present, increasing the gravitational tug on celestial bodies more than can be accounted for by visible matter. Further, based on the surprising discovered that the universe is expanding, not contracting under its own gravitational pull as expected, scientists have proposed that a strange, repulsive "dark energy" fills the cosmos countering gravity. The combined effect of these unseen entities, dark energy and dark matter, are so great, that they account for 96% of the matter and energy of the universe. In other words, the visible universe that we used to think was everything actually is only a tiny fraction of what is out there. What we can see in the "cosmic org chart" accounts for only 4% of what really influences the cosmos.

It's that way in the business world. too. Companies can create tidy org charts and draft neat process maps to describe how they work, but the unseen reality outside the visible systems may be what really dominates operations. Increasingly, experts in knowledge management are learning that easily overlooked and often invisible *intangibles* can dominate corporate value and performance. Numerous intangible transactions may be essential to the success of a company, including casual information sharing between trusted friends, helpful exchanges of tips and best practices between employees or between external partners and internal employees, or loyalty that is gained when people are included in decision making. The invisible linkages and hard-to-observe exchanges in a company's internal an external ecosystems may be the real engines of value creation, regardless of what is on a process map or workstream. By not understanding the value of such intangibles, corporations can easily break key linkages and crush subtle engines of value creation.

Many companies focus on their "value chains" - a term popularized by Michael Porter in his seminal 1985 work, *Competitive Advantage.* The value chain describes the linear chain of events as materials and products move from sourcing through manufacturing and out to the market. It is a highly useful paradigm for manufacturing and was highly applicable to much of the economy in the era when Porter was doing his research. But since that time, the explosion of the knowledge economy has changed the way we work and create value. Verna Allee, a revolutionary expert in knowledge management, has detailed the move from the value chain to modern ecosystems and Value Networks in her book, *The Future of Knowledge: Increasing Prosperity through Value Networks* (Burlington, MA: Elsevier Science, 2003). Verna Allee and Associates have introduced a clever, methodical tool called Value Network Analysis for analyzing and visualizing the transactions of intangibles and tangibles that affect a business. That's beyond the scope of this presentation, but it helps us realize that many of the intangibles that dominate the real world of business can be visualized, revealing bold new insights into how things work

This is Lesson 1 for engineers:

1. **Relationships are the dark energy and dark matter of the corporate universe. They matter more than org charts, titles, and budgets.**

Real success is not about brown-nosing or gaining the favor of one powerful person. It's about filling your work and your life with healthy relationships in which you build ties, communicate, gain loyalty and trust, do extra-mile favors for others, and nurture good will.. Sharing information and helping others is one way to do this.

Here is a corollary important for building the relationships that can bring success:

2. **Look past titles, salaries, and org charts to appreciate and work effectively with those around you.**

Learning to treat each person as important is the only safe and responsible path. Apart from the moral imperatives for kindness to all, there are also practical reasons why one should never put too much weight on the "visible mass" like titles that an individual has. Remember, official appearances and titles don't describe how things really get done. An employee must engage in some corporate astronomy to understand what really moves the planetary bodies and how to be effective. Understanding the agendas, needs, and, sadly, politics of various corporate groups and individuals is part of that work.

Many times key players have a lot of dark matter and thus a lot of influence that others can't see. These can include seemingly low-ranking people who might be easy for some people to ignore or even treat rudely. Administrative assistants, receptionists, sales staff, and many others without flashy titles can wield surprising influence. Learn how to work effectively with the people that can help you get things done. Always treat others, regardless of title, with respect. This includes not just those in the company, but your partners, vendors, suppliers, and other outsiders you work with. Again, these relationships can matter far more than you imagine.

Other Lessons

3. It's all about sales. You must sell.

Some engineers are tempted to look down on sales staff and consider their job of selling as something beneath the lofty ranks of engineering and science. That attitude will limit engineers, for in reality, success in a corporation is always about sales. No project moves forward unless it can be sold convincingly to those who must pay for it–management. Innovation does not happen without selling. A great invention does not become an innovation until it is adopted and used by others, and that requires selling at many levels.

In fact, almost every day engineers will have the opportunity or need to sell their work, their ideas, and their skills to others, whether it's their team, their supervisor, other teams in the corporation, or outsiders. It's especially important if you are trying to pursue innovation success.

As for the work of actually selling products to outside customers or partners, never look down on those who are exclusively engaged in the art of selling. They are the engines that drive your company. Smart engineers should learn from them to better understand markets, needs, opportunities, problems, etc., as well as just learning how to sell. Relationships with sales staff can only be helpful in your career.

4. Inoculate yourself against the corporate antibodies designed to kill innovation.

Enthusiastic engineers love to solve problems creatively and can be tremendous sources of innovation, a blessing that everyone needs and no one wants–or at least that's how it can seem when numerous internal barriers to innovation seem to stand in the way of innovators. This is a topic addressed more fully in *Conquering Innovation Fatigue* by Jeffrey Dean Lindsay, Cheryl Perkins, and Mukund Karanjikar (John Wiley & Sons, 2009). Whether it's politically-motivated "Not Invented Here" syndrome or well-intended corporate policies that stifle efforts to move innovative projects forward, most corporations seem to have large quantities of anti-innovation antibodies in their system that will strike when you are doing your best to take the company forward. Be intellectually prepared and understand that the reasons for the barriers are not likely to be mere stupidity or evil, but genuine concerns and risks that others see. Learning to see from the perspective of others can help you anticipate and overcome concerns more effectively. Again, building relationships is critical here for getting the information, help and perspectives you need. Innovation success is possible, but you must be prepared for the antibodies and have plans that help you go forward, when possible, in spite of these barriers.

5. Be able to be held accountable for every action.

The hidden networks of relationships entwined in an organization can result in swift punishment for stupid deeds or unkind words. They can result in others learning of gaps in behavior. For innovation success, you need to avoid these pitfalls and have the corporation and its people on your side as much as possible, not questioning your motives or accusing you of wrongdoing. There are problems enough for innovators as it is.

For every email you send, imagine a panel of critics later reviewing it to see if you were acting appropriately. When you browse the web at work, imagine you boss looking over your shoulder, because they may very well be doing that. When you chat on a corporate phone, never speak in ways that would embarrass you if someone were listening in. Corporations are not designed to be fair, and even the best of intentions and purest of actions may be misunderstood, but at all costs don't do stupid things that can harm your career. Be ready to be held accountable for every action and avoid the temptation to take moral shortcuts. For best results, strive for best behavior all the time. That's one step toward maintaining the healthy relationships you need to find innovation success in a corporation.

For further insights into innovation success, see http://Innovationedge.com/blog and http://InnovationFatigue.com.

(33c) New Chemical Introduction at Lubrizol

Frank van Lier, The Lubrizol Corporation, Wickliffe, OH

Lubrizol uses a staged New Chemical Introduction Process (NCIP) for managing new component development. This multi stage process has proven its value in many years of application and recently underwent an upgrade to refresh, refocus and revitalize the program. An explanation of the multi-stage process will be provided along with examples of the challenges and keys to success.

See more of this Session: Managing Innovation
See more of this Group/Topical: Management Division

(33d) Improvisation Techniques for Chemical Process Innovation

Peter J. Ludovice, School of Chemical & Biomolecular Engineering, Georgia Institute of Technology, Atlanta, GA, Lew Lefton, School of Mathematics, Georgia Institute of Technology, Atlanta, GA and Richard Catrambone, School of Psychology, Georgia Institute of Technology, Atlanta, GA

The importance of creativity in technical innovation has been strongly emphasized by commercial, educational and government organizations. Such organizations include the National Academy of Engineering and the National Science Foundation. While numerous protocols for design and trouble-shooting have been developed, they typically ignore the creative act within the innovation process. This is likely due to the systematic approach typically employed by engineers in developing such processes. While these protocols are effective at organizing the appropriate concepts, constraints and resources to tackle design and trouble-shooting challenges, they do little to catalyze the creative act in innovation. Recognizing that a more chaotic search through idea space can improve innovation, the application of techniques used in humorous improvisation has catalyzed innovation in non-technical areas. The equivalence of humor and innovation has long been recognized by Arthur Koestler and Edward de Bono. However, we hypothesize that the relative lack of success for this approach in technical areas is due to the additional constraints on feasible innovative technical ideas.

Our approach is based on the equivalence of humorous improvisation for searching idea space and stochastic simulation for searching physical state space.[1] Humorous improvisation is used as a random idea generator, just as random number generators can produce random moves in physical state space. However, the state distribution bias used in stochastic simulation is lacking in humorous improvisation and our recent efforts have focused on developing this equivalent biasing in the searching of idea space. Such a biasing would cause our improvisation protocol to slowly evolve to produce unique and unpredictable innovative technical ideas just as physical stochastic simulations evolve to a unique unpredictable state distribution. Such processes, in which simple stochastic elements produce unpredictable complex behavior, are called emergent processes. Simulations are valuable precisely because such unique complex behavior emerges from such simple simulation elements. Sawyer has recognized that creative ideas emerge from simple group improvisation elements in much the same way.[2] Recognizing the emergent nature innovation from improvisation helps in assessing creativity. Because true innovation is a rare event, it is difficult assess. However, the emergent nature of this process means that innovation is not a single random new idea, but an idea that emerges from a scaffold of ideas build by the group. We are planning to assess the effectiveness of this method by measuring the progress of such a scaffold.

This approach has been applied in various exercises with academic and professional engineering groups and feedback has been taken in the form of post-exercise questionnaires and videotaping of the improvisation sessions. The exercises were carried out by the authors who represent a unique combination of experience in science, engineering, humor, improvisation and educational psychology. We will discuss the results of these exercises and characterize what best practices have been derived from these results. The specific application of this improvisation technique to the chemical process industry will also be discussed.

[1] Ludovice, P.; Lefton, L.; Catrambone, R. "Improvisation for Engineering Innovation," Proc. ASEE National Meeting, Louisville, KY, Paper AC-2010-1650, 2010.

[2] Sawyer, R.K. "Group Genius: The Creative Power of Collaboration", Basic Books: New York, 2007.

See more of this Session: Managing Innovation
See more of this Group/Topical: Management Division

(68a) Chemical Product Design Course at Carnegie Mellon University

Nick Sahinidis and **Apurva Samudra**, Chemical Engineering, Carnegie Mellon University, Pittsburgh, PA

This presentation describes the chemical product design course at Carnegie Mellon University. The course introduces undergraduate students to the design of a chemical product. The lectures address aspects of the design of molecular structure, microstructure, and devices/processes that effect chemical change. Topics include market pull and technology push case studies; the design of refrigerants, polymers, and mixtures; techniques in the design of experiments, molecular property prediction, database searches; and intellectual property rights. This is a project-based course, for which an extensive report must be submitted. Recent course contents and projects worked on by students will be detailed in the presentation.

See more of this Session: Technology Transfer In (A) Product Design Via ChE Education and (B) Commercial Product Manufacturing
See more of this Group/Topical: Process Development Division

AIChE 2010 Annual Meeting, Paper 68b, November 8, 2010

A Product Design Capstone Course in the Chemical Engineering Undergraduate Curriculum at SUNY-Buffalo

Paschalis Alexandridis
Department of Chemical and Biological Engineering
University at Buffalo (UB) - The State University of New York (SUNY)
Buffalo, NY 14260-4200
palexand@buffalo.edu
http://www.cbe.buffalo.edu/alexandridis

Introduction

In addition to addressing the question "How should we make it?" (*process engineering*), several practicing chemical engineers are concerned nowadays with the question "What should we make?" (*product engineering*). Chemical engineers and the companies that employ them have traditionally focused in the efficient production of commodity chemicals and specialty chemicals, but are increasingly active in the development and production of (i) formulated products, (ii) consumer goods, (iii) bio-based concepts, and (iv) devices.

Formulated products are multi-component systems that have been designed and manufactured to meet end-use requirements. They are often multifunctional (i.e., they accomplish more than one function of value to the customer) and microstructured (i.e., their function and perceived value derives significantly from their internal structure in the range 0.1–100 μm). Formulated products, which include bulk and shaped solids, semisolids, liquids and gases, are also known as structured products, engineered products, dispersed systems, or chemical-based consumer products. Consumer goods of interest to chemical engineers offer functionality that is based on chemical/physical technology. Bio-based concepts include biomaterials, drugs, and technologies based on metabolic, cell, or tissue engineering. Devices include in their function physical or chemical transformations. (Costa et al., 2006)

The various chemical products may have little in common in terms of appearance or performance, however, they share common principles and practices in terms of development and manufacturing. Thus the need and the opportunity for a chemical product design course.

Product Design in the Chemical Engineering curriculum

The Product Design course has been developed for seniors majoring in chemical engineering, and is offered in the fall semester of the senior year. By that time, chemical engineering majors have completed successfully several math, physics, and chemistry courses, as well as a materials science and engineering course (which equips students with of knowledge of structure-processing-property for main classes of materials such as metals, ceramics and plastics). The students have also completed core courses in chemical engineering thermodynamics (including thermophysical properties, phase transitions), transport phenomena (including fluid mechanics, heat and mass transfer), and chemical kinetics / reactor design, and have some chemical engineering laboratory experience. Taken together, the above courses provide a broad foundation of knowledge that can be synthesized toward addressing problems of product design. The capstone Process Design course is offered to chemical engineering students in the spring semester of their senior year.

The Chemical Engineering Product Design course at SUNY-Buffalo encompasses the (1) general framework for product design and development (identify customer needs, convert needs to specifications, create ideas/concepts, select concept, formulate/test/manufacture product; also intellectual property, safety, environmental, marketing and financial considerations), and (2) structure-property relations that guide the search for materials with particular properties, as well as modifications of existing materials that would improve properties. These two main themes are reinforced by (3) case studies of successful products with well-documented history of invention, development, production, and marketing. The course material is integrated in (4) product design projects that synthesize (i) the market pull for innovations in the development of improved or new products, together with (ii) the technical push from advances in our fundamental understanding of structure-property relations at the atomic, molecular, nano-scale and continuum levels. The content covered in the course is rather broad, while the depth is limited by mostly time constrains.

Course organization

The class meets four hours a week throughout the semester, with the Instructor typically presenting new material or leading the discussion. Guest lecturers are invited to present information on their areas of expertise, e.g., library resources, intellectual property, product design practice. Several class periods are devoted to project presentations delivered by the students. Chemical engineering faculty members serve as consultants for the team projects. The teams meet with their consultants about once a week to review the progress and exchange ideas. Several books related to the course are placed on reserve for the students to use. Students have access to (through the university library) and are encouraged to utilize for their projects electronic books, journal articles, trade publications, and patents.

The course has been organized along several principles that are recognized as enhancing the student learning experience: (Davidson & Ambrose, 1994)

- Active learning is encouraged (the course material is relevant to every-day life and this makes it interesting to students; students are provided with real-life situations to analyze)
- Effective learning experiences are designed for students (problem-based learning is a significant component of the course; assignments apply class material to new contexts)
- Prompt feedback is provided (grading is done on the basis of detailed rubrics, and all results are communicated to the students; feedback is aimed to be corrective and supportive)
- Importance of time and effort spent learning is emphasized (instructor plans learning activities to permit best use of time; students are encouraged to make effective use of time)
- Student faculty-contact is encouraged (in and out of class; during meetings with consultants)
- Cooperation among students is encouraged (students work on several team projects throughout the semester; they also cooperate during in-class exercises; such cooperation can help students enhance self-esteem, collaborative skills, and personal responsibility)

• High expectations are communicated (the instructor sets high but attainable goals)
• Diverse talents and ways of thinking are respected (the course material is delivered in a way that varying learning styles are engaged; all students are encouraged to voice their views)

Assessment

Each student is assessed on the basis of quizzes and assignments that are completed individually, and projects that are carried out as team efforts (teams consist of preferably 3 members). Quizzes are held during class time and aim to test material presented in class and/or assigned for reading. Individual assignments include "TV/magazine Ad", "What's That Stuff?" and "Read the Label" (each of the these three assignments should address a different product). Team projects comprise "Project 1: Understanding Products", "Project 2: From Needs to Concepts" (problem statements for projects 1 and 2 are taken from the textbook, Wesselingh, Kiil, and Vigild: "Design & Development of Biological, Chemical, Food and Pharmaceutical Products"), and Project 3: "Find a Solution: Real-World Product Design and Development" (project statements have been obtained from www.ninesigma.com).

The overall course grade reflects performance in quizzes and in-class participation (counting for 10% of the overall grade), individual assignments (35% total: 10% for A.1, 10% for A.2, and 15% for A.3), and team projects (55% total: 20% for P.1, 10% for P.2, and 25% for P.3). Grading of assignments and projects is based on rubrics (that are provided to the students in advance). For each assignment/project, students submit a properly documented report document and a PowerPoint presentation (presented in class, as time permits).

The assignments/projects focus on synthesis (create, predict, design, improve, propose) and evaluation (select, decide, critique, justify, verify, debate, assess, recommend). For example, Project 3, that counts the most (25%) toward the course grade, is rather open-ended. Students are challenged with proposing a solution to a real-world problem articulated in a "Request for Proposals" (RFP) published by NineSigma, a company that facilitates solution-provider knowledge searches. RFPs identify critical innovation needs and seek a solution or access to knowledge and capabilities missing in the client's internal program. RFPs define the client's needs so that researchers ("providers") can understand what is needed and respond with the best possible solutions. Teams prepare a proposal for a solution to a NineSigma RFP using the RFP information as a springboard, and following the product design approach: (i) customer needs, (ii) product specifications, (iii) idea/concept generation, (iv) idea/concept selection, (v) product formulation and processing/manufacture, (vi) cost/market. The following RFP titles exemplify the types of problems the students have addressed: Development of Eco-Friendly Solvents; Enabling Freshness of Baked Cookie and Cracker Products; Peroxide-based Sterilants with High Sporicidal Activity; Seasonings or Ingredients to Impart the Taste of Fats to Non-Fried Potato Chips; Soil Conditioning Technologies to Enhance Nitrogen Use Efficiency; Natural Conditioners for Hair and Skin; Water Resistant, Recyclable Packaging Stock; Bio-degradable, Compostable Oxygen Barrier Film; Seeking Cleaning Formulations for Biofilm Removal from PMMA; Monomer Technologies for Imparting Strength in Hair; High Absorbent Desiccants for Coatings; Improving Gas Barrier Properties of Polylactic Acid.

Concluding remarks

A course on product design has been developed that is geared toward chemical engineering applications such as formulated, microstructured products and functional consumer goods. The course covers the general framework for product design and development (i.e., customer needs, product specifications, concept generation, concept selection, product formulation and manufacturing), and guides students to utilize structure-property relations in the search for new and improved products. Problem-based learning challenges students to analyze existing products and to propose solutions to real-world problems using their knowledge of chemical engineering principles. The students are supported in their projects by faculty consultants and are encouraged to utilize library and web-based resources. Students are assessed in multiple ways, based on work they do by themselves in and outside the class, and on team projects. The students appreciated the relevance of the course material to every-day life products and the opportunity for active learning and problem-based learning in a team setting. At the same time, the students viewed with some trepidation the open-ended nature of the course deliverables.

Bibliography

Journal articles advocating product design in chemical engineering:

- Do changes in the chemical industry imply changes in curriculum? Cussler, E. L. Chemical Engineering Education 1999, winter, 12-17.
- Molecular structure and property: Product engineering. Wei, J. Ind. Eng. Chem. Res. 2002, 41, 1917-1919.
- Chemical product engineering. Cussler, E. L.; Wei, J. AIChE Journal 2003, 49, 1072-1075.
- Chemical product engineering: An emerging paradigm within chemical engineering. Costa, R.; Moggridge, G. D.; Saraiva, P. M. AIChE Journal 2006, 52, 1976-1986.

Textbooks on product design with applications related to chemical engineering:

- Wesselingh, J. A.; Kiil, W.; Vigild, M. E. "Design & Development of Biological, Chemical, Food and Pharmaceutical Products"; J. Wiley & Sons, 2007 (ISBN 978-0-470-06154-1).
- Cussler, E. L.; Moggridge, G. D. "Chemical Product Design"; Cambridge University Press, 2001 (ISBN 0-521-79633-4).
- Wei, J. "Product Engineering: Molecular Structure and Properties"; Oxford University Press, 2007 (ISBN 9780195159172).
- Seider, W. D.; Seader, J. D.; Lewin, D. R.; Widagdo, S. "Product and Process Design Principles: Synthesis, Analysis, and Evaluation", 3rd ed.; J. Wiley & Sons, 2009 (ISBN 978-471-04895-5).

Textbooks on product design from the more common, mechanical engineering perspective:

- Ulrich, K. T.; Eppinger, S. D. "Product Design and Development", 4th ed.; McGraw-Hill, 2008 (ISBN 978-0-07-310142-2).
- Otto, K. N.; Wood, K. L. "Product Design (Techniques in Reverse Engineering and New Product Development)"; Prentice Hall, 2001 (ISBN 0-13-021271-7).

Textbooks on product design from a management/business perspective:

- Baker, M.; Hart, S. "Product Strategy and Management", 2nd ed.; Prentice Hall, 2007 (ISBN 978-0-273-69450-2).
- Rainey, D. L. "Product Innovation: Leading Change through Integrated Product Development"; Cambridge University Press, 2005 (ISBN 0-521-06601-8).
- Crawford, C. M.; Di Benedetto, C. A. "New Products Management", 9th ed.; McGraw-Hill, 2008 (ISBN 978-0-07-352988-2).

Books on teaching, helpful in developing a new course:

- Davidson, C. I.; Ambrose, S. A. "The New Professor's Handbook"; Anker Publishing Company Inc., 1994 (ISBN 1-882982-01-0).
- Wankat, P. C.; Oreovicz, F. S. "Teaching Engineering"; McGraw-Hill, 1993 (ISBN 0-07-068154-6).

(68c) Using Process Simulation for Technology Transfer and Process Facility Fit

Charles Siletti, INTELLIGEN, INC., Mount Laurel, NJ and Demetri Petrides, INTELLIGEN, INC., Scotch Plains, NJ

Technology transfer is, in large part, transfer of process knowledge from the development group to a manufacturing site. Because process simulation models represent a succinct, functional repository of process knowledge, they are a natural addition to the technology transfer process. This paper focuses on the capabilities and limitations of simulation models for technology transfer. Process-Facility fit is part of the technology-transfer process and is ultimately a determination of whether and how well a given process can be executed in a particular facility. Gap analysis is often the technique of choice for evaluating a facility. This presentation primarily focuses on the use of process simulation and scheduling tools to add quantitative information to gap analysis. A series of examples demonstrate how simulation can aid in determining gaps in primary equipment, buffer preparation equipment, purified water, utilities, labor and staffing. Techniques for managing uncertainties in process facility fit will also be covered.

See more of this Session: Technology Transfer In (A) Product Design Via ChE Education and (B) Commercial Product Manufacturing
See more of this Group/Topical: Process Development Division

(68d) Technology Transfer in Outsourcing Manufacturing of Specialty Polymers

Zhaoyang Ou, Zhen Lai and Chieh-Min Cheng, Xerox Corporation, Webster, NY

Outsourcing manufacturing of specialty chemicals has steadily grown over the years as a wider spectrum of business segments (biopharmaceutical, semiconductor) seek lower development and product costs, and faster product cycles. In contrast with commodity products with mature and widely applied technologies, specialty chemicals are often first developed and scaled-up internally in pilot scale before being transferred to external vendors for manufacturing. As such, it has presented unique challenges in technology transfer where not all critical manufacturing requirements are known and solutions are being developed alongside with vendor scale-up, which raises manufacturing risks and potential product costs. In this talk, some key lessons learned from the outsourcing manufacturing of specialty polymers will be discussed.

See more of this Session: Technology Transfer In (A) Product Design Via ChE Education and (B) Commercial Product Manufacturing
See more of this Group/Topical: Process Development Division

(68e) Technology Transfer of Pharmaceutical Freeze-Dried Sterile Liquid Manufacturing - A Case Study

John F. Peragine, Bristol-Myers Squibb Company, New Brunswick, NJ

A pharmaceutical freeze-dried sterile injectable liquid has been produced at the same manufacturing site for the past 40 years. This unique product has never been commercially manufactured at any other site since its inception. Since manufacturing at the current site is being discontinued, production needs to be transferred to a new site. Initial experimental attempts to transfer the current process to the new site have been unsuccessful. Although the manufacturing processes at both sites are very similar, transfer of the freeze-drying cycle has been problematic. The physical form of the freeze-dried product at the new site has not been comparable to the current site and has not met acceptance criteria. Since the new site will be the only site for worldwide production, the new manufacturing process needs to be harmonized (demonstrating that the new process yields final drug product equivalent to the currently marketed product). Process equivalence is also important to demonstrate for the required FDA and multinational filings before the new site can be approved. This paper discusses the key challenges of transferring the freeze-drying cycle and lessons learned during the technology transfer process.

See more of this Session: Technology Transfer In (A) Product Design Via ChE Education and (B) Commercial Product Manufacturing
See more of this Group/Topical: Process Development Division

(118a) Project Management in a Matrix Organization

Frank van Lier, The Lubrizol Corporation, Wickliffe, OH

Successful project management is a critical aspect for organizational success. It becomes even more challenging when the project manager and team members are part of a matrix organization. The ability to gain buy-in on goals and objectives, access the needed people and resources, and carry out projects requires deft people skills. Benefits can include higher quality outcomes and increased implementation speeds. This presentation will cover one industry veteran's perspective on how to accomplish project objectives in a highly matrixed organization.

See more of this Session: Project Management: What's New? What Works?
See more of this Group/Topical: Management Division

(118b) Environmental Law 101 Module

Chris A. Paul, Mcafee & Taft, Tulsa, OK

This tutorial, Environmental Law 101 Module, provides a summary of legal issues for engineering professionals and organizations most relevant for new project managers. An extended abstract will be submitted prior to the conference.

See more of this Session: Project Management: What's New? What Works?
See more of this Group/Topical: Management Division

Planning, Managing, and Best Practices for Project Procurement

Dr. Eldon R. Larsen, Marshall University Graduate College, South Charleston, WV

Paul Jason Dean, Kentucky Transportation, Ashland, KY

Abstract

Project procurement is all about acquiring materials, services, equipment, supplies, and anything else outside of the project team that is needed to complete a project satisfactorily. Procurement of goods and services can be one of the costliest portions of a project's budget. Procurement can also be a significant source of project risk and schedule delay. Yet, it is often underestimated in importance to the project. In this paper we discuss various aspects of project procurement, including contract management, and present some best practices.

Procurement Planning

Before a project team starts into actual project procurement, it is important to plan how procurement will take place, what procedures will be followed, what documentation will be needed, what levels of approval and signatures will be required, and how all of this affects the rest of the project planning, risks, resources, and so on. Additionally, the project team needs to decide what will be purchased, versus what the team will provide. Procurement involves both buyers and sellers. In fact, the project team could be a buyer, a seller, or both.

Often, procurement planning will include a determination of potential sellers and requirements for those sellers. Some organizations have lists of pre-qualified and/or preferred sellers to whom they specifically go for goods and services. Other organizations are required to openly solicit bids from anyone interested in providing the needed good or service. Thus, it is important to understand what policies will govern any given project in your organization.

When planning project procurements, there are many factors to consider:

- Project or product scope constraints related to performance, timing, and cost
- Health, safety, and environmental requirements
- Security concerns
- Intellectual property concerns
- Legal issues surrounding licenses, permits, regulations, safety, and employment
- Partnership arrangements or teaming agreements
- Organizational procurement policies and procedures
- Previous experience with sellers
- Lessons learned regarding procurements for previous projects
- Insurance needs
- Roles and responsibilities
- Past and present information about potential sellers

- Unique requirements for the project
- Previously established agreements and contracts
- Make-or-buy decisions
- How to manage multiple sellers, where applicable
- Important procurement lead times needed in the project plan
- Needed procurement completion dates
- The potential for needed insurance contracts and performance bonds
- Documentation that will be required of contractors by the buyer before, during, and at completion of contract work
- Seller selection criteria, including making these public when required
- Qualified sellers lists giving the names of potential sellers who have already been identified as qualified to provide a specific good or service

It is also important to look at whether it is better to make or to buy some goods needed for the project, and whether to supply internally, or to buy from outside, some needed services. Sometimes, rather than a make-or-buy decision, a project manager could also be faced with a *rent-or-buy* decision, similar to a make-or-buy decision, but where something is potentially rented rather than purchased.

Sometimes procurement planning will involve determining whether the buyer wants to use a *sole source* or a *single source*. Though these terms seem very similar, there is an important technical difference. A *sole source* is used when a specific seller is the only one who can provide exactly what the buyer needs, perhaps due to intellectual property protection like patents. In contrast, a *single source* might be used because of preferences of the buyer due to factors like negotiated long-term special pricing or special qualifications or performance history and a desire to maintain long-term relationships between the buyer and seller.

The result of procurement planning is a plan that guides how the procurement process will be managed. This plan can range from somewhat informal to very detailed and very formalized, depending upon the size and the needs of the project and according to organizational policies. Once procurement planning has occurred, it is important to go back an integrate this planning into the main plan for the project and into the work breakdown structure, network diagram, and Gantt chart as needed.

It is important in the procurement planning to clearly define roles and responsibilities for people involved in the procurement process. In larger projects it is highly recommended that an owner's technical representative be appointed with clearly defined responsibilities in monitoring project progress, quality, billing, and so forth as deemed appropriate for the project—many inexperienced project managers regret later on that they did not pay enough attention to this role.

Also, an important product of procurement planning should be a decision as to what kind of contracts, if any, will be used and how bidding on those contracts is to take place. Also, if independent estimates are to be obtained for the contract work, the procurement planning process should determine how this is to be obtained.

Because procurement often involves obtaining or providing project-needed goods and services through contracts and purchase orders, there is a need for project managers to understand contract administration and various purchasing terms and procedures.

Contracts

Contracts are legal, mutually binding agreements between buyers and sellers. When a buyer enters into a contract, he is legally bound to provide the monies agreed-upon for the sellers' products. Because contracts are legally binding, disagreements can be subject to resolution in the courts, if needed. Sometimes contracts can also take the form of *agreements, understandings,* or *purchase orders*.

Sometimes project managers decide to enter into contracts with sellers in order to share project risks with a contractor, when the project manager does not want to absorb the full risk, or when the project team is not adequately qualified or able to provide a needed good or service.

It is important to write project contracts so that they meet the specific needs of the project. Even if projects seem to be very similar, each project is still unique in some way. Thus, each contract can vary even for similar projects.

Who writes the contract—the buyer or the seller? In practice it could be either, depending upon the situation. However, if someone outside of the project team is going to write the contract, it is strongly advisable for the project manager to draft what he or she thinks should be in the contract, before seeing what the other party writes. This will help the project manager (1) to think about what is really needed and (2) not to become so focused only on what the other party writes that he forgets to include other important aspects that should be included in the contract. The party who writes the contract often will write it to be in his favor, so caution is important. Take time to get the contract right.

Contracts need to be approved and signed within an organization's policies, so the project manager needs to become familiar with those policies. Additionally, it is good practice to become very familiar with one's purchasing department and contract administrators in organizations where those will be called upon for help.

There are many types of contracts, some beyond the scope of this book. The more common, standard contract types are discussed below. Keep in mind that not all contract types are appropriate for all organizations. Organizational policies could greatly restrict what contract options are allowable. For example, in many governmental organizations there are laws that specify what types of contracts may be used.

Fixed-Price or Lump-Sum Contracts

Fixed-price or lump-sum contracts are those in which the total set price is agreed upon for a clearly defined good or service. The buyer must clearly define what good or service is desired. The seller is legally bound to provide the good or service for the fixed price. If there are mutually agreed-upon changes to the project scope, then the price can be adjusted accordingly. The simplest form of a fixed-price contract is a purchase order for a product to be provided by a certain date for a certain price. However, there are several variations of fixed-price contracts, and some of them can include incentives to meet specific project objectives. A few of these variations follow.

Firm Fixed Price Contracts (FFP). FFP contracts are one of the most commonly used project contracts. People like these contracts because the price is clearly agreed upon and understood. Any increased costs will be completely born by the seller, unless there is a mutually agreed-upon scope change, in which case the resulting price changes are born by the buyer.

Fixed Price Incentive Fee Contracts (FPIF). FPIF contracts are similar to FFP contracts, *except* that FPIF contracts allow for performance variations with incentives to meet pre-defined performance measures of success. These measures of success can and should be related to cost, schedule, and project and product performance. The final price is not known until the project is completed. There can be pre-defined price ceilings, above which the seller bears any additional cost and must still complete the contract obligations.

Fixed Price with Economic Price Adjustment (FP-EPA). FP-EPA contracts are often used when the contract spans a long period of time, like several years, which can affect the base cost of raw materials and labor costs. For example, over several years the prices for steel, concrete, and lumber can vary greatly, making it extremely difficult for a contractor to accurately estimate long-term costs of building materials. Therefore, FP-EPA contracts provide for pre-defined final price adjustments due to changes in factors like material costs, inflation, or regulatory changes. These contracts should use reliable financial indices and should be used to protect both the buyer and the seller.

Cost Reimbursable Contracts

Cost reimbursable contracts reimburse the seller for his costs, plus a fee to allow the seller to make a profit. The sellers costs often include both direct and indirect or overhead costs, and the indirect costs are usually but not always some percentage of the direct costs. Additionally, there can be incentives to meet specific project objectives. There are several common variations in cost reimbursable contracts, some of which follow.

Cost Plus Fixed Fee Contracts (CPFF), also called Cost Plus Percentage of Cost Contracts (CPPC). CPFF and CPPC contracts are used when the buyer and seller agree that the seller is to be reimbursed for allowable costs, which are not fixed in advance, plus the seller is to receive a fee based upon a percentage of the estimated cost when the contract is initially signed, which might not be the final actual cost. The fee is paid at the end of the job and does not normally depend on other factors like contractor performance. If the project scope changes through an approved change control process, then the fee could change.

Cost Plus Incentive Fee Contracts (CPIF). In some ways the CPIF contract is similar to the CPFF contract, except that for the CPIF contract the contractor or seller receives a fee called an incentive bonus, which bonus is based upon meeting certain specified performance criteria in the contract. The performance criteria could involve cost metrics, schedule metrics, and product and project scope quality and performance metrics. For these contracts, if the seller is able to provide the required product or service for a final actual cost that is less than originally estimated when the contract was signed, then there is often some provision for both the seller and the buy to share in the savings, and thus there is an incentive for the seller to pay careful attention to costs. There could also be monetary incentives for the seller if the contract is completed early, but not always. Every project's contract will be different in what incentives are specified.

Cost Plus Award Fee Contracts (CPAF). CPAF contracts are very similar to CPIF contracts. The buyer reimburses the seller for allowable costs. The seller's fee is based upon meeting agreed-upon subjective measures of performance specified in the contract, and is usually not subject to appeal.

Time and Materials (T & M) Contracts

Time and materials contracts tend to be a kind of hybrid between fixed priced contracts and cost reimbursable contracts. Similar to cost reimbursable contracts, time and materials contracts are open-ended in that the actual allowable costs to be reimbursed could be more or less than the original estimated cost. Sometimes these contracts specify fixed-prices per unit of a good or service that will be reimbursed. Often there are *not-to-exceed* limits on how much will be reimbursed.

Common Procurement Documents

There are a number of important documents frequently involved in procurement activities. Some of these documents have several different names, depending in part on the industry or organization using them. Depending on the needs of the project, the procurement documents should provide information in a manner that provides pertinent information and helps achieve the procurement objectives as easily and effectively as possible. Standardized formats, checklists, clear statements of work, clear expectations, complete legal provisions and constraints are all useful and important. Sometimes public disclosure and advertising are either required or recommended, in order to be fair to all participants. Generally sellers should follow buyers' document guidelines. Below, some of the more common documents are discussed.

Request for Quotation (RFQ). Requests for quotations often are used to ask for small-dollar amounts or off-the-shelf commodities not needing much or any customization. Usually RFQs do not require negotiation, unless it is minimal.

Request for Information (RFI). Requests for information ask for information about a potential seller that would help the buyer to understand that seller's qualifications to do the work needed in a procurement.

Request for Proposal (RFP). Requests for proposals are sometimes called *solicitations* or a *request for tender* (RFT). The seller's response would be a *proposal.* RFPs generally ask for proposals for customized goods or services which are sometimes high-priced or extensive in nature. The seller's proposal should include relatively detailed information about the approach the seller would use to provide the needed good or service. RFPs can result in negotiations with respect to schedule, costs, and performance or scope or the work to be accomplished.

Invitation for Bid (IFB). Invitations for bids or IFB are sometimes called *requests for bids* or RFB. These documents are similar to RFPs, but the IFB terminology is more common for government projects' sealed bidding.

Letter of Intent. A letter of intent is simply a document prepared by the buyer stating the intent of the buyer to purchase a good or service from a specific seller.

Procurement can involve many other documents. The project manager, buyers, and potential sellers should all become familiar with the specific documents and procedures required by the particular buyer's organization.

Bidder Conferences

For large project needs, where multiple potential sellers will be bidding on work to be performed for a buyer, the buy might hold a bidder conference. Bidder conferences might also be called contractor conferences, vendor conferences, or pre-bid conferences. These meetings occur will all the bidders interested in bidding, generally at the same time and location, *prior to* any actual bidding.

These conferences are used for the buyer to present pertinent information to all the interested potential bidders. The information would include clarification on the specific requirements, expectations, and constraints of the work to be provided to the buyer. These meetings provide an opportunity for bidders to ask clarifying questions and to better understand the buyer's needs. *All* interested bidders should be given exactly the same information and opportunities to obtain specifications and clarifying information.

It is highly recommended that the buyer provide each interested bidder with a complete and detailed record of the bidder conference, including answers to questions, specifications and drawings, handouts, presentation slides, and so forth. The buyer might even consider providing each interested bidder with a videotape of the meeting. Most important is that each bidder is treated equally.

It is also recommended that the buyer provide the bidders with a clear understanding of the evaluation criteria and procedures that will be used to select the final bidder which will provide the needed work to the seller.

Independent Estimates

For potentially large, expensive contracts, buyers often asked a qualified independent agent or organization to make an independent estimate of what the contract work *should cost*. Sometimes this independent estimate is done internally by the buyer. The independent estimate is used as a kind of check on the bidders' proposals. Significant differences between the independent estimate and bidders' proposals might indicate that there is a misunderstanding of what the buyer wants, and there was not sufficient information given to the bidders. If a long period of time has elapsed between the time of the independent estimate and the bidders' proposals, inflation or other economic factors might be affecting the discrepancies.

Contract Negotiations

For many contracts, especially those where the work is extensive and expensive, once a specific seller has been selected by the buyer, they will enter into contract negotiations. The objective is to reach mutual agreement on all the details of the contract between them: what will be provided, when, the price, payment arrangements, how conflicts and disputes will be resolved, roles and responsibilities, authorities, terms and laws, rights to intellectual property, and criteria and procedures for contract closure. Additionally, and very importantly, the contract needs to make clear the procedures to follow for any changes to contract scope, schedule, or costs.

For small contracts, there might not be any contract negation at all. However, it is advisable for the project manager to check with legal counsel or procurement experts to help determine when negotiations are needed and how they are to take place.

Contract Closeout

Contract closeout is done to formally signify that the work of the contract has been accepted by the buyer and that the terms of the contract have been achieved. This procedure verifies that all contract deliverables have been delivered in an acceptable manner. Additionally, all contract documents are finalized, recorded, and archived for future reference. Contracts can also be closed out early if for some reason it is necessary to end to contract before the contract work has been completed. This obviously might result in a contract dispute and a negotiated settlement of the contract, the procedure for which should have been detailed in the contract itself.

Green Project Procurement

Green Procurement is a new and emerging trend that focuses on the environmental sustainability of a product. The focus on green products is to reduce carbon dioxide emissions, increase water efficiency, increase energy savings, improve indoor air quality and achieve an overall smaller environmental footprint.

Many government entities and companies are enacting green procurement activities with a view of minimizing their overall environmental impact, and with the ultimate goal of specifying environmental preference for a product that truly has credibility as being green. A claim that an environmental product is green is simply not justification to mandate the product into a project. A green product has to be as good as or better than the traditional alternative. When specifying products many organizations and developers initially analyze the upfront or initial cost of acquiring a product. A new way of thinking must be utilized where the overall life cycle cost and the environmental footprint of the product is considered. The planning phase of project development offers the best opportunity to investigate and specify green products. This is where alternatives can be identified and where the environmental technical standards are outlined, along with performance and functional requirements.

The U.S. Green Building Council has a program called LEED (Leadership in Energy and Environmental Design) Green Building Rating System and accelerates the global adoption of sustainable green building and development practices through the creation and implementation of universally understood and accepted tools and performance criteria.

E-procurement

E-procurement is the utilization of the internet to facilitate purchase and sale of supplies, work and services through the Internet. E-procurement is a tool that allows a user to reach many more suppliers/contractors that a project sponsor would not be able to reach in traditional publications. Government, private and trade association web sites allow the buyers and sellers of goods and services to view all available opportunities for bid invitations. E-procurement is heavily utilized in large supply chain industries or just in time manufacturing facilities.

Many agencies, suppliers and contractors utilize their own e-procurement web pages or that of an industry trade group to promote services, proposal, and opportunities in a manner that allows them to reach more users. The federal government, states, and many large companies have web pages that outline procurement opportunities, policies, procedures, rules, regulations, and frequently asked questions tabs which outline material relevant to procurement in their organizations. This is all in an effort to disseminate information to their respective audiences.

Procurement Best Practices

Before completing this chapter on procurement, let's look at some of the *best practices* in the procurement field. The following best-practices summary is based mainly on work by William Gindlesperger (2009) and by Mark Smith (2003).

- Establish common procurement definitions, and the organization's standardized contract clauses that can be adapted as needed by specific project needs
- Create and deploy an integrated contract management system with a complete record and audit trail; where possible make it web-based
- Identify needs for new contracts and sources of alternative supply quickly
- Identify risks and potential exposure, due to contracts
- Create a pre-qualified supplier pool in which all suppliers are uniformly and objectively evaluated—know the suppliers in the pool and maintain relationships with fewer, yet higher performing suppliers
- Create purchase order templates pre-populated with approved suppliers and supplier profile data
- Maintain, update and utilize software packages (internal and external) that outline average unit bid prices broken down by quantity, county, state, region and the United States, to allow for more accurate pricing and competitive bidding
- Create thorough bid specifications with full details before an invitation for bid
- Create a fair bidding environment and match specifications with capable suppliers
- Streamline procurement procedures and create efficiencies to save time and money; facilitate access to approved suppliers
- Create mutual acceptance of the process that every supplier understands
- Call for, and award bids in a timely manner
- Teach suppliers that past *performance* is important to receive contract opportunities, and then *price* is the essential to win those contracts
- Maintain and nurture positive relationships, with contractors, trade organizations, government entities, utility companies, suppliers, etc, as a means to reach out and ask questions and utilize their expertise when a problem or question arises
- Be active and participate in partnering – make it a plan for success, as this allows for solutions and answers to problems in a timely manner and may resolve many issues where future litigation may arise
- Adhere to and practice a code of ethics
- Establish transparency by sending bid results to all suppliers that submitted a bid
- Monitor every step of the workflow process—ensure clear communications, attention to details, and a deliverable end product; flag contract expiration dates and confirm contractual milestones
- Unbundle or break down large project into smaller contracts to ensure greater participation of smaller contractors

- Have an owner's technical representative systematically ensure that work performed is per the contract—establish clear accountability, and correct problems before they grow
- Complete the project with accurate invoicing

Summary

Project procurement can be a very important and costly part of a project, and a very important part of project management. The upfront time invested in planning for procurements is extremely important in order to facilitate a procurement process that flows as smoothly and effectively as possible. Project managers should seek to establish procurement relationships between the buyer and sellers where both sides are winners; this is in the interest of everyone involved.

References

Gindlesperger, Willaim, *Supply Chain Management Review*, July 8, 2009.

Smith, Mark. *Supply & Demand-Chain Executive*, October 28, 2003.

About the Authors

Dr. Eldon R. Larsen is and AIChE Fellow and is a professor of engineering at the Marshall University Graduate College in South Charleston, WV. He serves as a Director in the AIChE Management Division board. He has taught project management courses in industrial and academic settings for over 16 years.

Paul Jason Dean works for the state of Kentucky in Transportation and manages projects constantly as part of his daily work.

(171a) Tools and Techniques for Managing Innovation From Beginning to End

Jack Hipple, Innovation-TRIZ, Tampa, FL, **Mukund Karanjikar**, Technology Holding LLC, Salt Lake City, UT and **Eldon R. Larsen**, Marshall University Graduate College, South Charleston, WV

Innovation and its management is the hot topic of the last decade in many industries, including the chemical industry. We have come to realize that Six Sigma and DFSS can eliminate variation in what we know how to do, or plan to do, but cannot by themselves produce innovation.

Innovation needs to start with an idea coupled to a need. Its flesh and bones need to be provided by people and their supporting organizations. Finally, the business and manufacturing concepts behind the innovation must be implemented by smart project management.

This session will review the necessary skills to make this journey successful, including proper idea identification, putting together innovation and project teams, and successful project management.

See more of this Session: R&D Management From Beginning to End
See more of this Group/Topical: Management Division

(171b) Finding the Right Ideas

Mukund Karanjikar, Technology Holding LLC, Salt Lake City, UT

This part of the Managing Innovation workshop will focus on techniques to identify the best opportunities for serious innovatin efforts.

See more of this Session: R&D Management From Beginning to End
See more of this Group/Topical: Management Division

Choosing the Right Team to Innovate

Jack Hipple
Innovation-TRIZ
Tampa, FL
8143-994-9999
www.innovation-triz.com

Abstract

Once focus areas and ideas are selected for an innovation effort, the people who will actually work on an innovation team are chosen. Frequently this is done through a "who's available" and "who knows the most" approach with little regard for personality styles and potential conflicts that can decrease the efficiency of the innovation effort. This portion of the workshop will focus on tools and assessments that are available to assist teams in improving their innovation efficiency.

1. Background and Introduction

The structuring of innovation teams is important. Everyone knows this or at least recognizes its importance. What kind of thought goes into assembling a team to work on a project or new product which has a major component of innovation within it? Well certainly representation from all the various functions within the organization is important. This would usually include research, marketing, manufacturing, legal and patent, as well as customer service and quality. The backgrounds and experiences of these people is important of course, but how often do we spend thinking about the *styles* of the people on the team. Now this is more than *liking* each other. This is about how people interact with each other and how they think about problems and challenges. Now this may in fact be important to the team leader who doesn't want to spend his or her time playing referee. But if it just mean that and nothing more, then it just means peace, not necessarily productivity.

2. Personality Styles

There are many personality style assessment measures that are used in industry to assist individuals in their career growth and in short term feedback, but seldom are such tools used pro-actively in team structuring and formation. In fact, many individuals, when asked what their "type" is (from a previous assessment) often produce a blank stare. They remember some of the "letters", but usually not all of them, nor what they mean. For the purposes of this presentation we will use the widely used Myers Briggs assessment and its four characteristics:

Extroversion/Introversion (E vs. I): This refers to the external communication style of a person. A person who loves to talk with strangers, glad hand at a function vs. someone who would prefer to talk on the phone, curl up with a book, and would not look forward to a large gathering of strangers

Sensing/Intuition (S vs. iN): This describes the difference between individuals who like to take in facts and data for decision making vs. an intuitor who might be just as comfortable with gut feelings and instincts or data that's not obvious at first glance.

Thinking/Feeling (T vs. F): Referring to the difference between some who makes decisions based solely on facts and data vs. concern for the human side and people impact of actions and decisions.

Judging/Perceiving (J vs. P): This refers to the difference between the desire to bring closure to a process and making final plans, critical paths, etc. vs. leaving the door open for possibilities not seen or readily obvious ("not putting all the eggs in one basket").

These styles of people would predict very different approaches to innovation. An ESTJ would dominate a team meeting from a verbal standpoint and only be interested in facts and data and who's going to do what when. An INFP would be relatively quiet (maybe their ideas would not be heard in the team meeting!), their concerns would be about the unsaid things by the customer and how the new process or product might affect the need for overtime on the assembly line and its operators' families. They might also be asking about whether all the ramifications had been thoroughly thought out. There is no right or wrong about any of these styles, but assembling a team without knowing the personalities and preferences is like putting together a baseball team and randomly assigning positions without tryouts and knowing who can pitch or catch or bunt.

3. Problem Solving Style

Another strongly embedded characteristic of people is their approach to problem solving. This can be measured via an assessment instrument known as the Kirton KAI™. This assessment returns a number between 32-160 with a 2σ deviation of 70-110. The lower the KAI™ profile, the more adaptive and structured the thinking and solution generations will be. Comfort within the existing paradigm is very high and there is high discomfort with ideas and solutions outside the existing paradigm. Adaptive KAI people love agendas, goals, objectives, and commitment dates. Their thought process is very externally visible—they explain it to you so you don't have to guess. You can spot one almost immediately by simply looking at someone's desk and seeing how orderly it is! At the other extreme will be individuals who, in a brainstorming session, will throw out ideas without evaluation or assessment (quantity is what is important), and who are not so concerned about plans and deadlines. Lastly, the manner in which they reach a conclusion or an answer to a problem is not so obvious and they usually need to backtrack and explain where they were coming from.

Let's now take a look at how there two primary characteristics of behavior and style preferences can affect team dynamics.

4. Team Dynamics

Let's think a minute about an innovation team that has been assembled to bring on stream a new product developed by a corporation's research department. What if we just assembled the team based on functional needs and paid no attention to the factors we've just reviewed. If half the team was extroverted and half introverted, what mechanism or team leadership is required to make sure that all ideas and concerns are heard? If the team is developing a new product requiring much testing and data analysis and the majority of the members are "N's", how likely is it that all the proper data will be gathered? If there's a contractual commitment data with penalties involved for late delivery, would it not be very important to make sure that there is good representation of "J's" on the planning team? If there is concern about broad patent coverage, it would be good if there were a few "N's" in the room to think about possibilities, not just today's products.

Let's think about a team that's challenged with a new competitive product. The adaptive KAI's on the team will be thinking about negotiating, cross licensing, and other "in the current structure" ideas. Highly innovative KAI™ personalities will be thinking about alternative products and patent circumvention. Are either of these "right" or "wrong"? Of course not! The point is that with only one type of person in the room, we only get one type of idea. And without knowing the profiles and openly discussing these in a team setting, it is not possible to be sure that we have all the right people in the room for the innovation challenge we have at hand.

™KAI is a registered treademark of M.J. Kirton

(171d) Manging Your Innovation Project

Eldon R. Larsen, Marshall University Graduate College, South Charleston, WV

Innovation projects are more complex than standard implementation projects as they frequently involve technology and partnerships outside the normal range of corporate activities. This portion of the workshop will focus on the necessary project management steps and skills to manage innovation related projects.

See more of this Session: R&D Management From Beginning to End
See more of this Group/Topical: Management Division

(171e) Discussion on Innovation Management

Jack Hipple, Innovation-TRIZ, Tampa, FL, **Mukund Karanjikar**, Technology Holding LLC, Salt Lake City, UT and **Eldon R. Larsen**, Marshall University Graduate College, South Charleston, WV

Participants will submit questions and discussion points for the presenters to respond to and comment upon from their experiences.

See more of this Session: R&D Management From Beginning to End
See more of this Group/Topical: Management Division

(229a) Why University-Industry Partnerships Matter

Anthony M. Boccanfuso, The National Academies, University-Industry Demonstration Partnership, Washington, DC

Increasing attention is being paid to university-industry research partnerships since they are perceived to be critical to advancing the discovery to innovation process and competitiveness.

While several advantages of these type of collaborations are generally agreed upon, there are numerous challenges and barriers and this session will discuss a number of topics:

* The advantages of engaging in university-industry partnerships

* The challenges faced in developing effective collaborations

* The various perspectives of companies, universities and the faculty and industry scientists and engineers who pursue these type of partnerships

* Suggestions for identifying appropriate partners

* The impact of government agencies (state and federal) and foundations on these partnerships.

and finally, an update on the activities of the National Academies' University-Industry Demonstration Partnership and the projects being pursued to achieve its mission.

See more of this Session: Managing Technology Collaborations I - Academia, Industry and Government
See more of this Group/Topical: Management Division

(229b) IP Challenges in University-Industry Collaborations

Ken Horton, Gore School of Business, Westminster College, Salt Lake City, UT

Universities and their Industry partners generally have the same business goals when they collaborate on research and development. One of the exceptions to this general rule, though, arises when the partners discuss the topic of the intellectual property (IP), who owns it, and how it is used. Some of these challenges are a result of the traditional business models that are used in the collaboration. These challenges can be overcome by utilizing new business models in the collaboration that treat the IP--and how it is used--differently.

See more of this Session: Managing Technology Collaborations I - Academia, Industry and Government
See more of this Group/Topical: Management Division

(229c) Managing Multilateral Technology Collaborations at the Institute for Collaborative Biotechnologies

David H. Gay, Institute for Collaborative Biotechnologies, University of California, Santa Barbara, Santa Barbara, CA

The Institute for Collaborative Biotechnologies is a uniquely interdisciplinary research and development alliance. Led by the University of California, Santa Barbara, in collaboration with MIT and Caltech, together with partners from the Army and industry, the ICB transforms biological inspiration into technological innovation. The presentation will address opportunities, challenges, progress achieved and lessons learned in managing multilateral technology collaborations between the Army, industry and academia.

See more of this Session: Managing Technology Collaborations I - Academia, Industry and Government
See more of this Group/Topical: Management Division

OPENING STEM CELL RESEARCH AND DEVELOPMENT: A POLICY PROPOSAL FOR THE MANAGEMENT OF DATA, INTELLECTUAL PROPERTY, AND ETHICS

Krishanu Saha[1*], David E. Winickoff[2*], and Gregory D. Graff[3*]

[1] Whitehead Institute for Biomedical Research, Cambridge, MA 02142

[2] Department of Environmental Science, Policy, & Management, Univ. of California, Berkeley, Berkeley, CA 94720

[3] Department of Agricultural and Resource Economics, Colorado State Univ., Fort Collins, CO 80523

[*]Authors contributed equally to this work.

ABSTRACT

The need to strike a better balance between free sharing and proprietary and regulatory restraint in the life sciences is becoming an important policy concern in the fields of health policy, law, and bioethics. Here we explore these issues in the field of stem cell research. Expanded funding for stem cell research and development holds unique promise for advancing medicine and human welfare. However, technical, proprietary, and regulatory conditions are not ideal: closed information, congested intellectual property, and regulatory complexity are likely to slow innovation, skew attention towards large markets, and prevent ethical settlement. Accordingly, we propose a new form of collective action among funding agencies, scientific journals, and research institutions to centralize data, cell lines, and technologies and to design and disseminate research tools. The goal is more efficient technology exchange within the research community, orientation of product development towards pressing public health needs, and promotion of 'best ethics' in research and development.

THREE PERVASIVE PROBLEMS IN STEM CELL R&D

Within each of three policy domains current conditions present serious problems for the pace of innovation, the distribution of resulting health benefits, and the public accountability of stem cell research. Moreover, these problems interact and are mutually compounding across domains.

Exchange of Data and Materials

Enhanced capacities to produce, manage, and disseminate data through new information technologies have fueled debates about the manner and timing of data release in the sciences [1]. Traditional norms around sharing research inputs have run headlong into the desire of scientists and their institutions to protect their interests in potentially valuable data, materials, and research results. Practices of delay and secrecy have resulted from the increased relevance of university life sciences to commercial markets and private investors [2, 3]. One mark of change is the proliferation of material transfer agreements between universities [2, 4]. Recent studies suggest that data withholding is becoming common and problematic [5].

Intellectual Property Rights

Following patenting trends elsewhere in the life sciences, scientists and their institutions have been claiming the essential building blocks of stem cell research as private assets. In the specific case of stem cell lines, the first and most obvious problem concerning materials arose from the combination of the Bush Administration's restrictive funding policies and the commanding patent position of the Wisconsin Alumni Research Foundation [6, 7]. Recent analyses show a significant rate of accumulation of new patents over stem cell technologies [8-11], with problematic implications for downstream innovation [12-14].

Bioethics and Regulations

As if technical and proprietary complexities were not enough, disagreement over ethics has given rise to two major problems for efficient and accountable governance of this work. First, in the U.S. federal moratorium on new human embryonic stem cell (hESC) lines has resulted in a vacuum both of funding and oversight. Even as private and state-funded research moves ahead, a national framework is lacking apart from some 2005 guidelines from the National Academy of Sciences. Second, the resulting state programs have proliferated a patchwork of state regulations that differ, sometimes only slightly, on a range of issues. To amplify matters, similar differences occur globally across national regulatory regimes.

TOWARDS AN INTEGRATIVE APPROACH

Scientists and policymakers have, of course, identified each of these issues, and important initiatives in stem cell research have been undertaken. These efforts should be applauded, but, across the board, they remain inadequate on two counts. First, none of them goes far enough to solve the particular problem it targets. Second, by targeting technical, IP, or regulatory problems in isolation these efforts neglect the important interconnections between these domains. Overlooking interaction between the different domains involves both a conceptual and a practical error. The conceptual error is to ignore the profound ways in which these domains are mutually constitutive categories. Norms and practices of sharing data and materials are simultaneously issues of property and of ethics. Questions of sharing data and materials raise important questions of property which, in turn, raise questions of ethics. To separate these questions is to perform a conceptual purification that, at a minimum, prevents optimal solutions. Yet, the overlap is not merely conceptual, it is practical. As the recently encountered problems with some of the NIH hESC lines illustrate, ethical accountability can be promoted only to the extent that provenance characteristics and data about cell lines are well known and shared [15, 16]. Efficient progress of ethically accountable stem cell research requires considerations that span the complexities and bottlenecks across all three domains. Not only would an integrative approach be more efficient, given the interdependencies, an integrative approach may be the only way to solve some of the individual problems.

THE LOGIC OF COLLECTIVE ACTION

The problems discussed above have not been adequately addressed largely because they are classic collective action problems, in which a diverse set of actors share common interests yet no one individual actor's incentives are sufficient to overcome the inertia of inaction. Mobilization requires leadership in the form of coordination and initial investment. Sufficient conditions for collective action have not quite yet emerged in any one of the three domains, let alone across them. And, given the currently decentralized pattern of funding and regulation in the U.S.—despite the articulation of clear interests among many parties—there is no clear leader. Under these circumstances, it is necessary to motivate *many, heterogeneous* potential actors through the use of various carrots and sticks. We argue that collective action with a push from major funding agencies can be a basis for opening up stem cell research in the face of multiple compounding constraints. It is precisely the major funding agencies, such as CIRM or the NIH, that have important roles to play in supporting and enhancing common-access resources in stem cell research. Such an "opening up" would result in a more efficient exchange of data, and with further action, sufficient access to research tools within the stem cell research community. We propose four specific actions for stem cell policy:

1. Lead the development of practical common norms

First, major funding agencies could convene interested segments of the research community to construct a practical normative framework for how stem cell materials and data should be managed and circulated within the community. Participation in the efforts above indicate that this group is heterogeneous and large—including scientists, lawyers, ethicists, state funding agencies, nonprofit agencies, universities, and industry. Leadership from major funding agencies is needed to structure this discussion in finding common ground. This commonly developed normative framework should seek to redress the negative effects spanning the technical, proprietary, and ethical domains.

2. Require access in research contracts

Second, major funding agencies should use funding contracts to require the sharing of research data and materials. By many estimates, stem cell lines and the fundamental protocols of cellular reprogramming and differentiation constitute common resources for the research community that are expected to drive broad progress in biomedical research. Thus, cell lines contain the hallmarks of what have previously been called "community resource projects" [17, 18]. Accordingly major funding agencies could classify the development of new human stem cell lines as a "community resource project". A key consequence of invoking this norm would be to include sharp stipulations in research contracts concerning access to stem cell lines and their associated technologies. The stipulations need not use a simple "open access" model, but could be structured in a variety of creative ways that would increase their flow while respecting the private rights and claims of inventors and their institutions [19].

3. Build a common data resource

Third, major funding agencies should support a common data resource that encompasses all of the necessary technical, IP, and ethics information that will be used by its funded scientists, and hopefully, by the community of stem cell scientists globally. The lack of success in this area to date suggests that generating and sustaining support and interest for such an initiative will require new carrots and sticks, and complementary efforts among major funders.

4. Foster further collection action

Fourth, major funding agencies should support further efforts to coordinate, design, build, and disseminate research tools and platform technologies to alleviate the most important technical, IP, and ethical bottlenecks facing the field. These first three actions can construct the foundation for further collective action in which research institutions themselves, and not the government, could to take the lead. With sufficient data available, consortia of interested research institutions could perform systematic analyses of key stem cell lines, research tools, or other platform technologies where some combination of technical, proprietary, and ethical bottlenecks have stymied work at the participating institutions and/or their commercial partners. Analysis of particular constraints for a particular cell line, research tool, or other platform technology is a time-intensive task that, if centralized and done once for many, would have significant efficiency gains for scientists, technology transfer offices, and ethics review boards alike. Informed by the results of those common analyses the consortia could go so far as to designing new technologies that work around the most important technical, IP, or ethical bottlenecks. These new technologies could then be developed and made available for both research use and commercial development of downstream applications through collective licensing or "protected commons" strategies—such as open source or patent pooling. The constituent components would be ethically vetted and chosen to meet regulatory standards across multiple jurisdictions. Such creative initiatives have taken root in the fields of agricultural biotechnology and synthetic biology [19-22], and could prove useful for stem cell research [23].

CONCLUSIONS

The field of stem cell research is full of promise. Yet, those who seek today to realize that promise must wade through a morass of data and material transfer issues, navigate or negotiate numerous intellectual property claims, and stand in peril of violating ethical or formal regulatory requirements. Major funding agencies can and should do more with its leadership than simply fund more research, taking a more active role in expanding the institutional underpinnings of a science commons that is under strain. Such leadership could catalyze fundamental normative and institutional changes that would expedite the exchange of technologies within the research community, orient investments over a wider range of public health needs, and promote a more ethically sound advance of stem cell R&D.

REFERENCES

1. Rai, A.K. and R.S. Eisenberg, *Harnessing and Sharing the Benefits of State Sponsored Research.* Berkeley Technology Law Journal, 2006. **21**: p. 1187-1213.

2. Merrill, S.A. and A.M. Mazza, *Reaping the Benefits of Genomic and Proteomic Research: Intellectual Property Rights, Innovation, and Public Health.* 2006: National Academy Press.

3. Merges, R.P., *Property Rights Theory and the Commons: the Case of Scientific Research.* Social Philosophy and Policy, 1996. **13**: p. 145-167.

4. Ku, K. and J. Henderson, *The MTA?rip it up and start again?* Nature Biotechnology, 2007. **25**: p. 721-721.

5. Caulfield, T., et al., *Patents, commercialization and the Canadian stem cell research community.* Regen Med, 2008. **3**(4): p. 483-96.

6. Wadman, M., *Licensing fees slow advance of stem cells.* Nature, 2005. **435**: p. 272-273.

7. Loring, J.F. and C. Campbell, *Intellectual Property and Human Embryonic Stem Cell Research.* Science, 2006: p. 1716-1717.

8. Esmond, R.W., R.A. Schwartzman, and T.J. Ebersole, *Stem Cells: The Patent Landscape.* Intellectual Property & Technology Law Journal, 2006. **18**(1): p. 1-4.

9. Glanzel, W., *Stem Cells - Analysis of an emerging domain of scientific and technological endeavour.* Steunpunt O&O Statistieken, K.U. Leuven, 2004.

10. Campbell, D., M. Noiseux, and G. Cote, *Potential for Stem Cells Science and Technology in Canada: Great Promises and Challenges.* Science-Metrix report, 2004: p. 63-63.

11. Holden, C., *Wisconsin Stem Cell Patents Upheld.* Science, 2008. **319**(5870): p. 1602b-1603-1602b-1603.

12. Scheinfeld, R.C. and P.H. Bagley, *The current state of embryonic stem cell patents.* New York Law Journal, 2001. **226**.

13. O'Connor, S.M., *Intellectual Property Rights and Stem Cell Research: Who Owns the Medical Breakthroughs?* New England Law Review, 2005. **39**.

14. Simkin, M., *How will intellectual property factors affect the commercialization of stem cell therapeutics.* 2005: BioEurope.

15. Streiffer, R., *Informed Consent and Federal Funding for Stem Cell Research.* Hastings Center Report, 2008. **May-June 2008**: p. 40-47(8)-40-47(8).

16. Baker, M., *Consent issues restrict stem-cell use.* Nature, 2008. **454**(7204): p. 556-556.

17. Wellcome Trust. *Sharing data from large-scale biological research projects: a system of tripartite responsibility. Wellcome Trust Meeting Report, January 14-15, 2003, Fort Lauderdale, FL.* 2003. Fort Lauderdale, FL.

18. Human Genome Program. *Summary of principles agreed at the first international strategy meeting on human genome sequencing.* 1996. Bermuda.

19. Winickoff, D.E., K. Saha, and G. Graff, *Opening stem cell research and development: A policy proposal for the management of data, intellectual property, and ethics.* Yale Journal of Health Policy, Law, and Ethics, 2009. **IX**(Winter).

20. Atkinson, R.C., et al., *Intellectual property rights. Public sector collaboration for agricultural IP management.* Science, 2003. **301**(5630): p. 174-5.

21. Rai, A. and J. Boyle, *Synthetic Biology: Caught between Property Rights, the Public Domain, and the Commons.* PLoS Biol, 2007. **5**(3): p. e58-e58.

22. Porceddu, M.B.C., et al., *Constructive Approaches to Intellectual Property Complexity in Today?s Agricultural Technology World.* Plant Mol. Breed, 2007. **5**: p. 294-295.

23. Winickoff, D.E., K. Saha, and G. Graff, *Opening stem cell research and development: A policy proposal for the management of data, intellectual property, and ethics.* Yale Journal of Health Policy, Law, and Ethics, 2008. **VIII**(Winter).

(238a) Property Prediction for Lipids Based Product Design and Analysis

Carlos Axel Diaz-Tovar[1], Rafiqul Gani[1] and Bent Sarup[2], (1)Chemical and Biochemical Engineering, Technical University of Denmark, Kgs. Lyngby, Denmark, (2)Alfa Laval Copenhagen A/S, Soeborg, Denmark

Vegetable oils and fats have an important role in human nutrition and in the chemical industry since they are a source of energy and fat-soluble vitamins, and also in the production of renewable sources of energy. Nowadays as the consumer preferences for natural products and healthier food increase along with growing interest in biofuels, the oleochemical industry faces in the upcoming years major challenges in terms of design and development of better products and more sustainable processes to make them. Fortunately, the unique condition of edible oil and fats of being mixtures of different chemicals provides a wide range of industrial uses and applications of these lipid-related compounds that, once they are purified, can be offered in the market as final products (edible oils and fats) or as raw materials for other processes (i.e. sterols as precursors of steroids). The aim of this work is to present the systematic development of computer-aided methods and tools related to the prediction of necessary physical properties suitable for lipid-based product design. The methods and tools include: a filled lipid-database of collected experimental data from the open literature and confidential data from industry, and, generated data from validated predictive property models; modeling tools for fast adoption-analysis of property prediction models and for fast development of process models relevant in product design and that are not available in process simulators. This was achieved by first identifying and classifying the lipid compounds found in the edible oils and their potential industrial uses and applications. Then creating a list of the physical-chemical properties needed for model-based product design. Next, collecting the available experimental data from different sources were collected for the identified lipid compounds and their corresponding properties. Finally, selecting and adopting the appropriate models to predict the necessary properties, to fill-out the lipid-database and to make it suitable for application with other computer-aided tools (such as commercial process simulators). For pure component property prediction, the GC-based method of Marrero-Gani and its extension to atom-connectivity methods were used to predict the single value properties like the normal melting point temperatures, the normal heats of formation as well as the critical properties. For temperature dependent pure component properties (vapor pressure, liquid heat capacity, viscosity and surface tension), published GC-based models have been adopted to obtain a general temperature dependant GC-based model. For liquid density, the modified Racket equation has been adopted to create a GC-based model fine-tuned for the lipid chemicals. The regression of the adopted GC-based model parameters needed experimental data. PC-SAFT EoS was used to generate pseudo-experimental data when real data was not available. With the development of the lipid-database and the associated property models, it has been possible to analyze the behavior of various lipids based products and design the conditions under which they can be recovered. The presentation will also show results of the recovery and analysis of lipids-based products illustrating the use of the developed property models and the lipids-database.

(238b) Quantitative Structure-Property Relationship Model for Prediction of IGC_{50} Toxicitiy Potentials

K. M. Yerramsetty, B. J. Neely and K. A. M. Gasem, School of Chemical Engineering, Oklahoma State University, Stillwater, OK

The effect of chemicals on animals can be multifold and requires laborious experimentation to test chemicals for their toxicity. Quantitative structure-property relationship (QSPR) models offer a substitute for experimentation and thereby save resources and avoid undue animal experimentation. However, this task is too complex to be addressed by a simple QSPR model. Therefore, as a first step, the toxicity of the chemicals on small microbial organisms is used as an indication of the effect on animals. The toxicity data used in this study involved the microbe *T. pyriformis* as the test organism. The 50% growth impairment concentration, which is denoted as IGC_{50}, was used as an indication of the toxicity potential of the chemicals.

This study was completed under a QSPR challenge conducted in 2009 by Case studies on the Development and Application of *in-Silico* Techniques for Environmental hazards and Risk assessment (CADASTER). The objective was to develop successfully a QSPR model on known IGC_{50} values and apply it to an external test set of unpublished IGC_{50} values. In this study, we present our new quantitative structure-property relationship (QSPR) model for predicting IGC_{50} values. Several descriptors were generated for the 1213 molecules in the database. A wrapper approach, which involves differential evolution combined with neural networks, was implemented to find the optimum set of 20 network descriptors and the network architecture, simultaneously. Our ensemble network resulted in an RMSE value of 0.817 for the external dataset considered.

See more of this Session: Physical Properties for Chemical Product Design
See more of this Group/Topical: Process Development Division

Pseudo-Component Viscoelastic Model of Soft Tissues

Upasana M. Sridhar, R. Russell Rhinehart & Sundar V. Madihally
upasana@okstate.edu, rrr@okstate.edu, sundar.madihally@okstate.edu
School of Chemical Engineering, Oklahoma State University, Stillwater, OK, USA

For Presentation at the 2010 AIChE Annual Meeting, November 7-12, 2010, Salt Lake City, UT, USA

1. Introduction (Motivation):

Understanding the response of human tissues to externally imposed stress and strain is critical to improving quality of life and healthcare in diverse aspects. Examples include: i) Improvements in automotive safety devices such as air bags and seat belts depend on better knowledge of how human parts, and the integrated whole, respond to the stresses of impact. ii) Development of biomedical imaging techniques, which observe *in vivo* tissue and organ responses to stress and strain, could be used to assess whether a tissue is functioning correctly or needs urgent care, if stress-strain patterns were characterized. iii) Development of synthetic prosthetic devices and *in vitro* regenerated tissue requires scaffolds that duplicate mechanical properties of native tissues. iv) Development of simulators that train next-generation physicians and development of robotic procedures both depend on models of tissue stress-strain response.

Significant effort in developing models of the stress-strain behavior of biological materials reveals that biological tissues show a more complex mechanical behavior than polymers, plastics, and metal films; and cannot use the conventional models like Quasi Linear Viscoelastic (QLV) model, introduced by Fung [1-7] developed for those substances. Most viscoelastic models of biological materials are grounded in classic approaches from these other disciplines. New modeling concepts are needed to appropriately describe mechanisms within biological structures.

Further, models are used in many ways. In one use, models are used to validate mechanistic understanding of materials. If the models are abstract modifications of those used for metals and polymers, the models do not reflect either mechanistic understanding or phenomenological fidelity [8-12] As a result, the models cannot be used to accept or reject hypotheses of molecular mechanisms within biological structures.

Finally, models of biological tissues are used for many situations including stress response to strain, stress response to cycling conditions, strain response to stress, etc. This also includes the response to conditions imposed after the material has been stressed or elongated from its natural at rest state. Current models derived to match one situation, and model parameters do not have transferability to models used for other situations. New modeling approaches are needed to have a universal model for all stress-strain treatments.

Partly, the models in current use are based on analytical solutions to either the calculus terms or differential equations. Analytical solutions include initial conditions or presumptions of uniform treatment in time. By contrast numerical solutions do not presume specific initial conditions, are not limited to a specific treatment history, and can be solved for either stress or strain as a response to the other. Numerical solutions permit the model to be transferrable to other situations.

This work demonstrates the modeling ability and transferability of a novel viscoelastic response modeling approach for biological tissues.

At this juncture it is important to mention that in order to develop a model to describe the behavior of biological materials two problems have to be addressed which are i) Modeling the viscoelastic behavior and ii) Model parameter optimization techniques.

2.Approach:

2.1 Modeling

Since tissue stress relaxes in time under constant strain, and since tissue shape progressively deforms under constant load, and since the internal structure of tissues gradually returns toward original structure when external stress or strain is relieved, time-dependent (viscoelastic) models are required. The time-dependent behavior also depends on the stress-strain history. Further, since tissues are comprised of several participating structures (cells, matrix, fibrils, etc.), each having individual mechanisms; a multi-component, viscoelastic model is required. Further, since tissue properties are not constant; nonlinear, multi-component, viscoelastic models are required. Finally, since many tissue components do not relax fully to the original internal structure, the commonly employed dashpot element (which lets the spring return to zero stress) is inappropriate. New constitutive relations are required for the nonlinear, multi-component, viscoelastic models.

This work will demonstrate the validity of appropriate nonlinear viscoelastic relations and a pseudo-component approach for modeling the complex tissue structures. Six composite models types developed using various combinations of the pseudo-components are used to examine the consistency of the composite model with the experimental data. The composite models are developed from a combination of a hyper-elastic spring or a spring-and-dashpot model along with total-relax-back-to-original structure or retain-the-as-stretched structure.

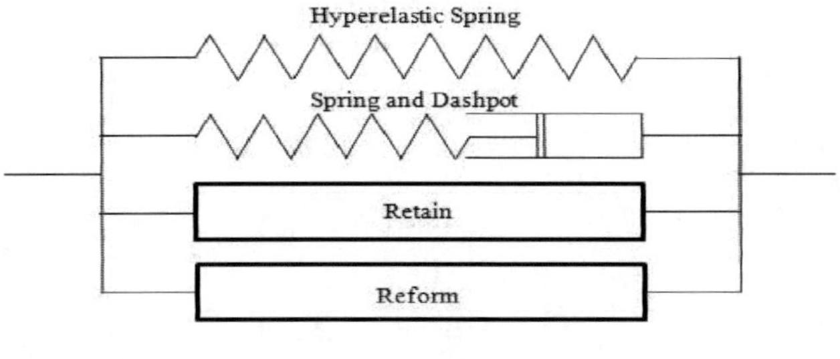

Figure 1: Pseudo-Components

The four pseudo-components used to develop the composite models are shown in Fig1. The 6 composite models are developed from various combination of 2 or 3 of these 4 components.

2.1.1.Pseudo-Component Model Components:

Although relaxation begins as soon as stress is applied, and continues until stress is relieved, the following illustration will separate the stages (externally applied stretching and internal relaxing) for convenience and communication clarity. The subsequent mathematical analysis will treat the two phenomena simultaneously.

Consider a material with zero internal stress at rest that is not stretched (Fig 2a). Let the length of the original film be L_0 with height H_0 and width W_0. The line segments that are randomly oriented demonstrate that the orientation is at a stress-free state. These line segments represent viscoelastic components. The line segments are termed as "components" which symbolizes a general category of functional relationships. For example, a line segment (component) in a macromolecule could represent folded sections, which in response to strain could be unfolded.

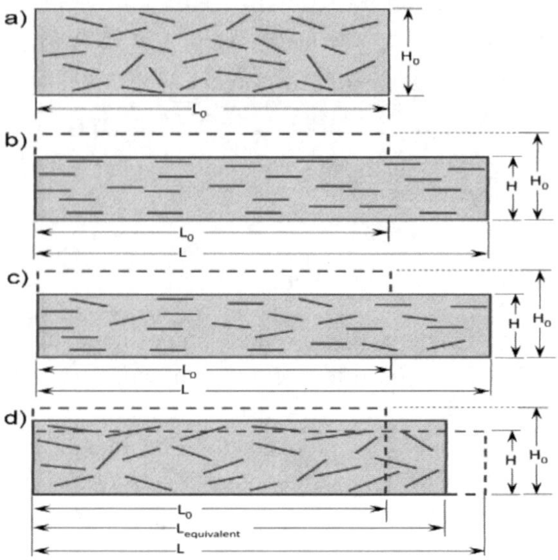

Figure 2: Viscoelastic stress relaxation

When strain stretches the folded section as a result of application of external deformation, internal stress is created. The stretch in the folded section is represented as aligned components (Fig 2b). The new length is now $L=L_0*(1+\varepsilon)$, and if the volume of the material remains the same, the lateral contraction creates new width and height of W and H. The stress determined is, $\sigma_0 = F/(H_0*W_0)$ where F is the tensile force while ε represents the strain. If held at length L, the stresses on the folded sections may cause them to unfold, partially alleviating the stress. The unfolding may happen over time, as thermal motion permits, providing a progressive relaxation to the new deformed state, as illustrated in Fig 2c.

Thermal processes may lead to refolding when the external deformation is removed. (Fig 2d). This leads to a permanent elongation in the length of the film represented as $L_{equilibrium}$ (equilibrium length) but with the original random, stress free structure.

The four pseudo-components shown in Fig 1 are discussed below.

Pseudo-component 1: Hyper-elastic Spring: This component characterizes the material in such a way that it does not relax internal stress. According to this component the material rebounds to its original structure and size on removal of external load. This component has a stress response to the strain that is nonlinear as shown in Fig 3, curve (a) or (b).

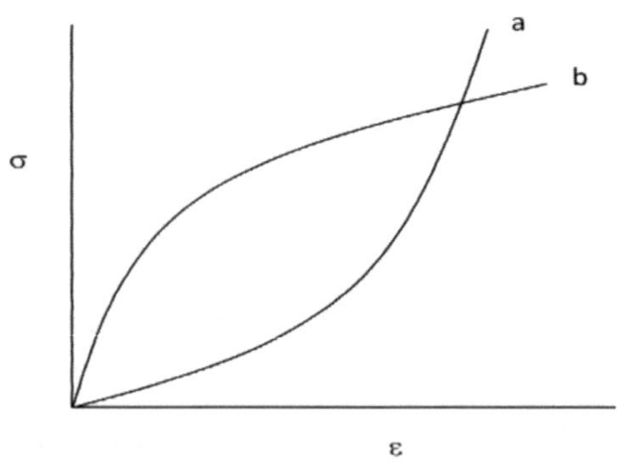

Figure 3: Strain to strain relation of a hyper-elastic spring

Pseudo-component 2: Spring and Dashpot: This component does not account for the necking in of the material i.e. it does not account of the changes in W and H of the material when the length changes due to elongation. When held in an elongated state, the dashpot elongates and the spring returns to its original stress-free state (length).

Pseudo-component 3: Reform: As opposed to spring and dashpot component this component counts for the necking in of the material i.e. it does account for the changes in W and H of the material when the length changes due to elongation. Partial reformation of the components in the material to the original orientation (Fig 2c) happens while the elongated material is held at length L. Prior to total relaxation if the external loading is removed, the material restores itself to an elongated state which is permanent (Fig 2d), which has the same internal structure as the original zero stress random orientation state. L_{equ} is the permanent elongated length. Since the volume of the film is a constant and since the film original length L_0 is lesser than L_{equ} ($L > L_{equ} > L_0$), the cross section area at this point is smaller than the original cross sectional area of the film. Yet, if the strain in Figure 1c when partial relaxation occurred, is not relieved, the expected stress would be that if the sample of original length, L_{equ} had been elongated to L. The strain would be $(L-L_{equ})/L_{equ}$. Although there would be a distribution of orientations and unfolding kinetics, the pseudo-component will be modeled as having the average behavior. In the fully relaxed state, these components are oriented in a random fashion similar to the original sample. Once again when strain is applied it leads to the same stress response identical to that of the original sample, but with a lower cross sectional area.

Pseudo-component 4: Retain: Alike reform, retain also counts for the necking in of the material because of reduction in cross-sectional area when the material is elongated. However, here, the material components retain the orientation that resembles a stretched orientation when the material is held at an elongated length, L. By arranging itself to a new structure orientation the material components relieve its stress. Even after the relaxation occurred, if the strain is continued, the pattern for the stress will persist from the newly oriented structure.

Combinations of these 4 pseudo-components are used to build composite models.

2.1.2. Composite Models:

Each of the composite models describe a structure in which the pseudo-components are acting in parallel (Figure 1). There are two aspects of the models needed to completely describe each of the composite models. One aspect describes the number of parallel pseudo-components that make up the composite model. The second aspect describes the type of each pseudo-component. Combined, the number of pseudo-components and the behavior of each is termed the model architecture. In initial investigations several basic architectures were investigated. Four types of Architecture are investigated.

Architecture A) 8 parameters – This structure has 4 elements set up parallel where in one of those is a nonlinear spring and the rest all are linear spring and dashpot components.

Architecture B) 9 parameters – This structure has 3 elements in parallel each of which is a retain pseudo-component.

Architecture C) 8 parameters – This structure has 3 elements in parallel out of which one is a nonlinear hyper-elastic spring while the other two are retain pseudo-components.

Architecture D) 11 parameters – This structure has 4 elements in parallel one of which is a hyper-elastic spring, and the others were reform pseudo-components.

Two criteria are used to select the best Architecture. They are 1) the sum of squared deviations of the model from the experimental data (SSD), and 2) simplicity of the model. Architecture A, even though it had 4 Pseudo-components, was limited by linear stress-strain relations and could not capture the experimental stress data as well as the other three

Architectures. Architecture B, regularly showed a very high τ value for one component, which signifies that out of the three one of the retain components did not relax, which in turn implies that it acts as a hyper-elastic spring. Architecture C explicitly uses a hyper-elastic spring as one pseudo-component, achieving identical best fit as Architecture B, but with one fewer parameters. Architecture D did not improve the fit to data relative to the undesired aspect of increasing the number of adjustable parameters. Accordingly, Architecture B is selected for this work.

Within Architecture B, one pseudo component is a hyper-elastic spring, and the other two would be the 6 permutations of Pesudo-Components 2, 3, or 4 as follows:

Composite Model 1: One hyper-elastic spring with two spring and dashpot components.

Composite Model 2: One hyper-elastic spring with two reform components.

Composite Model 3: One hyper-elastic spring with two retain components.

Composite Model 4: One hyper-elastic spring with one retain and one spring and dashpot component.

Composite Model 5: One hyper-elastic spring with one reform and one spring and dashpot component.

Composite model 6: One hyper-elastic spring with one reform and one retain component.

2.1.3. Mathematical Statement:

It is important that a mathematical statement is formulated for the composite models developed. Fig 3 symbolizes for a pseudo-component an instantaneous non-linear stress-strain relation modeled as,

$$\sigma_i = A_i(e^{B_i \varepsilon_i} - 1) \tag{1}$$

where "i" (subscript) represents the ith pseudo-component. In this equation the coefficients A and B should have the same sign. Fig 3 curve a and curve b illustrates the plot when A,B < 0 and A,B > 0 respectively.

Also, as far as this work is concerned each of the pseudo-component undergoing internal material deformation is modeled as one having a rate of internal stress relaxation of the order 1, which at infinite time tends towards complete relaxation of zero stress. The relaxation model with no strain-rate-induced stress is,

$$\tau_i \frac{d\sigma_i}{dt} + \sigma_i = 0, \sigma_i(t = 0) = \sigma_0 \tag{2}$$

As Fig 1 represents, for the material with internal deformation concept Eqn (1) is the stress model when instantaneous strain at one stage is applied, stress relaxation is revealed by Eqn (2) if for a period of time the elongation is held then. Yet, when the material is relaxing, if it is strained at a particular rate, then derivations done rigorously enumerates that Eqn (2) should also include the injection of internal stress at that particular rate, which happens as a result of rate of external strain.

$$\tau_i \frac{d\sigma_i}{dt} + \sigma_i = \tau_i \frac{\partial \tilde{\sigma}_i}{\partial \varepsilon} \frac{d\varepsilon_i}{dt} = \tau_i A_i B_i \exp\left(B_i \frac{Lequ_i}{L} \varepsilon\right) \frac{Lequ_i}{L} \frac{d\varepsilon_i}{dt}, \ \sigma_i(t = 0) = \sigma_0 \tag{3a}$$

$$\tau_i \frac{d\sigma_i}{dt} + \sigma_i = \tau_i \frac{\partial \tilde{\sigma}_i}{\partial \varepsilon} \frac{d\varepsilon_i}{dt} = \tau_i A_i B_i \exp\left(B_i \varepsilon\right) \frac{d\varepsilon_i}{dt}, \ \sigma_i(t = 0) = \sigma_0 \tag{3b}$$

Eqn (3a) is for the reform part of the recovery of the internal structure, Eqn (3b) is for the retain part which allows a stress relief. Eqn(3a) is analogous to spring-and-dashpot model since the reformation of the spring is allowed whereas Eqn (3b) is not similar to spring-and-dashpot model, because of the fact that the "spring" cannot return to a state of zero elongation.

Numerical methods can be used to solve Eqn (3a) and (3b). Any stress or strain history can be modeled and can be solved with the help of numerical methods.

Given from Eqn (3a) the stress of pseudo-component "i" which is partially relaxed, at the completion of a Δt time increment, the strain equivalent on the strain component can be determined by the inverse of Eqn (1)

$$\varepsilon_{equ,i}\,(t + \Delta t) = \ln\left(\frac{\sigma_i(t+\Delta t)}{A_i} + 1\right)/B_i \tag{4}$$

Eqn (4) gives the value of equivalent strain from which for a pseudo-component the equivalent length is calculated as

$$L_{equ,i}(t + \Delta t) = L_0(1 + \varepsilon_i(t + \Delta t)) \tag{5}$$

The sample length at the next time instant, t, can be calculated from experimental strain, ε_0 as

$$L(t) = L_0(1 + \varepsilon_0(t)) \tag{6}$$

and hence the new strain that is effective on the pseudo-component is

$$\varepsilon_i(t) = \frac{[L(t) - Le_{equ,i}(t)]}{L_{equ,i}(t)} \tag{7}$$

For each pseudo-component the stress vs time is modeled by sequential application of Eqns (6), (7), (3a) or (3b), 4 and then (5). As numerical methods are used to solve equations Δt should be small relative to the time-constants and the time periods taken for the corresponding changes in strain rate.

Here, it should be noted that the measurement during experiment in a tensile test is tensile force but the tensile tests report stress. The tensile force obtained here is represented as the sum of all forces attributed to each of the pseudo-components.

Assuming that each of the pseudo-components retain a volume that is constant upon deformation, and the contraction in each of the dimensions is uniform, and the length of the material L is long enough that the geometric end effects due to the clamping of the sample could be eliminated, the extension in length due to strain causes the cross-sectional are to reduce to as

$$A = \frac{A_0}{(1+\varepsilon_0)} \tag{8}$$

If for a ith pseudo-component having a representation as Eqn (3a) and if it is relaxed to the equivalent length, the "original" area that is unstressed is

$$A = \frac{(1+\varepsilon_i)A_0}{(1+\varepsilon_0)} \tag{9}$$

Hence the force due to the stress on i^{th} pseudo-component of Eqn (3a) will be,

$$F_i = \sigma_i A_i = \frac{(1+\varepsilon_i)\sigma_i A_0}{(1+\varepsilon_0)} \tag{10}$$

The measured stress is determined by volume weighting, by $\emptyset_i$, the volume fraction, ratio of sum of the forces due to each of the pseudo-components to the original area A_0,

$$\sigma_{composite}(t) = \sum_{i=1}^{N} \frac{\emptyset_i \sigma_i(t)(1+\varepsilon_i(t))}{(1+\varepsilon_0(t))} \tag{11}$$

where "N" stands for the total number of pseudo-components. For the fluid rearrangement components, internal volume does not remain constant and there is no material deformation, making $\varepsilon_i = \varepsilon_0$

It would seem that the volume fraction, $\emptyset_i$, and the stress-amplitude factor, A_i, from Eqn (1) are independent model parameters. However, when Eqn (11) is written in terms of strain only, the two coefficients appear as a product.

$$\sigma_{composite}(t) = \sum_{i=1}^{N} \frac{\emptyset_i A_i (e^{B_i \varepsilon_i(t)} - 1)(1+\varepsilon_i(t))}{(1+\varepsilon_0(t))} \tag{12}$$

Experimental stress-strain-time data cannot separate the functionality of φ_i and A_i. Accordingly, the model coefficients that characterize the viscoelastic properties of each pseudo-component that are adjustable in the "AB" material will be the φA_i product, B_i, and τ_i. If the pseudo-components are identified and volume fractions are known, then τ_i , A_i , and B_i would be the adjustable model coefficients. However, the pseudo-component is not meant to represent one particular component, but the collective behavior of the integrated whole. This work will propose τ_i , the φA_i product, and B_i as adjustable model coefficients.

At each time step, Δt, Eq. (11) will provide the value of the superficial stress measured on the composite. One major asset of this time-incremental, numerical model vs. an analytical integrated model, is that for the time-dependent behavior rate of strain there are no assumptions. The computational algorithm is not changed even when the strain rate is started, stopped, changed, or re-started. Further, it is a simple task to replace the instantaneous stress-strain relation of Eqn. (1) and its inverse of Eqn. (4) with another relation, and also not violate the convolution integral requirements in often-used viscoelastic models. In order to determine values of the model parameters that is adjustable least squares regression using $\sigma(t)$ and $\varepsilon(t)$ data and the pseudo-component model that is numerically-solved to predict the measured stress of the composite models will be used.

2.2 Optimization:

The second problem is associated with model optimization. Research reports difficulty in optimizing viscoelastic model parameter values to best match the model with experimental stress-strain-time data. There are multiple local optima, and approaches to the minima are often exasperatingly slow (even with classic best practice nonlinear optimizers such as Levenberg Marquardt). Further, optimization must include constraints on the parameter values, suggesting that direct search methods may be more appropriate than gradient based methods. Accordingly, techniques such as best-of-N starts or direct search techniques have been investigated for determining the probable global optimum subject to multiple constraints. The applicability of emerging optimization techniques for model parameter adjustment is explored along with new regression approaches and the quality of the model.

A new optimization technique called leapfrogging is also explored [13]. This technique starts with a set of players (trial solution) located randomly in the DV space. This technique relocates the position of players by reflecting the player with worst OF value across the player with the best OF value at each iteration. Test cases on this technique revealed that this technique gave better optimized values when compared to other techniques with fewer number of function evaluations[13].

The model parameters were adjusted by the optimizer so that the sum of squared deviations of the model from the experimental data for the entire range with 4 or 5 stages of sequential stretch-and-hold process. Since there are multiple optima in this high dimensional nonlinear search, the optimizer is run for "N" times from the random set of initial model coefficient values. The value of "N" should be chosen properly in such a way that it is not too small that the optimizer is not able to capture the best model coefficient values within that range, not too big that the optimizer takes more time to get the best model coefficient values increasing the computational burden. Hence an appropriate value for the "N" should be calculated. The value of "N" is calculated [14] so that the best set out of the N results within the user defined probability stands as one of the best set of model parameter values out of all possible solutions. During each trial out of the "N" trials the optimizer stops when there is no significant improvement statistically in the objective function, i.e. minimizing the sum of squared deviation of the model from the experimental data, relative to variance in the data [15,16].

3.Experimental [17,18]:

3.1. Sources of Material:

For the purpose of generating the polymer samples of Lactic and Glycolic Acid (PLGA) composite scaffold the 50:50 PLGA polymer pellets, ester terminated (nominal) with 90-120 kDa Mw were obtained from LACTEL absorbable polymers (Pelham, AL), the chitosan (200-300 kDa molecular weight, Mw, 85% DD) and gelatin type – A (300 Bloom) were obtained from Sigma Aldrich Chemical (St. Louis, MO), Apper Ethyl Alcohol, 200 proof, anhydrous and chloroform were obtained from Pharmaco.

For the purpose of generating Polycaprolactone (PCL) of 47 kDa (Mn) was purchased from Polysciences (Warrington, PA). Low molecular weight Chitosan (50 kDa based on viscosity), type A porcine skin gelatin (approximately 300 bloom), 500 kDa DS (contains 0.5-2.0% phosphate buffer salts) and Toluidine Blue O (with 90% dye content, approximately) were obtained from Sigma Aldrich Chemical Co (St. Louis, MO). Glacial acetic acid was purchased from Pharmco Products Inc (Brookfield, CN). Pure ethanol was from AAPER (Shelbyville, KY).

3.2.Method:

The PLGA composite scaffolds were prepared by 4 steps which is explained briefly [19]. First a thin PLGA film by air drying PLGA solution (4 % 21 wt/v) prepared by dissolving the polymer pellets in 5mL chloroform, overnight in a chemical fume hood on a Teflon sheet (United States Plastic, Lima, OH) (8cm×6cm) fixed to a flat aluminum plate is formed. The formed PLGA film as a result of air drying has a smooth and hydrophobic surface. Secondly in order to perform the etching process formed PLGA film was completely submerged in 1N NaOH solution for 10 minutes. This helps to create surface roughness on the smooth surface of PLGA film and also makes it hydrophilic. Thirdly, the etched PLGA film are washed with excess water and punching holes on the surface in a square pitch 1 cm apart from each other using a stainless steel needle and a hammer. Finally, to layer the the PLGA film with a 3 mL mixture of 0.5% (wt/v) chitosan and 0.5% (wt/v) gelatin solution dissolved in 0.7% (v/v) acetic acid, on both sides and freeze drying it in a lyophilizer to form the porous layer.

For preparing PCL scaffolds, first, 3-4 mL of 10% (wt/v) PCL solution prepared in glacial acetic acid, was gently dropped on the surface of water in an two inches in diameter Teflon dish and air dried until the matrix was completely formed. The matrix was neutralized in ethanol for ten minutes and, then, washed with water. The side that did not touch the water when formed is referred as "top side" and other side is referred as "bottom side". Since majority of the published studies use chloroform as a solvent, PCL matrices were formed by air drying 2 mL PCL solution (10% w/v) in chloroform and were also used to compare obtained results.

In order to do the tensile testing, from freshly prepared samples, 12 mm 30 mm rectangular strip was cut and utilized for tensile testing in an INSTRON 5842 (INSTRON Inc., Canton, MA) mechanical testing machine at 10 mm/min, as described previously [20]. In brief, tests were performed at room temperature (25°C) for the dry condition and at 37°C for the wet condition (in phosphate-buffered saline (PBS) solution [21]). Before performing the tensile test in wet condition, the samples were incubated in 100% ethanol for ten minutes before testing. The first and second elastic modulii were calculated from the slopes of the linear portion (0 - 3% tensile strain for the first elastic modulus and 40 – 90% tensile strain for the second elastic modulus) of the stress-strain curve.

To measure the thickness of the matrices, digital micrographs were obtained at various locations through an inverted microscope (Nikon TE2000U, Melville, NY) equipped with a CCD camera, as described previously. These images were quantified for the thickness using image analysis software Sigma Scan Pro (SPSS Science, Chicago, IL), calibrated using a micrograph of a hemocytometer at the same magnification. Four to five images were obtained per sample with at least ten points per image. The calculated minimum thicknesses were used for determining the stress values in each sample.

SIS also was utilized. SIS is a dense connective tissue obtained after removing the mucosa, serosa, and muscle layers from the small intestine. It is a natural matrix rich in Type-1 collagen and the physicochemical and mechanical characteristics of the matrix, such as permeability, thickness, mechanical properties, and ultrastructural properties have been well characterized [22]. Similar, ramp-&-hold experiments were conducted using commercially available SIS.

4. Results and Discussion:

The models developed from the composite models were compared with the experimental stress strain relationship with time. Experimental tensile test data gave the stress time relationship. From the data the strain data was calculated by knowing the strain rate. Strain increased with ramp up in stress and remained constant when the stress was released and the material was relaxed. The stretching and relaxing of the stress helped in generate a ramp up and hold strain data against time.

PCL and PLGA stress time data were obtained and from them strain data was calculated. The strain rate for PCL in AA was 1%. With the help of optimizer the fitting of the experimental data with the 6 different composite models was tested. It was found that composite Model 1 gave the lowest sum of squared deviation (SSD) of 0.395 MPa^2. Composite Model 1 is a combination of hyper-elastic spring with two spring and dashpot components. Table 1 shows the lowest SSD values out of 34 trials for different composite models along with the model parameters value. With 99% confidence that the best-of-N random starts will result in one of the best 13% of all possible values for the SSD gives the N value as [14]

$$N = \frac{\ln(1-0.99)}{\ln(1-0.13)} = 34 \tag{13}$$

Table 1: SSD and parameter values for various composite models for PCL

Model	SSD (MPa^2)	ØAx1 (MPa)	B1	ØAx2 (MPa)	B2	Tau2(s)	ØAx3 (MPa)	B3	Tau3(s)
1	0.39	-4.80	-0.36	-0.91	-58.9	2.06	-1.25	-63.1	90.2
2	3.79	3.88	0.29	-1.35	-562	5.35	-1.14	-6.52	61.0E3
3	1.62	-1.19	-63.0	169	0.20	1.71	4.42	0.32	303
4	0.49	-4.94	-0.35	-1.30	-79.6	78.6	-17.0E2	-0.02	1.39
5	0.62	-5.37	-0.43	-0.96	-73.2	5.23	-0.88	-69.3	90.8E5
6	0.81	-4.20	-0.64	41.3	0.50	1.45	-1.43	-98.6	52.6

Fig 4 shows the fitting of the composite Model 1 with the obtained experimental data having the generated strain vs. time data.

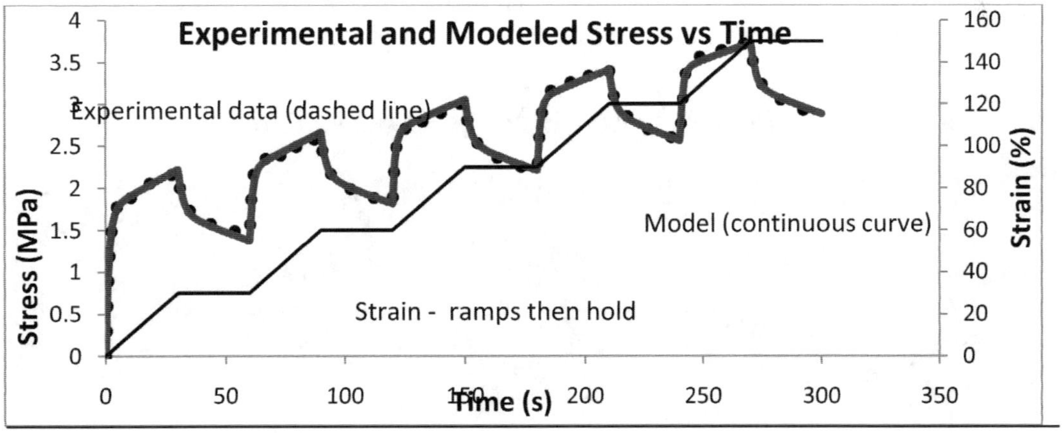

Figure 4: Experimental and Modeled stress vs. time for PCL (composite Model 1)

In Fig 1 the dotted line represents the experimental stress time data. The solid line represents the generated strain time relationship. The continuous curve on the dotted line represents the fitting of the model data with the experimental data. It can be seen that the model data provides a close fit to the experimental data.

Note that most modeling work only reveals a model fit to one strain period followed by one relax period. Partly, other work is limited by the analytical derivations within the model that presume a constant strain period initialized at the totally relaxed state, followed by the zero strain rate period, with the initial value for the hold period as the ending stress of the initial strain period. They do not develop subsequent equations for subsequent strain, hold, strain, hold, etc. periods. In this work, one model is used for all stages. The number of model parameters, 8, which is fully adequate to model 5 successive strain-and-hold stages is equivalent to the number needed for analytical models to model one strain and hold event.

Similarly, various composite models were tested for PLGA. It was found that composite model 4 did a better job compared to other models. Composite Model 4 is a combination of hyper-elastic spring with one retain component and a spring and dashpot component. Composite Model 4 gave a least SSD of 0.542 MPa^2. Table 2 lists the SSD values along with the various model parameter values obtained for various composite models.

Table 2: SSD and parameter values for various composite models for PLGA

Model	SSD (MPa^2)	∅Ax1 (MPa)	B1	∅Ax2 (MPa)	B2	Tau2(s)	∅Ax3 (MPa)	B3	Tau3(s)
1	1.77	-1.37	-0.99	0.01	4.49	74.7	-1.47	-6.14	2.77
2	5.69	1.29	0.32	2.28	0.36	240	-0.77	-164	4.15
3	1.12	-1.52	-0.81	0.18	1.04	62.6	19.5	0.34	1.58
4	0.54	-5.02	-0.14	-0.39	-26.5	120	2.26	0.79	3.07
5	2.01	0.19	0.86	-3.67	-2.11	2.49	-2.73	-0.45	481
6	0.93	-3.31	-0.27	1.35	0.92	3.96	-0.49	-160	67.9

Fig 5 shows the fitting of composite Model 4 data with the experimental data. A good approximation of the model data with the experimental data could be seen.

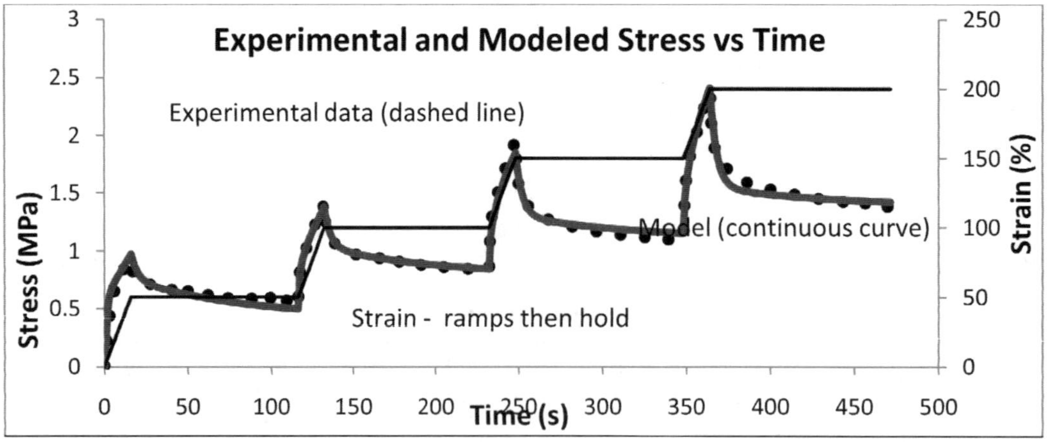

Figure 5: Experimental and Modeled stress vs. time for PLGA (composite Model 4)

Also, the same was modeling was done for SIS data. Composite Model 4, a combination of hyper-elastic spring with one retain component and a spring and dashpot component did the best job by giving a SSD value of 0.48 MPa^2. Table 3 shows the SSD and parameter values for various composite models for the SIS data.

Table 3: SSD and parameter values for various composite models for SIS

Model	SSD (MPa^2)	ØAx1 (MPa)	B1	ØAx2 (MPa)	B2	Tau2(s)	ØAx3 (MPa)	B3	Tau3(s)
1	1.77	0.22	3.91	4.43	2.58	0.71	-0.87	-3.61	251
2	2.68	0.4	3.18	3.94	0.46	144	-0.39	-160	1.55
3	0.67	0.22	4	0.28	4.69	1.18	-0.83	-4.19	309
4	0.48	0.098	4.96	-1.19	-3.13	379	0.38	5.05	0.55
5	1.69	0.14	4.51	324	0.04	0.67	-1.63	-1.8	394
6	0.65	0.27	3.68	0.19	4.57	2.08	-0.58	-6.29	364

Fig 6 shows the fitting of the composite 4 model data with the experimental data.

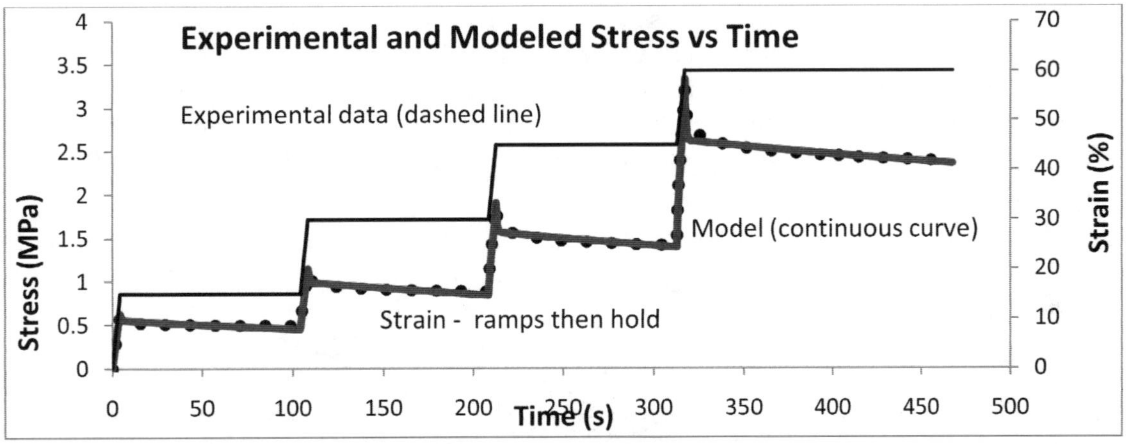

Figure 6: Experimental and Modeled stress vs. time for SIS (composite Model 4)

A useful model will be able to translate to other stress-strain situations. As a demonstration of the transferability of this model, Figure 6 shows the model response for a cyclic strain application. Strain is steadily increased until stress reaches an upper limit, decreased at the same rate until stress reduces to a lower limit, and subsequently repeated. Fig 6 shows the cycling data for the PCL model Architecture B, Model 1.

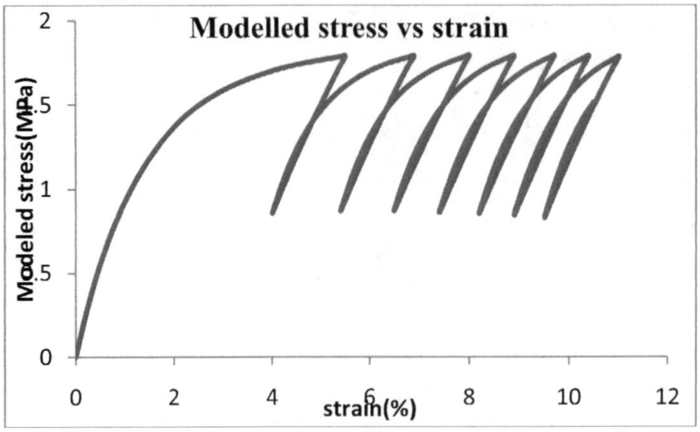

Figure 7: Cycling curve for PCL Model 1.

The upper limit for the modeled stress was kept as 1.8 MPa whereas the lower limit was at 0.8 MPa. Fig 6 shows that the strain progressively increases with each cycle, indicating that one of the component is reforming to the new conditions. However note that the strain increase between two consecutive cycles progressively decreases. Eventually, not shown, subsequent cycles overlap each other. This behavior is similar to that we have obtained experimentally, further revealing transferability of the models to other situations.

The same demonstration of transferability of the model that fit the experimental data was tested for PLGA and SIS data. Fig 8 and fig 9 show the cycling nature of the composite model 4 for PLGA and SIS.

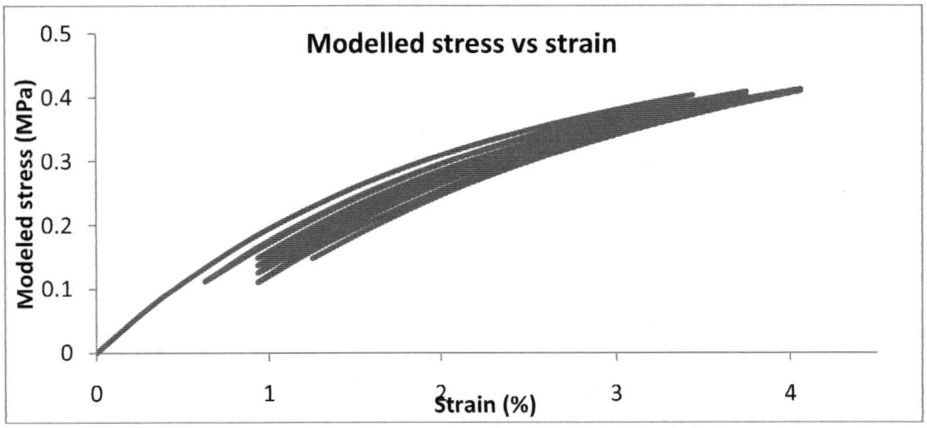

Figure 8: Cycling curve for PLGA Model 4.

The limits were 0.40MPa and 0.13MPa for the PLGA data. Transferability of composite model 4 for the SIS data could be better seen for the PLGA data with the cycles overlapping each other.

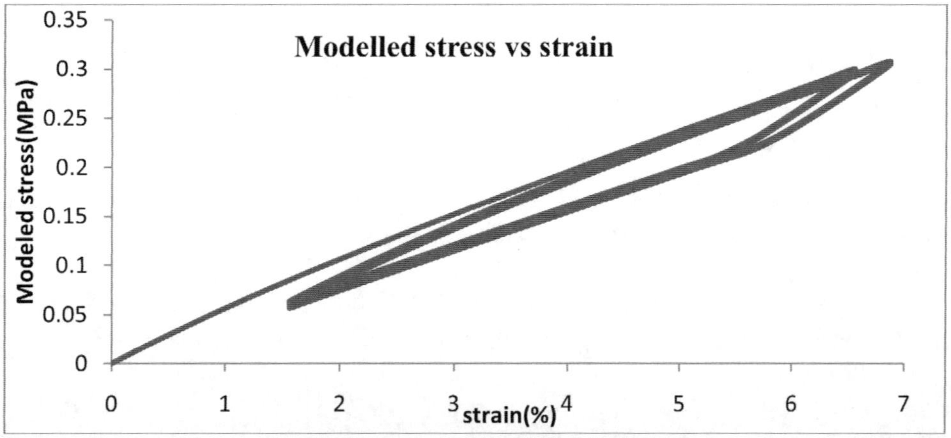

Figure 9: Cycling curve for SIS Model 4.

The upper and lower limits were respectively 0.30 MPa and 0.63 MPa for the SIS data. Here fig 9 shows that the cycles overlap with each other as required and expected showing the transferability of the composite model 4.

5. Conclusion:

It was demonstrated that pseudo-component models provide an excellent fit to the stress relaxation data over 4 or 5 sequential stages of strain and hold influence. Also demonstrated was the transferability of the model to predict material behavior under very different (cyclic) treatments.

References :

[1] Fung YC. Elasticity of soft tissues in simple elongation. Am J Physiol 1967;213(6):1532-1544.

[2] Nekouzadeh A, Pryse KM, Elson EL, Genin GM. A simplified approach to quasi-linear viscoelastic modeling. J Biomech 2007;40(14):3070-3078.

[3] Duling RR, Dupaix RB, Katsube N, Lannutti J. Mechanical characterization of electrospun polycaprolactone (PCL): a potential scaffold for tissue engineering. J Biomech Eng 2008;130(1):011006.

[4] Sarver JJ, Robinson PS, Elliott DM. Methods for quasi-linear viscoelastic modeling of soft tissue: application to incremental stress-relaxation experiments. J Biomech Eng 2003;125(5):754-758.

[5] Craiem D, Rojo FJ, Atienza JM, Armentano RL, Guinea GV. Fractional-order viscoelasticity applied to describe uniaxial stress relaxation of human arteries. Phys Med Biol 2008;53(17):4543-4554.

[6] Boyce BL, Jones RE, Nguyen TD, Grazier JM. Stress-controlled viscoelastic tensile response of bovine cornea. J Biomech 2007;40(11):2367-2376.

[7] Defrate LE, Li G. The prediction of stress-relaxation of ligaments and tendons using the quasi-linear viscoelastic model. Biomech Model Mechanobiol 2007;6(4):245-251.

[8] Wineman A. Nonlinear Viscoelastic Solids--A Review. Mathematics and Mechanics of Solids 2009;14(3):300-366.

[9] Zhong C. A Nonlinear Viscoelastic Model for Describing the Deformation Behavior of Braided Fiber Seals. Textile Research Journal 1995;65(8):461-470.

[10] Provenzano P, Lakes R, Keenan T, Vanderby R, Jr. Nonlinear ligament viscoelasticity. Ann Biomed Eng 2001;29(10):908-914.

[11] Marklund E, Eitzenberger J, Varna J. Nonlinear viscoelastic viscoplastic material model including stiffness degradation for hemp/lignin composites. Composites Science and Technology 2008;68(9):2156-2162.

[12] Hongbin L, Noonan DP, Zweiri YH, Althoefer KA, Seneviratne LD. The development of nonlinear viscoelastic model for the application of soft tissue identification.

[13] Rhinehart, R. R., M. Su, and U. M. Sridhar, "Leapfrogging: a novel optimization approach", Submitted to the European Journal of Optimizations Research, March 23, 2010.

[14] Iyer, M. S., and R. R. Rhinehart, "A Method to Determine the Required Number of Neural Network Training Repetitions," IEEE Transactions on Neural Networks, Vol. 10, No. 2, March, 1999, pp. 427-432.

[15] Padmanabhan, V., Rhinehart,R.R. (2005). "A novel termination criterion for optimization." American Control Conference ,vol.2284, pp. 2281-2286.

[16] Cao, S., and R. R. Rhinehart, "A Self-Tuning Filter," Journal of Process Control, Vol. 7, No. 2, pp. 139-148, 1997.

[17] Makornkaewkeyoon, K. (May, 2010). Polycaprolactone Matrices Generated in Aqueous Media: Natural Polymers Immobilization and Stress Relaxation Behaviour Chemical Engineering Department. Stillwater, Oklahoma State University. **Master of science:** 64.

[18] Mirani, R. (December 2009). Stress Relaxation Modelling of Composite Porous Scaffolds. Chemical Engineering Department. Stillwater, Oklahoma State University. **Master of Science:** 85.

[19] Lawrence BJ, Maase EL, Lin HK, Madihally SV. Multilayer composite scaffolds with mechanical properties similar to small intestinal submucosa. J Biomed Mater Res Part A 2009;88(3):634-643.

[20] Sarasam, A. and S.V. Madihally, Characterization of chitosan-polycaprolactone blends for tissue engineering applications. Biomaterials, 2005. **26**(27): p. 5500-8.

[21] Mirani RD, Pratt J, Iyer P, Madihally SV. The stress relaxation characteristics of composite matrices etched to produce nanoscale surface features. Biomaterials 2009;30(5):703-710.

[22] Kropp BP. Small-intestinal submucosa for bladder augmentation: a review of preclinical studies. World J Urol 1998;16(4):262-267.

(238d) Design of Biofuel Additives Using Molecular Design Techniques

Subin Hada[1], Charles C. Solvason[2] and Mario Richard Eden[2], (1)Department of Chemical Engineering, Auburn University, Auburn, AL, (2)Department of Chemical Engineering, Auburn University, Auburn University, AL

Biofuels blended with gasoline are one of the few alternatives that has not required significant new infrastructure or change on the part of consumers or auto manufacturers. As a result, biofuels are today the most widely deployed substitute for conventional fossil fuels in transportation. In the U.S., biofuels can currently be blended up to 10% (ethanol) and 20% (biodiesel) in every gallon of fuel.

The two most common renewable feedstock sources for making biofuels are plants rich in sugars and bio-derived oils. Ethanol can be produced from a number of feedstock; in addition to corn and sugarcane, other feedstocks include switchgrass, woody biomass, agricultural residue, wood residue, and municipal solid waste. However, different feedstocks produce different fuel characteristics. For example, the type of feedstock impacts the cold flow properties of the fuel, engine performances and efficiency, and fuel degradation (DARPA, 2010). It has been shown that the deviation in properties is a result of the variations in the residuals left in the blend from different sources of feedstock. Therefore, it is desired to molecularly design biofuel additives to account for the effect of residuals in the blend in order to achieve the performance properties of the petroleum based fuel. In this way, biofuels can be achieved that are adaptable to a range or blend of feedstock and the desirable fuel characteristics like higher energy density and wide operating temperature range. Moreover, such a sustainable biofuel must also meet the specifications required by the transportation and the aviation industry.

To meet this objective, fuel characterization data would be combined with property clustering techniques in a reverse problem formulation (Solvason, 2010). The characterization data consists of multitude of property values specified by aviation industry to ensure an adequate performance. To facilitate an efficient design we propose consolidating these various properties into a latent property domain using multivariate statistical techniques. Characterization based molecular design is then used to build additives that match the fuel specifications in the latent property space. Sustainability is controlled by environmental and ecological constraints on the design of the additives. Additives found can then be used to offset the impact the residuals on the property of the biofuel blend.

References:

Defense Advanced Research Projects Agency (DARPA), "Fact Sheet," March 2010.

C.C. Solvason, N.G. Chemmangattuvalappil, M.R. Eden (2010), "Multi-Scale Chemical Product Design using the Reverse Problem Formulation," 20th European Symposium on Computer Aided Process Engineering – ESCAPE20.

See more of this Session: Physical Properties for Chemical Product Design
See more of this Group/Topical: Process Development Division

Thermostatic Self-Heating Coffee Cup

Michael Bolmer
Tempra Technology
Bradenton, FL

Most of the papers presented this week are about large industrial processes or microscopic phenomena, but engineering is also applied to everyday household objects. There is evidently a large potential market for coffee, soup, and other beverage in self-heating cups. Previous attempts to produce such a product have resulted in cups that provide lukewarm coffee in cold weather and scalding hot coffee in hot weather. The problem with these cups is that they provide the same temperature rise no matter when or where they are used. This paper describes the development of a cup which will provide the desired final drinking temperature over a wide range of operating conditions.

In 2001 Nestle launched the Nescafé "Hot When You Want It" Coffee Can in the UK. It heated using the reaction of calcium oxide with water. When it was activated on a cold day the coffee temperature rarely rose above 38°C (100°F). When it was activated on a hot day the customer stood a possibility of getting scalded. It was withdrawn from the market due to lack of temperature control after selling over six million units. OnTech launched another self-heating coffee cup in the U.S. with another calcium oxide heater and coffee recipes from Wolfgang Puck. They were on the market for several years, but again withdrew because of too hot or too cold coffee. These two ventures show that there is a market for portable self-heating beverages, but the market demands that the beverage be delivered at a consistently safe and palatable drinkable temperature. Tempra Technology has developed methods to control the final drinking temperature of beverages in self-heating cups.

Market surveys show that customers prefer to drink coffee at temperatures between 58°C and 70°C. Therefore, on cold days we need enough energy to heat the beverage from 5°C to 58°C, or a maximum required temperature rise of 53°C. Second, we need to prevent all of this energy from overheating the beverage when it starts at a hot temperature. On a hot day (35°C) the required minimum temperature rise would be 70°C – 35°C = 35°C. Therefore, we need to be able to eliminate up to (53°C – 35°C) / 53°C, or 34% of our available energy.

We have developed a method to stop the reaction before all the reactants are consumed. Possible methods to stop the reaction include adding:

- a catalyst poison
- a complexing agent
- a crystallizing agent
- a defoaming agent
- a gelling agent
- a precipitating agent

Next, we will show how this temperature control works for two different reaction chemistries. The first heating reaction is between potassium permanganate and glycerol:

$KMnO_4 + C_3H_8O_3 \rightarrow$ heat

The suppressant works by complexing one of the reactants (glycerol), making it much less reactive. The complex is reversible, so that over time glycerol is slowly released to the reaction mixture, and is ultimately all consumed. This is desirable for two reasons. First, a small amount of reaction keeps the beverage hot over an extended time. Second, consuming all the reactants at the time of use means there are none left when the cup is disposed.

A simple model illustrates the effect of a complexing agent on a reaction:

Heating Reaction 1	$A + B \rightarrow E$	$dA/dt = -k_1AB$
Complex Equilibrium 2	$B + C \leftrightarrow D$	$K_2 = BC/D$

reactant A = $A_0(1 - x_1)$
reactant B = $A_0R_1 - A_0x_1 - A_0R_1x_2$
complexing agent C = $A_0R_1R_2 - A_0R_1x_2$
complex D = $A_0R_1x_2$
product E = A_0x_1
$R_1 = B_0/A_0$
$R_2 = C_0/B_0$
x_1 = extent of reaction 1
x_2 = extent of reaction 2

Heating Reaction 1	$-A_0\Delta x_1/\Delta t = -k_1A_0\,(1 - x_1)\,(A_0R_1 - A_0x_1 - A_0R_1x_2)$
Complex Equilibrium 2	$K_2 = (A_0R_1 - A_0x_1 - A_0R_1x_2)\,(A_0R_1R_2 - A_0R_1x_2)\,/\,(A_0R_1x_2)$

The equations were solved over small intervals using Microsoft Excel. The results are shown in Figure 1. The three curves show the effects of different levels of complexing agent (R_2). At any given conversion (x_1), the rate of reaction is lower as the amount of complexing agent is increased. If the complexing agent is added part way through the reaction, the reaction rate will switch from one curve to another with a slower reaction rate (Figure 2). We seek to add the amount of complexing agent that would reduce the reaction rate to just balance the rate of heat loss out of the cup, so that the beverage temperature would remain constant for an extended time.

The second heating reaction does not occur unless a catalyst is present. The suppressant works by precipitating the catalyst. Here, the precipitation is irreversible, and the reaction stops completely. Figure 3 shows two heating curves measured in a calorimeter, one with the stopping agent added half way through the reaction. The overshoot on this curve shows that the reaction continued near the thermocouple until the stopping agent was mixed to that point, then the liquid from where the reaction stopped sooner cooled the thermocouple to the bulk average temperature. Because the catalyst is at a low concentration compared to the reactants, much less stopping agent is required, which reduces the cost and size of the cup.

How much you can slow down a reaction with a reversible complexing agent is limited by the amount of complexing agent you can add, and therefore this is a limit to the controllable operating temperature range. With an irreversible stopping agent the potential controllable operating temperature range is larger.

The chemical reaction stopper is stored in a separate compartment from the reaction chamber. The reaction stopper is added to the reaction mixture when the beverage reaches the desired temperature. The chamber is in thermal contact with the beverage, but thermally insulated from the reaction mixture, which is at 100°C as soon as the reaction gets going, and is therefore no help in determining when to release the reaction stopper. The chamber is sealed by a material that releases at the desired drinking temperature, and then the reaction stopper is delivered to the reaction when the beverage is at the desired drinking temperature, no matter how long it took the beverage to reach that temperature. On a cold day the reaction stopper is never released, and the entire heating potential is used.

Finally, Figure 4 shows actual measured heating curves for coffee cups operating at three different temperatures. The black lines show what the maximum and minimum heating curves that produced coffee at the desired drinking temperature would be with no temperature control. The temperature control extended the useable operating temperatures from between 15°C and 23°C to between 7°C and 35°C. This takes the self-heating cup from being only useable indoors to being useable outdoors.

Relevant Published Patent Applications

US2007/0261692	Thermostatic Temperature Control for Self-Heating Containers
US2008/017184	Self Heating or Cooling Container
US2008/245358	Slow Cooking Heating Formula
EP2179229	Chemical Heating Compositions and Methods
WO2007/147108	Portable Heating Device Ventilation
WO2008/058062	Method for Adding a Fusible Material to a Container Wall

Figure 1

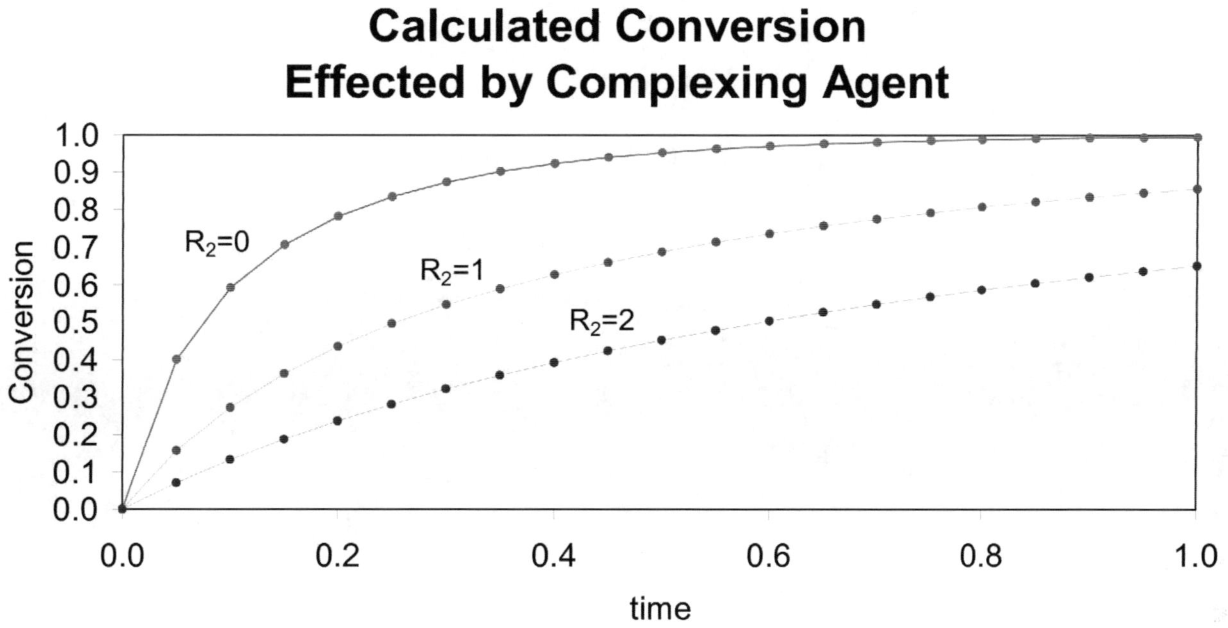

Figure 2

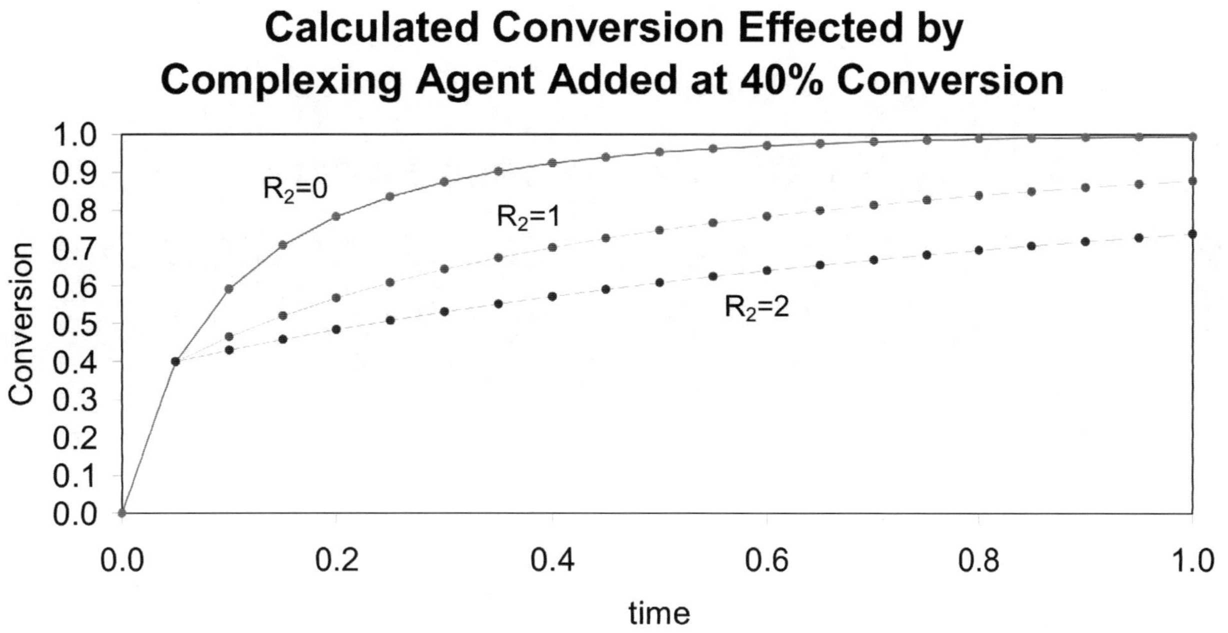

Figure 3

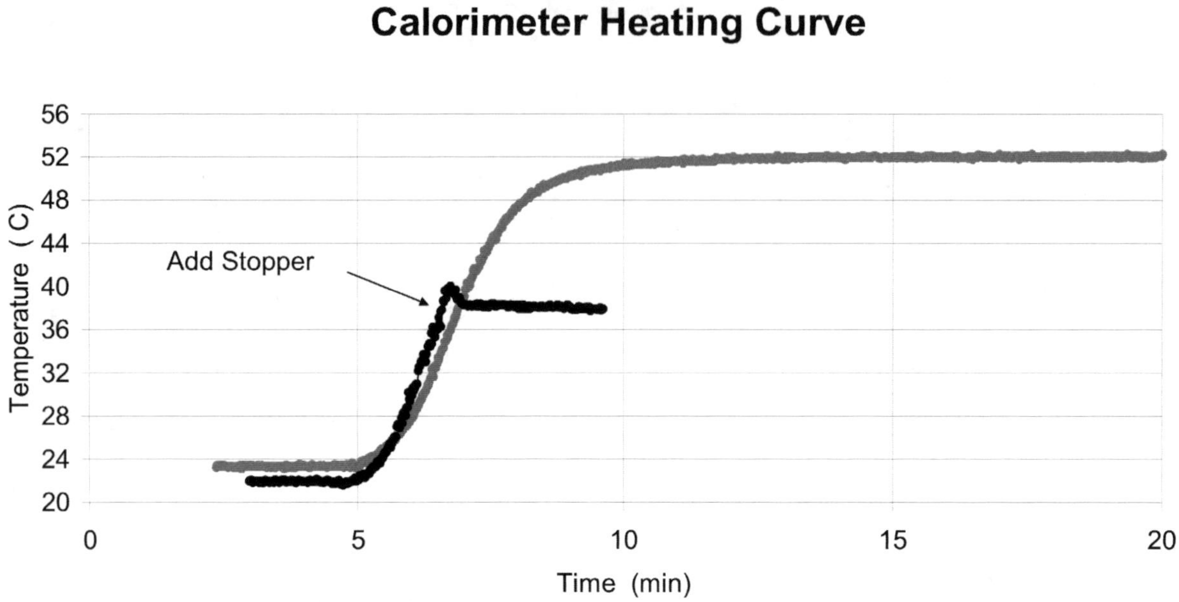

Figure 4

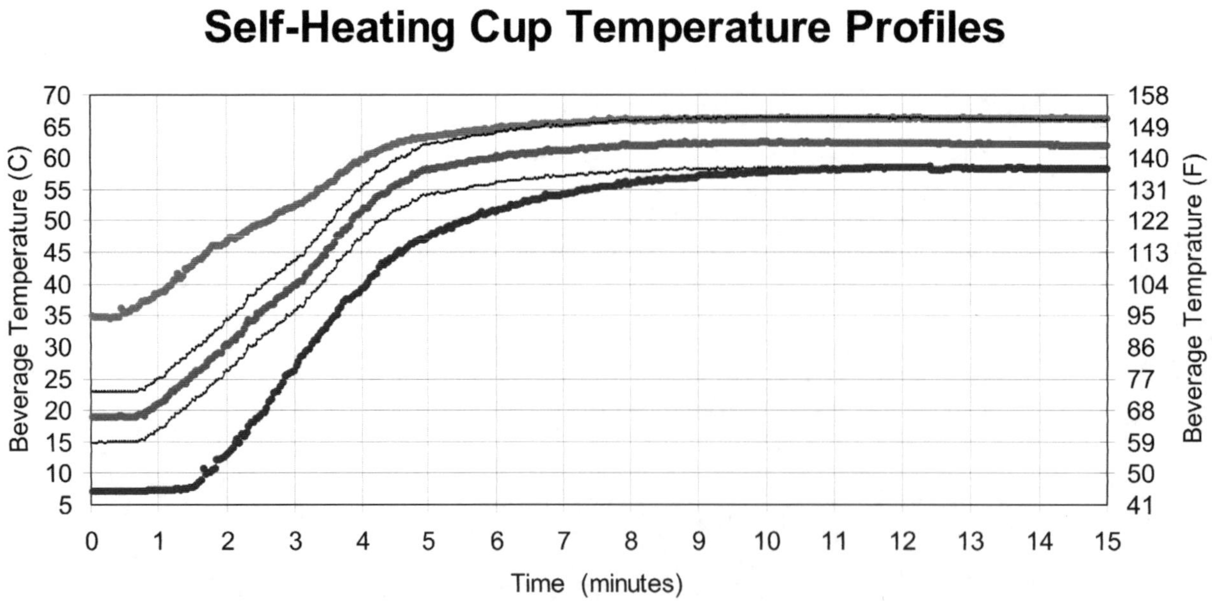

(240a) Agro-, Bio-, Chemical- and Energy Conversion - Process Intensifications and Their Applications in Intensified Integrated Plants and Engineered Ecosystems

G. Akay, School of Chemical Engineering and Advanced Materials, Newcastle University, Newcastle upon Tyne, United Kingdom

Process Intensification (PI) initially emerged as a novel process design philosophy in which the processing volume was deliberately reduced while the processing fields were enhanced in order to achieve the same levels of production rate. The resulting processes have lower capital and operating costs and in some cases novel products are obtained. PI is principally driven by the enhancement and control of momentum, heat and mass transfer and catalytic processes which in turn imply dependence on advances in materials science and fabrication technology for reactors. Further enhancement in PI is achieved through the synergistic superimposition of several processing fields within a reactor, integration of intensified unit processes within a production platform (plant), facilitating multi-component interactions especially in biological systems and sensing and controlling of the processes. This incorporation of these process conditions within a reactor require a new approach to reactor and process design philosophy which, nevertheless, also provides novel processes with a smaller number of steps, higher energy and conversion efficiencies and high selectivity as well as novel products.

We can assume that such intensified integrated plants which have been well analysed, already exist in the form of the human body. The mimicking of the human body in vitro (tissue engineering and biomedical engineering) does also help to understand the biological processes of the human body while providing a model for intensified reactors through which generic BioProcesses Intensification has been recently achieved, with intensification levels of ca. 10-100 fold compared with present technology. Furthermore, in chemical processes, the inherent restrictions of bioprocesses do not apply and hence the human body model of chemical plants will have higher degrees of freedom, only restricted by the availability of support technologies such as materials, sensors and control systems.

The human body can also be used for Energy (conversion)-PI deviating significantly from the well established conversion processes. Here the principle target is the establishment of Intensified Integrated Bio-refineries (IIBR) based on the conversion of biomass to power and chemicals where the process is integrated with agriculture. Such generators need to be distributed plants due to the distributed nature of the biomass feedstock. Intensified unit processes in IIBR include combustion, separation, syngas-to-power/fuel conversion and gas and process water clean-up.

Although we propose that the human body and vital organs can be a model for intensified integrated

Biochemical or Chemical plants and unit processes, nature's processing strategy is not always intensified. This is particularly true and equally very important in agriculture. Hence the application of the PI- principles, now commonly accepted in Chemical-PI, can enhance output from ecosystems in terms of plant/crop growth and maintenance. This is particularly important in cultivation in marginal land under water and nutrient stress. It is also possible to intensify plant nitrogen fixation from air thus reducing fertilizer demand. Such methods can be deemed to be Agro-Process Intensification which can open new fields for main stream chemical engineering leading to engineered ecosystems. Observed current level of intensification is ca. 3 fold, far in excess of that achieved by genetic modification.

In this review, after a summary of nature's processing strategy in the human body and plant/crop growth, recent examples of Agro- Bio- Chemical and Energy-Process Intensifications will be given. Advances in materials science (especially in nano-structured macro-porous materials and catalysts) relevant to PI will be summarised with examples. Performance and cost data of a recent integrated intensified energy plant will be provided to illustrate the benefits of intensified-integrated plants.

See more of this Session: Process Intensification Plenary Session
See more of this Group/Topical: Process Development Division

(240b) Energy Challenges and the Role of Process Intensification

Laurence R. Weatherley, Chemical & Petroleum Engineering, University of Kansas, Lawrence, KS

The presentation will comprise an overview of some of the key issues affecting process engineering research and development in relation to energy. The impact of the current forecasts of energy supply and demand, the desire for energy independence, and the drive towards the use of renewable resources will be discussed. Some of the challenges offered by the use of renewable resources for transportation fuels production will be summarized, showing the possibilities offered by a systems approach to the production of alternative fuels. Some developments in process engineering and the development of process intensification techniques which impact on energy production and energy efficient processing will be briefly discussed. The presentation will conclude with a case study of process intensification which illustrates the potential for new hardware technologies in achieving access to other sources of energy and in improving energy efficiency. A range of technologies being developed will be reviewed briefly covering some of the novel reactor types and relevant coupled reaction/separation processes. The presentation will conclude with a case study comparing process intensification technologies for continuous biodiesel production based on the criteria of residence time, energy efficiency, operating and capital cost, temperature control and the current status of development.

See more of this Session: Process Intensification Plenary Session
See more of this Group/Topical: Process Development Division

(240c) Chemical Looping Processes: An Example for Process Intensification

L.-S. Fan, Chemical and Biomolecular Engineering, The Ohio State University, Columbus, OH

The concept of chemical looping reactions has been widely applied in chemical industries, e.g., the production of hydrogen peroxide (H_2O_2) from hydrogen and oxygen using 9,10-anthraquinone as the looping intermediate. Fundamental research on chemical looping reactions has also been applied to energy systems, e.g., the splitting of water (H_2O) to produce oxygen and hydrogen using ZnO as the looping intermediate. Fossil fuel chemical looping applications had been used commercially with the steam-iron process for coal from the 1900s to the 1940s and had been demonstrated at a pilot scale with the carbon dioxide acceptor process in the 1960s and 1970s. There are presently no chemical looping processes using fossil fuels in commercial operation. A key factor that hampered the continued use of these earlier processes for fossil energy operation was the inadequacy of the reactivity and recycleability of the looping particles. This factor led to higher product costs for using the chemical looping processes, compared to the other processes that were petroleum or natural gas based. With CO_2 emission control now being considered as a requirement, interest in chemical looping technology has resurfaced. In particular, chemical looping processes are appealing due to their unique ability to generate a sequestration-ready CO_2 stream while yielding high energy conversion efficiency. Renewed fundamental and applied research since the early 1980s has emphasized improvement over the earlier shortcomings. New techniques have been developed for direct processing of coal or other solid carbonaceous feedstock in chemical looping reactors. Significant progress is underway in particle design, reactor development, and looping system integration, as demonstrated by the operation of several pilot or sub-pilot scale units worldwide, making it possible that chemical looping technology may be commercially viable in the future for processing carbonaceous fuels.

This presentation will describe the fundamental and applied aspects of modern chemical looping technology that utilizes fossil and biomass as feedstock and involve both the metal – metal oxide and metal oxide – metal carbonate as reactive looping systems. The presentation will focus on looping reactor and system engineering and its feature of process intensification schemes in the context of the conventional processes. Comparisons of the process efficiencies for the intensified system versus the conventional system will be presented and illustrated.

See more of this Session: Process Intensification Plenary Session
See more of this Group/Topical: Process Development Division

Process Intensification – An Industrial Perspective

Ravi Dixit
John Pendergast

Abstract

There are many challenges facing the chemical industry such as high energy costs, rapidly escalating prices due to materials cost, developing alternative feedstocks strategies, and more. All of these problems challenge the sustainability of the chemical process industries, and particularly challenge the U.S chemical industry. The implementation of process intensification is seen as a powerful and enabling tool for developing process options that are more sustainable, having both reduced energy and capital requirements. This talk will highlight the motivations and some success stories regarding the development of process intensification options. The barriers to implementation of process intensification from an industrial perspective will also be presented. This talk will focus on how various process intensification methodologies are applied at Dow to drive product and process research. Examples will be given covering areas such as multifunctional reactors and enhanced separations.

There are many challenges faced by the chemical industry particularly by the U.S. chemical industry. The first and most obvious challenge is the globalization of the industry. Although globalization has been going on for some time, its pace is accelerating. Individual markets in North America, Europe, Asia, and South America have merged into one global market. The implications are enormous. On the one hand it means, better trade, better utilization of labor and capital, and perhaps most important, a better living standards for millions of people worldwide. However, it also means tougher competition as companies like Dow are competing globally. It means more demanding customers that have more choices and access to suppliers world wide.

The second challenge faced by our industry is how best to manage the cyclical nature of the chemical industry. This is mainly an issue for commodities and basic products. The third challenge is commoditization. It is a basic fact of our business that every specialty product or technology eventually becomes a commodity. As one of Dow's Corporate Fellows phrased the problem, "What was best in the past, is only good or average today, and it may even be irrelevant tomorrow".

The biggest challenge we face as an industry in the U.S. is rising energy and feedstock costs. Energy is important for Dow for two reasons. First, we use electricity and steam to power our chemical transformations. Secondly, we need a source of carbon as raw materials. What most of the world calls fuel, we call feedstock. Figure 1 shows the spectrum of products Dow produces.

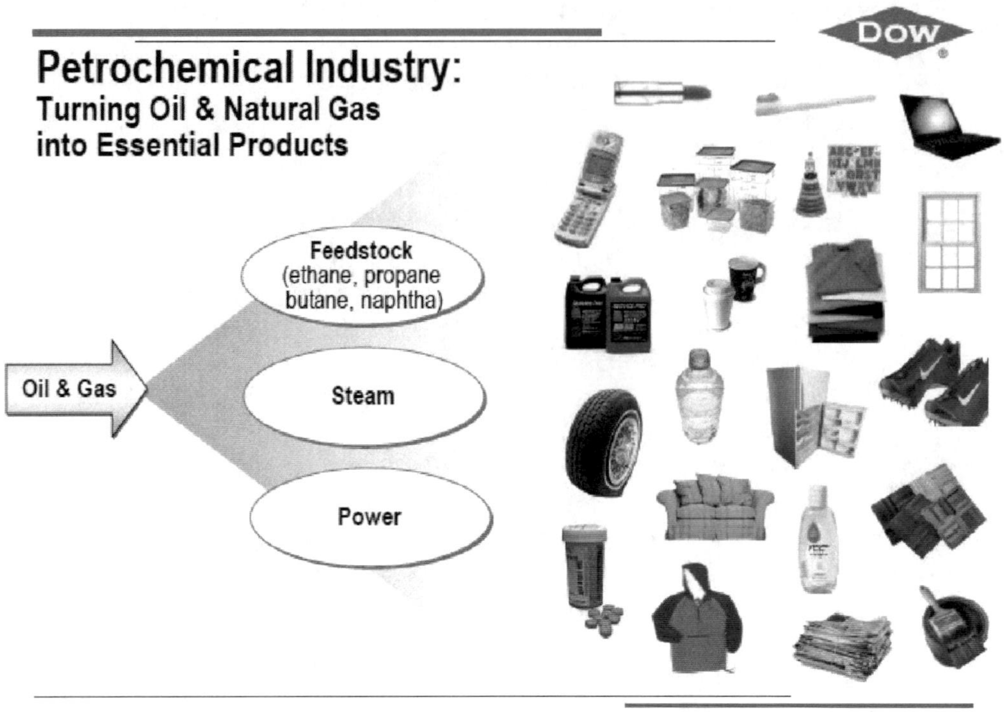

Figure 1. Dow's Business – Turning Oil & Natural Gas into Essential Products.

Dow is the world's largest (non-utility) producer and consumer of power and steam. We require 3700 MW of electricity to operate our plants. This is equivalent to the combined energy usage of San Francisco, San Diego and Oakland. This is largely produced internally by combined-cycle co-generation of electricity from natural gas and utilization of the steam to power our process plants at large integrated sites.

Dow feedstock demand is equivalent to 800,000 barrels per day of crude oil equivalents. In 2002 Dow's combined fuel and feedstock bill was $8 billion. This represented 29% of Dow total costs that were related to energy. In 2006, this cost increased to $22 billion, or nearly half of our $50 billion revenue. For every 10% variation in that $22 billion, the impact on Dow's earnings is about 45 cents per share.

Dow uses 600 trillion BTUs of energy annually. This is equivalent to 2/3 of the residential energy use in the state of California. In 1995 we established a set of aggressive goals on energy, the environment, health and safety. The energy goal was to reduce our energy intensity by 20% in 10 years. The results of this effort are shown in Figure 2.We are committed to improving energy efficiency of our plants because not only it is the right thing to do but also because it makes financial sense.

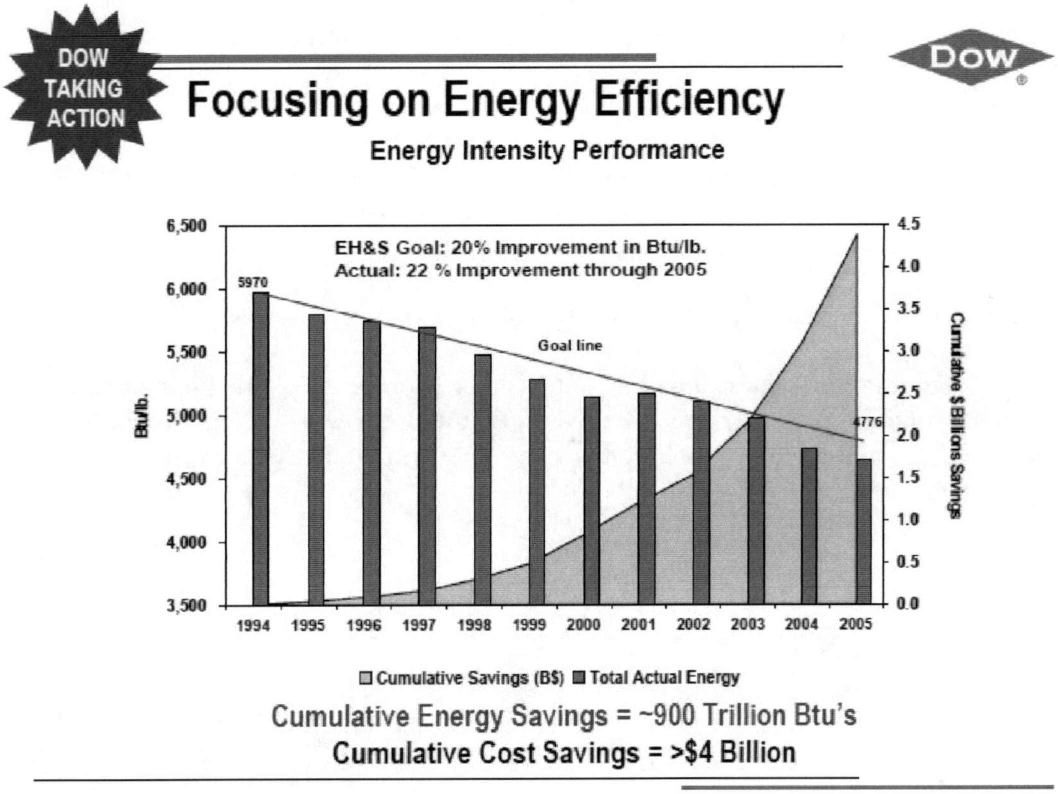

Figure 2. Reducing Energy Intensity of our Processes

From 1995 to 2005 Dow reduced its energy use per pound product by more than 20%. This resulted in cumulative energy savings of 900 trillion BTUs, enough to power all the homes in the entire state of California for a year. We have far exceeded our energy goal we set in 1995. We now have set even more challenging goal of further reducing our energy intensity by 25% by 2015.

We are addressing the energy and feedstock challenge in many ways.

1. We have made an effort to integrate upstream. We have formed JV's with oil producers in and the middle –East and with ethanol producers in South America. This gives us access to low cost hydrocarbon feedstocks.
2. Achieving economy of scale. As every engineer knows, scaling up reduces the average cost per unit of product. Our large integrated site models continue to set the standard for efficiency and low cost production
3. Exploring and evaluating alternative feedstocks such as coal, biomass and renewable green feedstocks and alternative energy sources such as wind and solar.
4. Biomass and renewables. Dow's scientists are pushing the boundaries of technology to make bio based processes more competitive. Dow's sugar cane to polyethylene plant in Brazil is a good example.

Another key and significant component to address the energy challenge and is our strategy of increasing innovation, both in our products and our processes. However, the most important component of innovation is people. Having the right people with the right competencies working on the right projects can change the technology landscape.

Dow's efforts in process intensification are aligned with this innovation initiative. We define Process Intensification as technology that substantially and dramatically reduces the number, size and complexity of the unit operations. Our emphasis is on making major step changes in technology, not incremental or small changes. The focus is on reducing both the capital and energy intensity of our existing and new processes.

One example of Process Intensification that has undergone successful commercialized is our Epichlorohydrin process. This is shown in Figure 3. Dow developed Rotating Packed Bed technology to produce HOCl, a key intermediate in the production of epichlorohydrin.

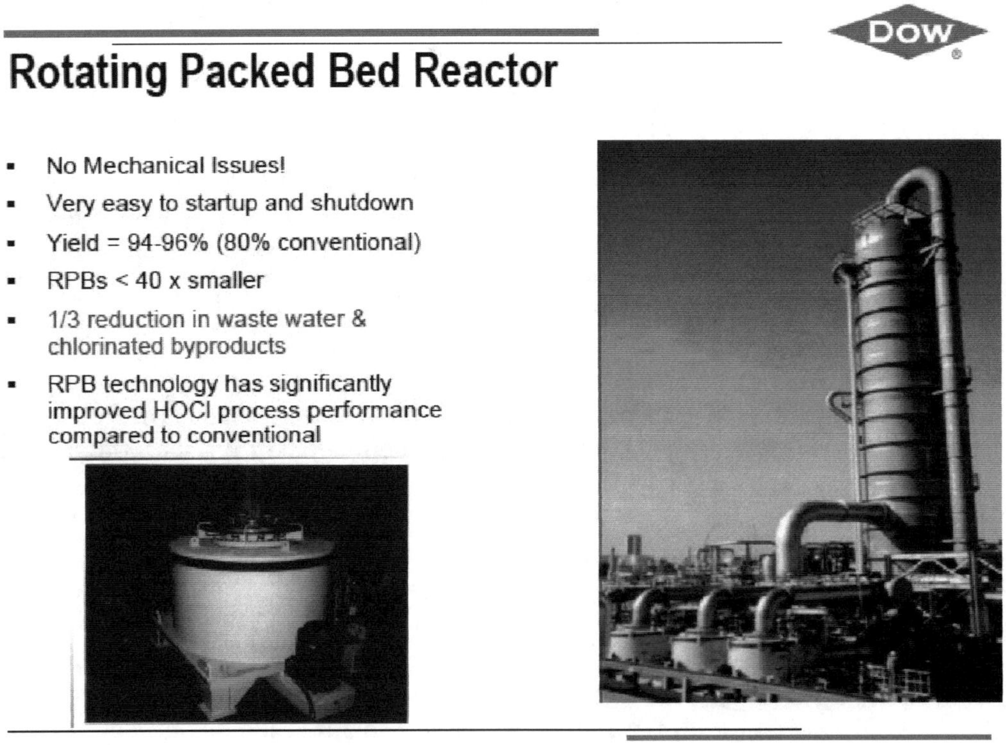

Figure 3. Rotating Packed Bed Reactor for HOCl.

We have applied the concepts of process intensification and reactive separations to chemistries that can be represented by a series-parallel reaction network. There are many process chemistries that follow this broad scheme; production of EO/EG, PO/PG, ethanol amines, glycol ethers, and chlorinated products. Processes to produce these products are greatly impacted by application of reactive distillation and diving wall column to significantly reduce capital and energy intensity. This talk will highlight these process intensification applications by a few examples. We have started applying process intensification principles to site-wide and even to enterprise-wide optimization. This will be illustrated with an example.

Process Intensification and Process Synthesis are not ends in themselves. These approaches and methods are sound business practices because the methods provide ways to reduce energy consumption and reduce capital. However, process intensification implementation in a large company like Dow is not without hurdles. Barriers to implementation will be addressed.

The talk will conclude with remarks on how process intensification efforts are integrated and aligned with those of Sustainability. Process Intensification and Process Synthesis are good for business and good for the environment at the same time.

(241a) Quantitative Structure-Property Relationship Models – A Comparison of Different Methodologies

K. M. Yerramsetty, **B. J. Neely** and K. A. M. Gasem, School of Chemical Engineering, Oklahoma State University, Stillwater, OK

Increasingly, investigators are using quantitative structure-property relationship (QSPR) models to describe various thermophysical properties; as such, improved implementation strategies and guidelines of this modeling technique are needed.

A designed experiment is used to evaluate both linear and nonlinear QSPR methodologies and linear, non-linear, and hybrid descriptor feature selection strategies for two different case studies involving melting point and acentric factor data. In all case studies, the same initial set of descriptors generated from different software packages are reduced to a final modeling set. The descriptors in these final sets are compared and contrasted, and differences in the non-linear QSPR model predictions using these sets are calculated. Efficacy of several feature selection methods such as differential evolution, genetic programming, mutual information and support vector machines are compared. This information provides useful insight for both novel modeling efforts and improved predictive capabilities.

See more of this Session: Product and Molecular Design
See more of this Group/Topical: Computing and Systems Technology Division

(241b) Characterization Based Chemical Product Design for the Nanoscale

Charles C. Solvason, Nishanth Chemmangattuvalappil and Mario Richard Eden, Department of Chemical Engineering, Auburn University, Auburn University, AL

The National Research Council (NRC) has recently recognized the importance of developing integrated formulated product design tools. In response, it created the Committee on Integrated Computational Materials Engineering (CICME) which developed and published a road map for what it termed a "Grand Challenge" [1]. In the report, the committee noted that in order to alleviate strain put on U.S. manufacturers from the swiftly changing and increasingly global marketplace, integrated design closely coupling computational models with manufacturing processes would be required. The term 'integrated' recognizes that the properties of products are controlled by a multitude of separate and often competing mechanisms that operate over a wide range of length and time scales. It is the linkage of the scales that remains the 'Grand Challenge' [1]. In response to this call from CICME, the research community has shifted focus from developing only the physical form, function, and aesthetics of assembled products to the design of chemically formulated products [2]. Chemically formulated products are products designed at the molecular level to deliver a specific desired attribute that may exist at multiples scales such as the quantum-, nano-, meso-, macro-, and mega-scales. Examples of formulated products include pharmaceuticals [3], proteins [4], organic semi-conductors [5], nano-structured materials [6] and many more product types.

Using today's technology, designing these optimized molecules for a specific end-use has never more attainable. Recent achievements owe much of their successes to increased computational speeds and the advent of novel algorithms that improved the transfer of information between the formulated scales [6]. One type of promising new algorithms utilize the reverse problem formulation which has been shown to significantly reduce the complexity and required computational time of product design problems. For example, Eden et. al. [7] demonstrated the ability of the reverse problem formulation to decouple product design problems from constitutive equations so that molecules can be designed prior to verification at the various scales, effectively reducing the computational complexity. Furthermore, Eljack et al. [8] extended the reverse problem formulation to include information from processes operating at the macro-scale. Additionally, Satyanarayana et al. [9] recently demonstrated the use of the reverse problem formulation combined with grid technology for polymer design with long range order approaching the meso-scale. While each of these applications of the reverse problem formulation has incorporated elements of multi-scale product design, the integration of the reverse problem formulation into a multi-scale framework that includes the nano-scale has yet to be achieved.

The main objective for this research is to extend the reverse problem formulation algorithm to include aspects of each of the length scales, including the nano-scale, thereby creating a framework where product synthesis, design, and optimization can be achieved with significantly reduced computational time. The research plan to achieve the objective entails the following: (1) the development of a centralized framework, (2) the development of an algorithm that chooses the appropriate properties to bridge each of the length scales, (3) the successful implementation of a property clustering algorithm to load the property descriptions into the framework. Independent validation steps will also be built in to the method and verified using published experimental results because of the "profound importance of experimental results to calibrate and validate computational methods and fill gaps in the theoretical understanding" [1]. Due to the size of this project, proof-of-concept case studies will initially focus on experimentally derived parameters derived from characterization techniques. Information from the molecular scale on short range order, such as group structure, conformation, and stereoregularity, will be combined with information from the nano-scale on long range order, such as the size, shape, and aspect ratio of the particle and the molecular weight, structure, and polarity of the polymer matrix. In particular, functional groups will be designed for attachment to single walled carbon nano-tubes in order to improve dispersion, which has been shown to significantly improve the mechanical, electrical, and optical properties of nano-composite products.

References

[1] Committee on Integrated Computational Materials Engineering, National Research Council (2008). Integrated Computational Materials Engineering: A Transformational Discipline for Improved Competitiveness and National Security. The National Academies Press, USA.

[2] Hill, M. (2004) Product and Process Design for Structured Products. AIChE Journal, 50(8), pp. 1656-1661. [3] Solvason C.C., Chemmangattuvalappil N.G., Eden M.R. (2009) Decomposition Techniques for Multi-Scale Structured Product Design: Molecular Synthesis. Computer Aided Chemical Engineering (In Press).

[4] Floudas C.A., Fung H.K., McAllister S.R., Monningmann, M., Rajgaria R. (2006) Advances in protein structure prediction and de novo protein design: A review. Chemical Engineering Science, 61(3), pp.966-988.

[5] Farhang A.R and Deeter T. (1999) Theoretical characterization of phenylene-based oligomers, polymers, and dendrimers. Synthetic Metals, 100(1), pp. 141-162.

[6] Fermeglia M. and Pricl S. (2009) Multiscale molecular modeling in nanostructured material design and process system engineering. Computers & Chemical Engineering (In Press).

[7] Eden M.R., Jergensen S.B., Gani R., and El-Halwagi M.M. (2003) Reverse problem formulation based techniques for process and product synthesis and design. Computer Aided Chemical Engineering, 15(1), pp. 451-456. [8] Eljack F.T., Eden M.R., Vasiliki K., Kazantzi Q., and El-Hawagi M.M. (2007) Simultaneous process and molecular design - A property based approach. AIChE Journal, 53(5), pp. 1232-1239.

[9] Satyanarayana K.C., Abildskov J., and Gani, R. (2009) Computer-aided polymer design using group contribution plus property models. Computers & Chemical Engineering, 33, pp. 1004-1013.

See more of this Session: Product and Molecular Design
See more of this Group/Topical: Computing and Systems Technology Division

(241c) Managing the Complexity in Liquid Formulation Design

Elisa Conte and Rafiqul Gani, CAPEC, Chemical and Biochemical Engineering Dept., Technical University of Denmark, Kgs. Lyngby, Denmark

The present day chemical industry is evolving beyond commodities towards specialty chemicals and consumer oriented chemical based products. Formulated products such as shampoos, paints, sunscreens, hair sprays enter in this last category of consumer oriented products. They are constituted by Active Ingredients (AIs), solvents and additives and in order to be commercially successful they need to meet many and often conflicting requirements of the market. This shift in the chemical industry is not simply a change from commodities to specialties [1], but a substantial innovation from material valued for their purity, to materials sold for their performance behaviour [2]. This evolution in the market has to be followed by an evolution in the product design discipline. Involving very different chemicals, formulations are complex products which make the modelling effort challenging. The 'scientific part' of chemical engineering consists of breaking down such complex systems into subsystems and to understanding the fundamental concepts associated to the behaviour of chemicals in solution. The 'engineering part' of chemical engineering consists in applying the gained knowledge in the design of a formulated product which matches the desired characteristics, and a process which is capable of producing it [3], although in liquid formulations, the processing part may not be as demanding as the structured chemical products. The development of systematic methodologies, tools and strategies is crucial in order to organize, systematize and improve the design and evaluation of complex systems for the production of chemical products [3]. A common practice is still the experiment-based trial-and-error approach. Model-based computer-aided tools which perform a fast screening of numerous candidates and alternatives could drastically reduce time and resources by providing "virtual and focused" experimentation possibilities. The virtual Product and Process Design laboratory (virtual PPD-lab) [4-5] provides a hybrid approach to product and process design, where focused virtual experimentations coupled with systematic generation and screening of chemical product alternatives can be performed. Through the virtual PPD-lab the product (formulation) developer can ask for the generation of a short list of the most promising feasible candidates (generation of alternatives) and also test some of his/her ideas (candidate formulations) through validated model-based computer-aided tools. The experimental resources, in this way, are reserved for the final selection and verification step, after a set of promising candidates have been identified, thereby achieving faster to market at reduced cost for any potential chemicals based product (formulation). Tailor made work-flows for design of specific products have already been developed and incorporated in the virtual PPD-lab software. A work-flow for the design of liquid formulated products has been developed and tested for the systematic generation and screening of alternatives in the design of paints [4], insect repellents and sunscreens [5]. The aim of this paper is to illustrate the power of the virtual PPD-lab in performing virtual experiments in relation to an industrial case study on hair spray. Given the AIs, a list of feasible solvents and the additives, a suitable solvent mixture matching the a priori defined targets needs to be identified. However, before the design problem can be solved appropriate property-mixing models need to be generated. This is because the AIs involved are polymers and appropriate property models helping to understand their phase behavior with solvents first needs to be developed and implemented into the virtual PPD-lab software. Once the models have been added to the model library of the virtual PPD-lab, the established work-flow (algorithm) for liquid formulation design is applied to determine the optimal product. The case study is an excellent example to illustrate how the virtual PPD-lab can help to manage the complexities associated with the design of industrial products as well as to reducing the experimental costs. The complexities involved consist of molecular design (candidates for solvents and additives), mixture design (liquid formulations) and testing of complex multicomponent phase behavior.

[1] Hill M. AIChE J. 2004, 8(50), 1656-1661. [2] Villadsen J. Chem. Eng. Sc. 1997, 52(17), 2857-2864. [3] Wintermantel K. Trans IChemE-Part A. 1999, 77, 175-188. [4] Conte, E., Morales-Rodriguez, R., Gani, R. Proceedings of 19th European Symposium on Computer Aided Process Engineering. 2009, 26, 249-254. [5] Conte, E., Morales-Rodriguez, R., Gani, R. Proceedings of 10th International Symposium on Process System Engineering. 2009, 249-254.

See more of this Session: Product and Molecular Design
See more of this Group/Topical: Computing and Systems Technology Division

(241d) Use of Aggregation Predictions in the Design of Protein Formulations

Brock C. Roughton[1], Anthony I. Pokphanh[1], T. Steele Reynolds[1], J. Laurence[2], E. M. Topp[3] and K. V. Camarda[1], (1)Department of Chemical and Petroleum Engineering, University of Kansas, Lawrence, KS, (2)Department of Pharmaceutical Chemistry, University of Kansas, Lawrence, KS, (3)Industrial and Physical Pharmacy, Purdue University, IN

A major concern in the development of therapeutic protein formulations is protein aggregation, and excipients are often used to increase the stability of such formulations. An ideal excipient binds with aggregation prone regions on the protein to limit interaction of that region with other protein molecules. The goal of this project is to predict aggregation-prone regions and use this information within a computational molecular design formulation to develop excipients which interact favorably with these regions. To build a predictive model based both on the protein's structural characteristics and on the molecular structure of an excipient, aggregation was evaluated for a set of model proteins and excipients using computational estimation techniques, and these results were compared to experimental values in the literature. Two algorithms were evaluated for the prediction of protein aggregation: Aggrescan (http://bioinf.uab.es/aggrescan/) and SAP (Spatial Aggregation Potential). Aggrescan assigns each amino acid an aggregation propensity score. An aggregation prone region, or hot spot, is identified by a sequence of amino acids with high propensities. SAP uses molecular simulation to determine hot spot regions that are hydrophobic and solvent accessible. In this algorithm, each residue is scored and the results are mapped to the three-dimensional protein structure. Both SAP and Aggrescan predicted similar hot spot regions for each protein, with SAP being slightly more accurate when compared to experimental data. However, the computational effort required to use the SAP algorithm is not compatible with a computational screening tool. A simple correlation was also developed linking the percent hydrophobic surface area of each protein to experimental aggregation values. These results demonstrate that aggregation potential can be estimated effectively using structural parameters of a given protein which are easily accessed. The effect of excipient structure also needs to be quantified in order to design excipients to stabilize specific protein systems. A docking simulation algorithm was employed using model proteins together with the excipients trehalose, poly(vinylpyrrolidone) and guanadine hydrochloride. The SAP algorithm was used to identify hot spot regions, and the docking algorithm was used to measure binding energies for each excipient-hot spot pair. Smaller binding energies suggest more favorable protein-excipient interactions. Correlations were then developed relating topological indices to the binding energy for each protein-excipient pair. The correlations are used within a molecular design framework to determine novel excipient structures that exhibit minimized binding energies. Results are analyzed using hydrogen-deuterium exchange and FTIR analysis to verify the extent of protection by the excipients.

See more of this Session: Product and Molecular Design
See more of this Group/Topical: Computing and Systems Technology Division

(241e) Design of Base Fluids for High Pressure/ High Temperature Drilling

Apurva Samudra and Nick Sahinidis, Chemical Engineering, Carnegie Mellon University, Pittsburgh, PA

One of the important factors that influence drilling costs in oil and gas production, is the drilling fluid used. Drilling fluids, also known as drilling muds, are suspensions with many soluble and insoluble additives in a base fluid like water or diesel. These fluids perform many functions and can be effective in optimizing the drilling operations. Although under high pressure/high temperature (HPHT) conditions, oil-based or synthetic fluid-based drilling muds have better performance than water-based fluids, environmental restrictions, and prohibitive costs prevent their use in such wells, particularly in offshore drilling operations. At the same time, increasing demand for oil and gas has lead to exploration of deep reservoirs with extreme temperature and pressure conditions and designing base fluids to withstand such demanding conditions is a challenging task.

Computer-aided molecular design (CAMD) is an effective technique used for product design. CAMD techniques based on group contribution have been successfully applied to problems like refrigerant design, solvent replacement design, and polymer design. The design framework developed by Samudra and Sahinidis [1], based on the GC+ model [2], is efficient, versatile, and allows integration of different property models. Although the original GC+ model includes some environmental metrics, the framework can incorporate other molecular modeling methods to predict properties like LC50. Another advantage of the framework is its modular structure and use of mixed-integer optimization in each module to increase the calculation speed by orders of magnitude compared other enumeration-based methods. This paper presents the use of the CAMD framework to design the base fluid for HPHT drilling. The properties that dominate the choice of drilling fluid are included in the design model. These include boiling and melting points, flash point, viscosity, and surface tension. One of the objectives in this work has been to include reliable property models to predict environmental metrics and use them in the design process. Properties like octanol-water coefficient, aqueous solubility, and LC50 are used to guide the molecular design. The targets for molecular design are formulated after an extensive review of deep well conditions and previously used synthetic fluids. We discuss the results of the design problem and compare them to current synthetic fluids like esters, mineral oils, and polyolefins.

Finally, we report on the use of computer-aided mixture/blend design using known fluids and mixture property models.

Bibliography:

[1] Samudra, A. and N. V. Sahinidis, Design of secondary refrigerants: A combined optimization-enumeration approach, in M. M. El-Halwagi and A. A. Linninger (eds.): Proceedings of the 7th

International Conference on the Foundations of Computer-Aided Process Design, CRC Press, pp. 879-886, 2009.

[2] R. Gani, P. M. Harper, and M. Hostrup. Automatic Creation of Missing Groups through Connectivity Index for Pure-Component Property Prediction. Industrial & Engineering Chemistry Research, 44:7262–7269, 2005.

See more of this Session: Product and Molecular Design
See more of this Group/Topical: Computing and Systems Technology Division

(241f) Structure-Property Relations for the Design of Organic Glasses for Organic Electronics and Nanolithography

Richard A. Lawson and Clifford L. Henderson, School of Chemical & Biomolecular Engineering, Georgia Institute of Technology, Atlanta, GA

In the last decade, there has been an increasing interest in the use of small molecule organic glasses instead of polymers for a wide variety of high technology applications, especially organic electronics and as photoresist materials for next generation high volume integrated circuit fabrication. These small molecule organic glasses are attractive compared to polymers for a variety of reasons. They can have unique electronic and photoactive properties for use as electrically conducting materials, photovoltaic devices, electroluminescent devices, and a variety of other applications. In photoresist applications, they are favorable because they provide molecules that are much smaller than polymers, effectively reducing the molecular pixel size of the resist. One major advantage that molecular glass materials have over polymers is that they can easily and repeatably be made as completely monodisperse materials. Polymers can have significant variations in chain-to-chain molecular weight (polydispersity) and variations in chain compositional uniformity both along an individual chain and from chain-to-chain, especially in complex ter- and tetra-polymers that are commonly used in high technology applications. Since molecular glasses have a well-defined structure compared to these polymers, it is possible to generate accurate structure-property relations for a number of important material properties based purely on the chemical structure of the small molecules.

We have developed bond and group additivity models that accurately predict the glass transition temperature (Tg) of both organic electronic and molecular photoresist materials. The correlation between glass transition and conventionally predicted properties such as boiling and melting point have been examined to extend glass transition predictions to new material classes. Another important material property, especially in molecular photoresists, is solubility and dissolution rate characteristics. To that end, we have extended the conventionally calculated partition coefficient (log P) and distribution coefficient (log D) to predict a priori molecular resist solubility. We have determined that that dissolution rate can be broken down into a kinetically controlled and a thermodynamically controlled regime. Application of log D and another structure-predictor that can be extracted from the log D calculation allows accurate prediction of dissolution rate across both regimes. These and other new areas of property prediction for the design of organic molecular glass materials will be discussed.

See more of this Session: Product and Molecular Design
See more of this Group/Topical: Computing and Systems Technology Division

(280a) Analysis of An Air Separation Unit as Part of An IGCC Power Plant

Dustin D. Jones[1], Debangsu Bhattacharyya[2], Richard Turton[1] and Stephen E. Zitney[3], (1)Chemical Engineering, West Virginia University, Morgantown, WV, (2)Department of Chemical Engg., West Virginia University, Morgantown, WV, (3)Collaboratory for Process and Dynamic Systems Research, National Energy Technology Laboratory, U.S. Department of Energy, Morgantown, WV

The air separation unit (ASU) plays a key role in improving the efficiency, availability, and operability of an oxygen fed integrated gasification combined cycle (IGCC) power plant. An improved integration between the ASU and the balance of the plant, especially the gasifier, and the gas turbine (GT) has significant potential for enhancing the overall plant efficiency. Considering the higher operating pressure of the GT, an elevated-pressure air separation unit (EP ASU) is usually favored instead of the conventional low-pressure air separation units (LP ASU). This process is generally chosen when most, or all, of the nitrogen produced is used as a fuel diluent for the gas turbine. In addition, if air is extracted from the GT to be used as an air feed to the ASU, it is generally desired to have the operating pressure of the ASU close to that of the GT in order to maximize the benefit of the high-pressure air. Due to the decrease in separation efficiency with increasing pressure and because of stringent purity requirements, for both oxygen and nitrogen, considerable modifications to the existing LP ASU technology is required. A detailed study on various possible configurations of an EP ASU and their impacts on the overall process efficiency are scarce in the open literature. Some configurations have been patented which are found to be very power intensive. In this study, a number of possible configurations has been considered with an objective of reducing the auxiliary power requirement. Plant operability is also considered with each of the configurations. It has been suggested in the open literature that in order to maximize the benefit of the high-pressure air available by air extraction an EP ASU should be used so as to increase the savings from the decreased load on the main air compressors (MAC). No study has been found within the open literature investigating whether equivalent savings are possible from an LP ASU with a pumped liquid oxygen cycle at varying degrees of integration. Studies carried out here have shown that an LP ASU with a pumped liquid oxygen cycle has nearly equivalent power savings as that of an EP ASU up to approximately 60 - 70% integration. A study was also completed for determining the optimal operating pressure of the ASU considering the degree of integration, and fraction of nitrogen injection. Results of this study will also be discussed.

See more of this Session: Gasification Process Development
See more of this Group/Topical: Process Development Division

(280b) Application of Graphical Techniques to Coal Gasification: Analysis of Partial Pressure and Target $CO:H_2$ and $CO_2:H_2$ Ratios

Lwazi Ngubevana, School of Chemical and Metallurgical Engineering, University of the Witwatersrand, Johannesburg, South Africa, **Diane Hildebrandt**, Center of Material and Process Synthesis, University of the Witwatersrand, Johannesburg, South Africa, Johannesburg, South Africa and David Glasser, Centre of Material and Process Synthesis, University of the Witwatersrand, Johannesburg, South Africa, Johannesburg, South Africa

The technique of graphically representing coal gasification has been proposed by Ngubevana et al[1].This method allows one to make simple assumptions and graphically represent the mass and energy balances around a gasifier. Using this method, one is also able to graphically represent the reaction equilibria. Different gasifier feeds and operating conditions can also be explored using this method. This paper looks at the application of this technique. Changes in possible gasifier product will be explored by looking at how the partial pressures of components change through the gasifier given certain operating conditions. Using this technique, one is also able to map out and determine the partial pressures of the components without actually calculating them; by simply knowing the gasifier product, one can easily determine these. The second and probably most important analysis that is done here is on the possible syngas composition ($CO:H_2$ and $CO_2:H_2$ ratios) from the gasifier. The traditional requirement of a $CO:H_2$ ratio of 2 for Fischer Tropsch (FT) synthesis is explored. Of biggest interest is a new FT chemistry as proposed by Patel et al[2]. This chemistry proposes setting a $CO_2:H_2$ ratio of 3 as the target for FT synthesis. Some optimisation work is also done on the possible syngas composition for both the traditional and proposed new chemistry routes. The mass and energy balance are of course a very important part of all the analysis done here.

1. Ngubevana L, Hildebrandt D, Glasser D. *AICHE National Spring Meeting 2007, Topical 6: 7th Natural Gas Utilization (T6), #30 – Coal and Gas to Liquids I (T6001)* .

2. Patel B, Hildebrandt D, Glasser D, Hausberger B.*Thermodynamics Analysis of Processes. 1. Implication of Work Integration. Ind. Eng. Chem. Res. 44 (2005); 3529-3537.*

See more of this Session: Gasification Process Development
See more of this Group/Topical: Process Development Division

BIOMASS CATALYTIC STEAM GASIFICATION

Thermodynamics Analysis and Reaction Experiments in a CREC Riser Simulator

Enrique Salaices[a], Benito Serrano[b], Hugo deLasa[a].

a Chemical Reactor Engineering Centre, the University of Western Ontario, London, Ontario, Canada N6A5B8.
b Chemical Engineering Department, Universidad Autonoma de Zacatecas, Zacatecas, Mexico.

Abstract:

A thermodynamic equilibrium model based on evaluations involving C, H and O element balances and various product species up to C_6 hydrocarbons is reported in this study. This model establishes the effects of biomass composition, temperature, and steam on the various gas product molar fractions. On this basis, most significant parameters determining the chemical interconversions and distribution of chemical species are identified. Based on the proposed equilibrium model and using glucose as model biomass species [$C_6 H_{12} O_6$], an optimum gasification temperature close to 800 °C and a steam/biomass ratio between 0.5 and 0.7 g/g is established.

This study has the special value of comparing thermodynamic equilibrium predictions with experimental data obtained in a CREC Riser Simulator using a fluidizable Ni-Al_2O_3 catalyst. Results are relevant for scaled-up gasifiers. They show that for reaction times longer than 30 seconds chemical species are essentially equilibrated and that the proposed model does provide adequate description of various product fractions. Data obtained also demonstrate the shortcomings of equilibrium models for gasifiers with reaction times shorter than 10 seconds and the need of non-equilibrium models to describe gasifier performance at such conditions.

1. EXPERIMENTAL SET UP AND METHODOLOGY

The catalytic gasification experimental runs were conducted in the CREC Riser Simulator [Error! Reference source not found.]. This reactor is a bench scale internal recycle batch reactor with a capacity of 53 cm^3, and allows for the loading of 1 g of catalyst. A series of sampling valves (Figure 1) complete the assembly of the CREC Riser Simulator and allow the withdrawal of reactants and products in short periods of time. In addition, a gas chromatograph allows the analysis of reaction products using both, gas chromatographer with thermal conductivity detector and mass spectrometer.

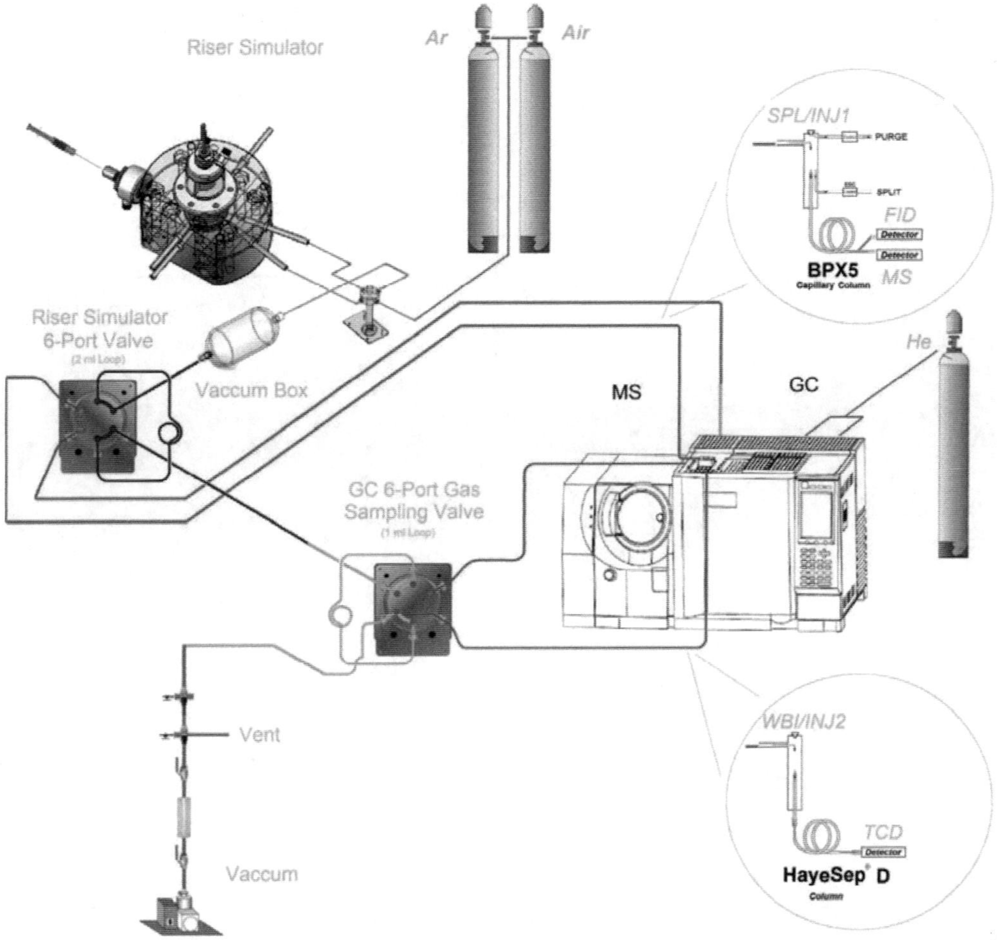

Figure 1. Schematic description of the CREC Riser Simulator and its accessories

2. THERMODYNAMIC EQUILIBRIUM MODEL FOR STEAM GASIFICATION

The product species considered for the equilibrium calculations are the following: H_2, CO, $C0_2$, H_2O, CH_4, C, C_2H_4, C_2H_6, C_3H_8, C_4H_{10}, C_5H_{12}, and C_6H_{14}.

From the elemental analysis of different biomasses, it can be proven that the contribution of nitrogen and sulphur species evolving from the reactor are negligible in terms of the equilibrium calculations, with this being is in agreement with previous reported models [2, 3]. The formation of tar is neglected in the thermodynamic calculations given its low concentration, although, it has to be considered in gasifier plant operation.

Thus, the simplified overall mass balance for the gasification reaction can be written as follows:

$$C_xH_yO_z + \omega\, H_2O \longleftrightarrow \alpha\, H_2 + \beta\, CO + \gamma\, C0_2 + \psi\, H_2O + \zeta\, CH_4 + \Omega\, C + \\ + \varepsilon C_2H_4 + \eta C_2H_6 + \theta C_3H_8 + \lambda C_4H_{10} + \mu C_5H_{12} + \nu C_6H_{14} \qquad (9)$$

This proposed equilibrium model considers the following gas phase reactions occurring after volatilization of the biomass, which are considered to be at equilibrium:

Table 1 Chemical reaction in the steam gasification of biomass

Name of reaction	Chemical equation	$\Delta G^o_{f(298)}$ [kJ/ mol]	$\Delta H^o_{f(298)}$ [kJ/ mol]	K (800 C)	Eq.
Dry reforming of methane	$CH_4 + CO_2 \leftrightarrow 2CO + 2H_2$	168.635	123.760	132.013	(10)
Steam reforming of methane	$CH_4 + H_2O \leftrightarrow CO + 3H_2$	140.098	205.310	169.182	
Water-gas shift reaction	$CO + H_2O \leftrightarrow H_2 + CO_2$	-28.538	-42.200	1.0051	(11)
Heterogeneous water-gas shift reaction	$C + H_2O \leftrightarrow H_2 + CO$	89.824	130.414	7.0401	
Boudouard equilibrium	$C + CO_2 \leftrightarrow 2CO$	118.362	172.615	6.499	(12)
Hydrogenating gasification	$C + 2H_2 \leftrightarrow CH_4$	-50.273	-74.900	0.049	
Ethylene	$2CO + 4H_2 \leftrightarrow C_2H_4 + 2H_2O$	-111.651	-104.256	1.738e-08	(13)
Ethane	$2CO + 5H_2 \leftrightarrow C_2H_6 + 2H_2O$	-212.787	-172.779	1.475e-08	(14)
Propane	$3CO + 7H_2 \leftrightarrow C_3H_8 + 3H_2O$	-293.149	-165.051	8.743e-14	(15)
Butane	$4CO + 9H_2 \leftrightarrow C_4H_{10} + 4H_2O$	-376.793	-161.968	7.669e-19	(16)
Pentane	$5CO + 11H_2 \leftrightarrow C_5H_{12} + 5H_2O$	-457.916	-159.719	4.321e-24	(17)
Hexane	$6CO + 13H_2 \leftrightarrow C_6H_{14} + 6H_2O$	-539.699	-158.303	2.785e-29	(18)

Nine independent nonlinear equations with nine unknowns (as described in Appendix I) result from algebraic manipulation of this system of twelve variables (α, β, γ, ψ, ζ, Ω, ε, η, θ, λ, μ, ν). The Newton-Raphson (NR) model is used to solve the nonlinear system of equations containing constrained variables. The NR method uses a truncated Taylor series estimate of the function values to obtain better estimates of the unknowns.

2.1. Equilibrium Constants.

Effect of Temperature on the Equilibrium Constant. Since the standard-state temperature is that of the equilibrium mixtures, the standard property changes of reaction, such as ΔG^o and ΔH^o, vary with the equilibrium temperature.

The dependence of ΔG^o on T is given by:

$$\frac{d(\Delta G^o / RT)}{dT} = \frac{-\Delta H^o}{RT^2} \qquad (19)$$

According with the definition of the *standard Gibbs energy change of reaction* **Error! Reference source not found.**

$$\frac{d\Delta G^o}{RT} = -\ln K \tag{20}$$

$$\text{with: } \Delta G^o = \Delta H_0^o - \frac{T}{T_o}(\Delta H_0^o - \Delta G_0^o) + R\int_{T_0}^{T}\frac{\Delta Cp^o}{R}dT - RT\int_{T_0}^{T}\frac{\Delta Cp^o}{R}\frac{dT}{T} \tag{21}$$

Therefore

$$\frac{d\ln K}{dT} = \frac{-\Delta H^o}{RT^2} \tag{22}$$

Eq. 22 establishes the effect of temperature on the equilibrium constant, and hence on the equilibrium conversion.

Effect of Pressure on the equilibrium gas composition. At reaction equilibrium, the gas species composition is a function of the temperature and pressure. An analysis of this composition can be made applying Le Chatelier's principle to the reaction network.

$$\frac{K}{\left[\Pi_i(\phi_i)^{v_i}\right]} = \left[\frac{(y_{A3})^{v3}(y_{A4})^{v4}}{(y_{A1})^{v1}(y_{A2})^{v2}}\right]P^{v_T} \tag{23}$$

3. THERMODYNAMIC EQUILIBRIUM MODEL PREDICTIONS

3.1. Fuel Composition Effect on the Steam Gasification of Biomass.

The influence of different fuel compositions on the product gas composition is examined in the present study, by varying on one hand the carbon to hydrogen content (in wt%) on the other hand the carbon to the oxygen content. This variation is carried out over a wide range of fuel compositions: from carbon to hydrogen content of 1:2.11 (Jute stick) to carbon to hydrogen content of 1:0.69 (Coal), and carbon to oxygen content of 1:1 (Glucose) to carbon to oxygen content of 1:0.111 (Heterotrophic). In order to obtain comparable results, the temperature and pressure is set constant at 800 °C and 1 atm, respectively.

Figure 2 reports the H_2 mole fraction of the product gas as a function of steam/biomass ratio for different biomass composition (on a dry basis). The hydrogen concentration in the product gas is shown both, to increase with increasing steam/biomass ratio from 0.0 to 1.2, and to be proportional with the carbon to hydrogen content in the biomass considered. It is predicted that increasing the steam/biomass ratio beyond 1.0 g/g does not lead to a significant higher concentration of H_2 in the product gas.

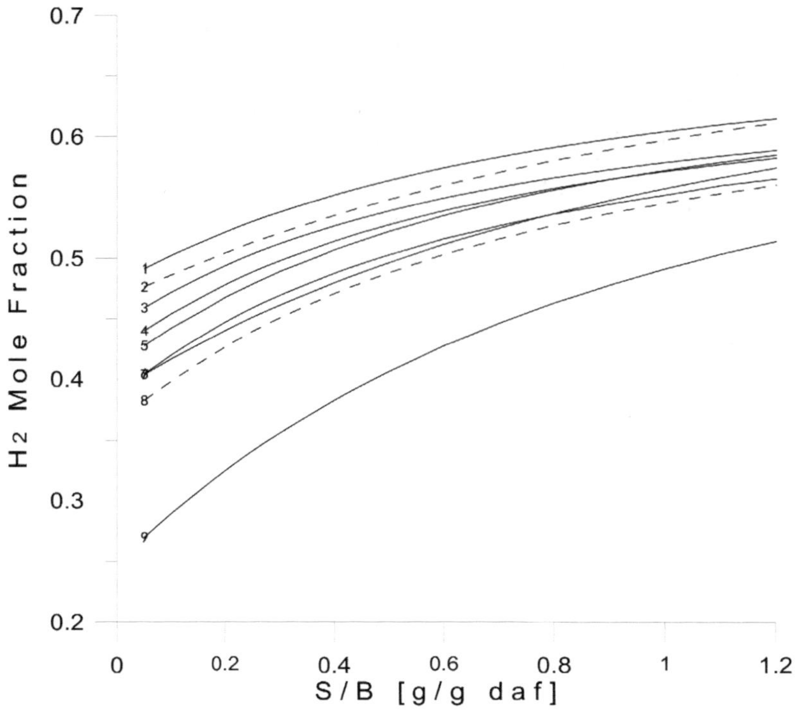

Figure 2. Hydrogen mole fraction versus steam/biomass ratio at 800 °C and 1 atm. Equilibrium model predictions of different biomass composition. From higher to lower H_2 mole fraction (1) jute stick, (2) glucose, (3) heterotrophic, (4) potato starch, (5) pine sawdust, (6) legume straw, (7) softwood bark, (8) wasted wood and (9) coal.

3.2. Steam/Biomass Ratio and Temperature Effect on the Steam Gasification of Biomass.

Previous experimental investigations [2, 5-10] have demonstrated the effect of the steam/biomass ratio on the product gas composition. An increase in the steam/biomass ratio is expected to have an increase in the hydrogen output according to both, the water gas shift reaction, and the steam reforming reaction. In addition, excess steam is often used to drive the cracking of higher hydrocarbons and reforming reactions.

However, an upper limit is expected due to an increase in the steam in the product gas, and the energy associated with the generation of steam needs to be considered when excess steam is introduced to the system. Such consideration demonstrates the importance of selecting the optimal steam/biomass ratio in the steam gasification of biomass from the point of view of process efficiency. Figure 4 reports the fractional distribution of hydrogen in the product gas as a function of the steam/biomass ratio and temperature (on a dry basis).

On the basis of this analysis, a steam/biomass ratio of 0.5 to 0.7 g/g and temperatures around 800 °C have being identified as optimum conditions for the process given it is at these conditions that the equilibrium model predicts a highest H_2/CO ratio of 2.

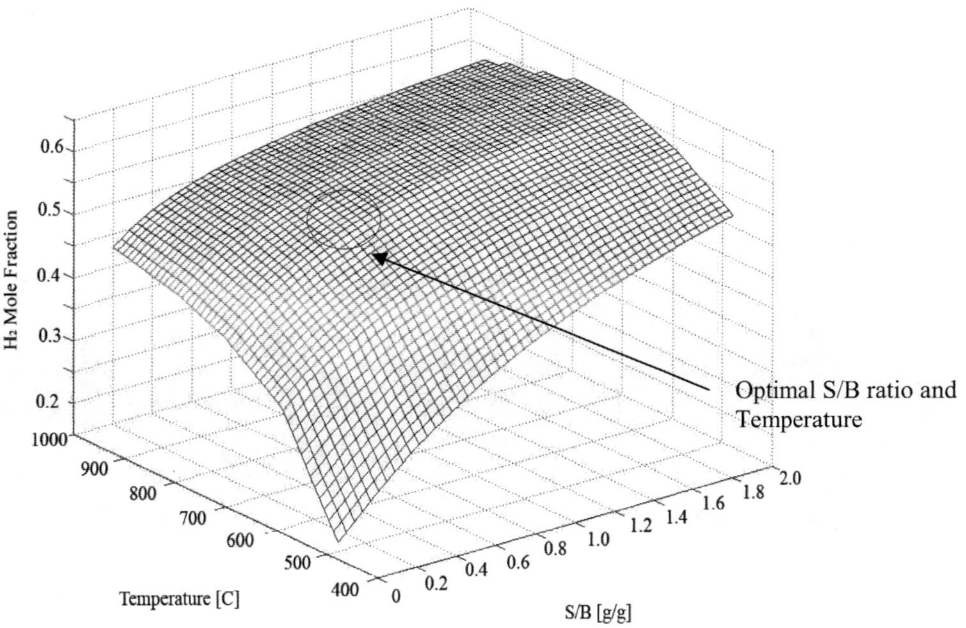

Figure 3. Equilibrium fractional distribution of H_2 on a dry basis at P = 1 atm and various Steam/Glucose (S/B) ratios and Temperatures.

4. EQUILIBRIUM MODEL PREDICTION VERSUS EXPERIMENTAL RESULTS

Equilibrium studies can be used to predict the maximum possible conversion and product distribution in a chemical reacting system. By comparing experimental results with equilibrium calculations one can establish the relation between thermodynamics and chemical kinetics of the process. In the case of this study, the thermodynamic equilibrium model proved to be useful in analyzing and understanding the experimental results.

Using glucose, as a model compound of biomass, steam gasification experimental runs were carried out at the following conditions: a) non-catalytic steam gasification runs at 700°C and 30 seconds of reaction times, and b) catalytic steam gasification runs at 700°C and reaction times of 5, 10, 20 and 30 seconds.

Figure 5 a-d reports the molar fraction compositions of the major product species on a dry basis, including H_2, CO, CO_2 and CH_4 as a function of steam/biomass ratio for: a) equilibrium model prediction (——), b) catalytic steam gasification using 2.5% Ni supported catalyst at 5 seconds (⊕), 10 seconds (✥), 20 seconds (▲), 30 seconds (◊) and c) non-catalytic steam gasification runs (✱).

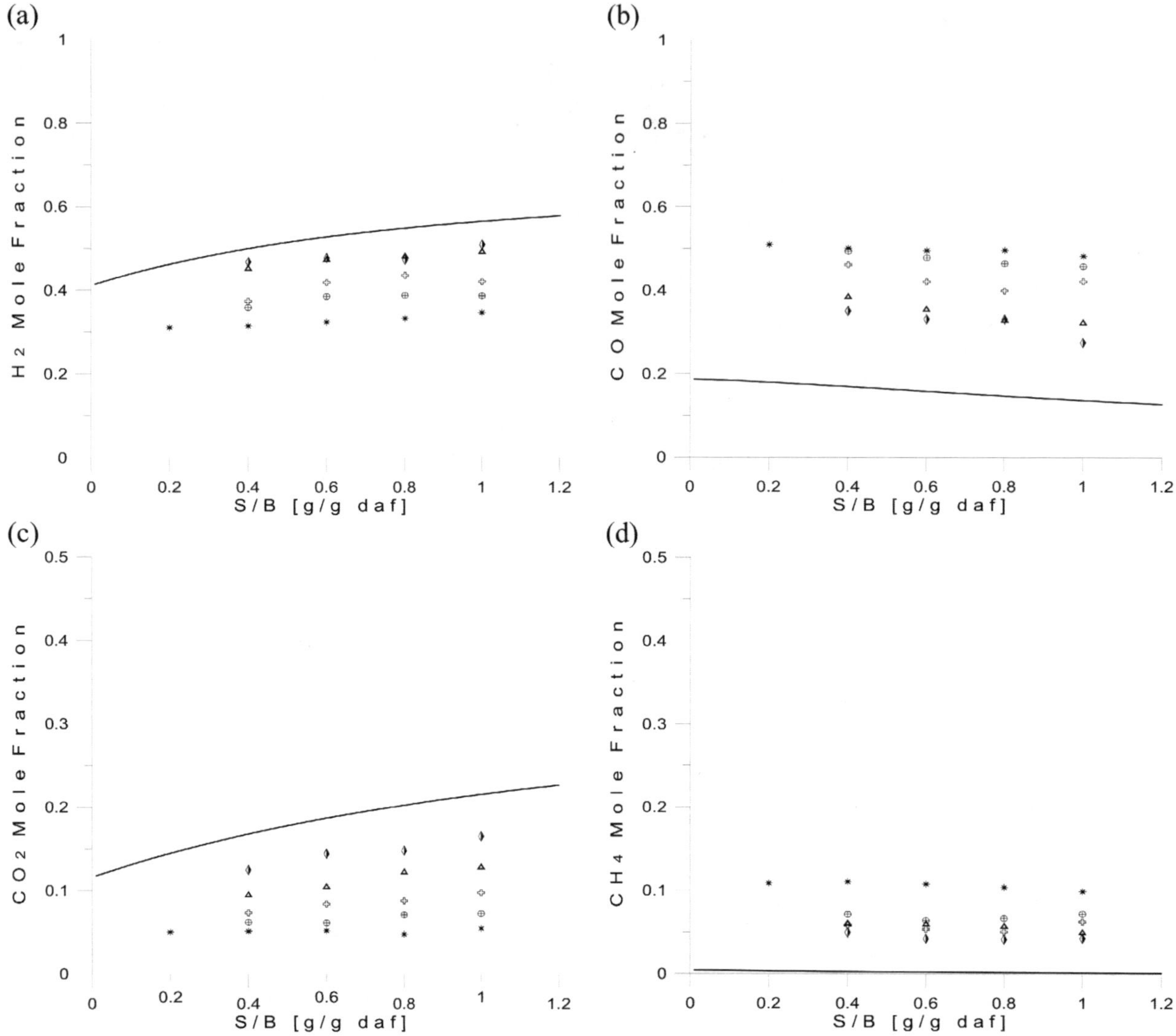

Figure 5. Mole fraction of product gas for gasification of glucose at different steam/glucose ratios, 700°C and P = 1 atm for a) H_2, b) CO, c) CO_2, and d) CH_4. Solid line (—) equilibrium model, (✱) non-catalytic gasification runs, experimental data using 2.5% Ni supported catalyst at (⊕) 5 seconds, (✚) 10 seconds, (▲) 20 seconds and (♦) 30 seconds of reaction time.

The H_2 concentration in the product gas for glucose is shown in **Error! Reference source not found.**5a. By augmenting the steam/biomass ratio from 0.2 - 1.0 g/g, the hydrogen fraction increases from 31.08 - 34.68%-mol for non-catalytic and from 46.86 - 51.02%-mol for catalytic experimental gasification runs at 30 second of reaction time. The equilibrium model over predicts the H_2 mole fraction by around 18.5%-mol for non-catalytic and 4.5%-mol for catalytic experimental runs for longer reaction times (> 20 sec) over the span of experimental steam/biomass ratios. This over predicted value is consistent with the previous non-catalytic experimental data analyzed (15, 18 and

21 %-mol for Legume straw **Error! Reference source not found.** [$C_{1.00}$ $H_{1.55}$ $O_{0.87}$], pine sawdust **Error! Reference source not found.** [$C_{1.00}$ $H_{1.59}$ $O_{0.64}$] and pine **Error! Reference source not found.** [$C_{1.00}$ $H_{1.13}$ $O_{0.62}$], respectively).

At thermodynamic equilibrium conditions, the CO mole fraction composition in the product gas decreases with an increase in the steam/biomass ratio according to the water-gas shift reaction, Eq. 11, when excess of gasification agent is present in the system. In the other hand, at low steam/biomass ratios (no excess of gasification agent) solid carbon remains in the equilibrium state and must be considered according to Eq. 12.

Error! Reference source not found. displays the CO concentration in the product gas as a function of the steam/biomass ratio. The CO mole fraction decreases from 51.01 – 48.23%-mol for non-catalytic and from 38.42 – 30.20%-mol for catalytic experimental gasification runs at 30 second of reaction time. It can be seen in **Error! Reference source not found.**5b that the equilibrium model follows the trend of the experimental data showing 33 %-mol less CO for non-catalytic and 15 %-mol less CO for catalytic gasification runs for longer reaction times (> 20 sec) trough the entire steam/biomass ratio analyzed.

All this suggests that a combination of reactions govern the overall gasification reaction network, with the water-gas-shift reaction being a major influence on the observed product yields. The cause for this observed behaviour may be assigned to the role of the nickel catalyst. This metal supported catalyst enhances the water-gas-shift reaction, and thus increases CO_2 and H_2 production, while a decrease in CO occurs.

An increasing trend in the CO_2 was also observed with increasing steam/biomass ratio. **Error! Reference source not found.**5c reports that the equilibrium model over predict the CO_2 by 13.8%-mol and 4.6%-mol for non-catalytic and catalytic experimental runs at 30 second of reaction time, respectively.

The equilibrium model, **Error! Reference source not found.**, shows that almost no methane (~ 0.2%-mol) is produced at 700°C and 1 atm. The equilibrium prediction differ from the experimental data showing around 10.3%-mol more CH_4 for non-catalytic and 4.3%-mol more CH_4 for catalytic experimental runs for longer reaction times (> 20 sec) over the span of experimental steam/biomass ratios. This indicates that equilibrium state is not reached during experiments; devolatilization of biomass gives high contents of methane and higher hydrocarbons which do not react completely leading to equilibrium concentrations of H_2, CO, CO_2, and H_2O according to the reforming of methane reaction, Eq. 10 and the higher hydrocarbons cracking reactions, Eq. 13 to Eq. 18.

5. CONCLUSIONS

The following are the main conclusions of the present study:

a) It is shown that the thermodynamic equilibrium model involving a set of nine independent reactions is able to represent the molar fraction of various product species in a CREC Riser catalytic gasification unit for 30 seconds of contact time and a wide range of steam/glucose ratios and temperatures. The CREC Riser Simulator unit closely mimic the expected fluidization conditions in a scale-up catalytic gasifier.

b) It is proven that at the operating conditions of the CREC Riser Simulator and longer contact times (e.g. 30 seconds) the catalytic biomass gasification process is not a kinetic rate controlled chemical transformation. It is also shown that thermodynamics do provide an effective tool for assessing product distribution with 800 °C and steam/biomass ratio between 0.5 and 0.7 g/g being identified as an optimum value for this process. Practical considerations to prevent ash agglomeration have shown that 700 °C is a preferred temperature for operation and it is at this temperature level where all experiments are developed in this study. Mass balances were performed for each one of the experiments developed in the CREC Riser Simulator. Mass balances closed with an average value of 95%-103%.

c) It is demonstrated that a shorter contact times (e.g. 5 seconds) the catalytic gasification becomes a kinetically limited process and that further kinetic studies are required to establish needed rate equations for product distribution predictions at various contact times, temperatures and steam/biomass ratios. The data gathered strongly suggest that the water-gas shift, reforming of methane, solid carbon and higher hydrocarbons reactions (C_2 +) are the ones limiting the approaching to thermodynamic equilibrium.

REFERENCES

1 de Lasa H.I. Riser simulator for catalytic cracking studies. Canadian Patent #1,284,017. 1991. U.S. Patent # 5,102,678. 1992.

2 Ginsburg J., de Lasa H. Catalytic gasification of biomass in CREC fluidized riser simulator. International Journal of Chemical Reactor Engineering. 2005, 3, A38.

3 Schuster, G., Loffler, G., Weigl, K., and Hofbauer, H. Biomass steam gasification – An extensive parametric modeling study. Bioresource Technology. 2001, 77, 71-79.

4 Smith J. M., Van Ness H. C. Abbott M. M. Introduction to Chemical Engineering Thermodynamics, fifth Edition. The McGraw-Hill Companies, Inc. 1996.

5 Ligang Wei, Shaoping Xu, Li Zhang, Changhou Liu, Hui Zhu, Shuqin Liu. Steam gasification of biomass for hydrogen-rich gas in a free-fall reactor. International Journal of Hydrogen Energy. 2007, 32, 24-31.

6 Franco, C., Pinto, F., Gulyurtlu, I., and Cabrita, I. The study of reactions influencing the biomass steam gasification process. Fuel. 2003, 82, 835-842.

7 Walawender WP, Hoveland DA, Fan LT. Fundamentals of thermochemical biomass conversion. Amsterdam, Elsevier. 1985, 813-817.

8 Singh SK, Walawender WP, Fan LT, Geyer WA. Wood and fiber science. New York: Pergamon Press. 1986, 6, 53-119.

9 Prasad BVRK, Kuester JL. Process analysis of a dual fluidized-bed biomass gasification system. Industrial & Engineering Chemistry Research. 1988, 27, 304-310.

10 Herguido J, Corella J, Gonzalez-Saiz J. Steam gasification of lignocellulosic residues in a fluidized bed at a small pilot scale. Effect of the type of feedstock. Ind Eng Chem Res. 1992, 31, 1274-1282.

Kinetics of High Pressure Biomass Gasification

Pradeep K. Agrawal[1], Kristiina Iisa[3], Anna Elgqvist[1], Steven Lien[2], and Scott A. Sinquefield[2]

[1]School of Chemical & Biomolecular Engineering,[2]Institute of Paper Science & Technology, Georgia Institute of Technology, Atlanta, Georgia 30332; [3]National Renewable Energy Laboratory, Boulder, Co

A major advantage of biomass gasification in thermochemical platform for biomass utilization is that any type of biomass can be converted to syngas, making it attractive for unmerchantable wood, harvest residue, and waste from forest and paper industry. The mechanism of biomass gasification is complicated and can be considered as the net result of two processes in series - pyrolysis and char gasification. Several variables are important in the kinetics of biomass gasification, including ash content and composition, gasification temperature and pressures, and heating rates. A major goal of present study is to develop kinetic models for high pressure biomass gasification. A combination of two complementary rate measurement techniques is being utilized in this study: (i) pressurized entrained flow reactor (PEFR) provides high heating rates ($\geq 10^3$ °C/sec) and short residence times (1-20 sec), and (ii) pressurized thermo-gravimetric analyzer (PTGA) provides lower heating rates (0.1-1 °C/sec), but larger residence times (5-100 min).

The work presented here describes the results obtained with loblolly pine using both techniques. PTGA equipped with a mass-spectrometer provides important information on the evolution of different gaseous species during volatilization. The char yield is adversely affected by the pyrolysis pressure; slight increase in char yield is observed with increasing pyrolysis pressure. The BET surface area of the char generated in the pressurized entrained flow reactor increases with increasing pyrolysis temperature and increasing pyrolysis pressure. Scanning electron microscope (SEM) pictures of the char generated in the PEFR indicate nearly spherical morphology of the particles obtained from pine pyrolysis. However, a combination of spherical and long fibrous particles is observed for the switch grass feed. The spherical char particles generated from the PEFR pyrolysis are hollow, which is in contrast with the char particles generated from TGA (lower heating rates).

Figure 1.SEM image of switchgrass char generated in PEFR pyrolysis at 5 bars and 1000 C.

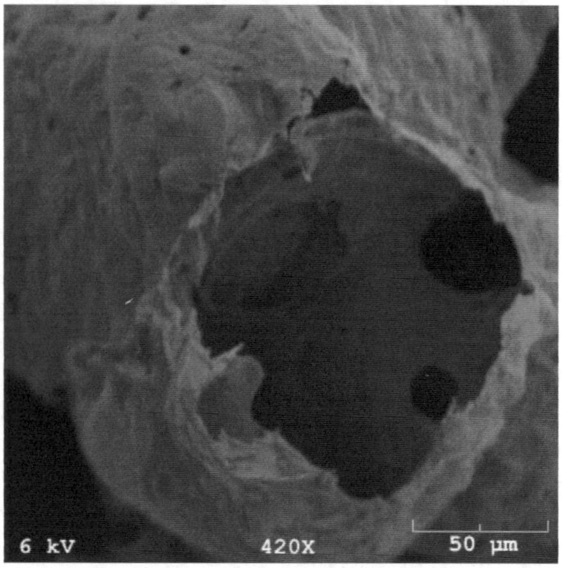

Figure 2. SEM image of switchgrass char generated in PEFR pyrolysis at 5 bars and 1000 C.

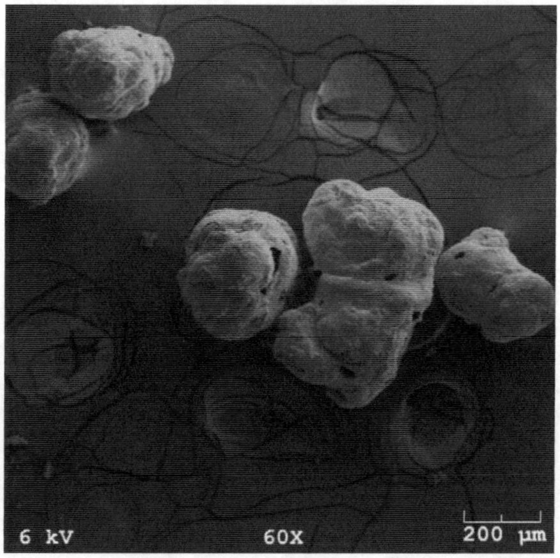

Figure 3. SEM image of loblolly pine char generated in PEFR pyrolysis at 5 bars and 1000 C.

The following observations can be made based on these results from the BET surface area measurements and the SEM images.

(i) Higher surface area is obtained if the pyrolysis is carried out at a higher temperature and/or higher pressure. It remains to be seen if this higher surface area translates into higher char reactivity.

(ii) The char generated from loblolly pine pyrolysis consists of nearly spherical particles, ranging in size 150–300 µm.

(iii) The char generated from switchgrass consists of two different types of particles: nearly spherical shape (150-200 µm),and elongated fiber-like particles (80-500 µm).

(iv) Nearly spherical particles appear hollow and would easily crumble. That does not appear to be the case with the fiber-like particles.

CO_2 gasification of pyrolysis-generated char was studied at different pressures and temperatures. The reactivity of char (towards CO_2) is influenced by the heating rates; in general the chars generated in PEFR (high heating rates) are more active than those generated in PTGA.

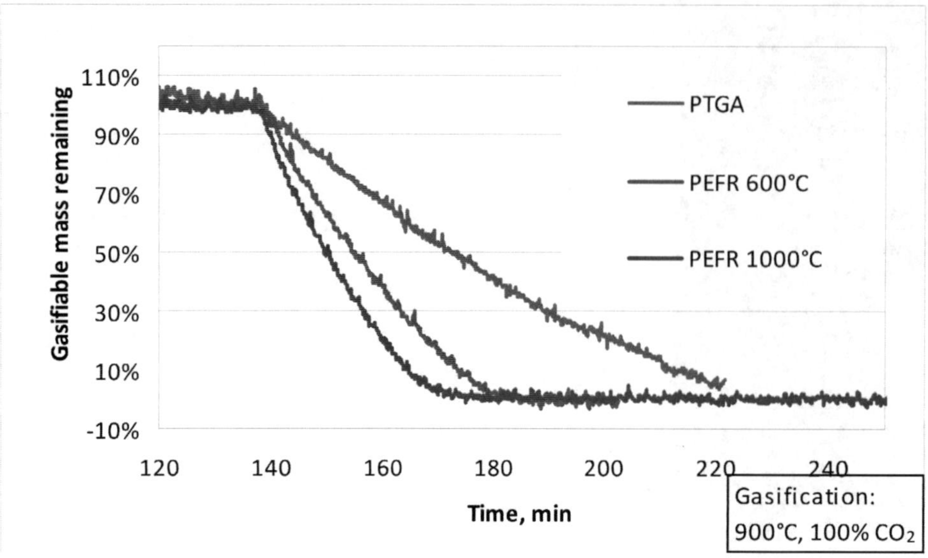

Figure 4. PTGA Gasification of pine char obtained from pyrolysis (effect of heating rate on char reactivity)

The char gasification data are coupled with transport effects. However, at lower temperatures, it is possible to obtain intrinsic kinetics. Attempts to build mathematical models will be presented.

Gasification Fluidized Bed Reactor Modeling

Alireza Abbasi[1], Hugo I. deLasa[1], Masahiro Kawaji[2], Paul E. Ege[3*]

(1) Department of Chemical and Biochemical Engineering, University of Western Ontario, London, ON, Canada
(2) Department of Chemical Engineering & Applied Chemistry, University of Toronto, Toronto, ON, Canada
(3) Reactech Process Development Inc., Markham, ON, Canada
*Corresponding author (paul.ege@reactech.com)

1 1. Introduction

Gasification reactions in fluidized beds depend strongly on the flow conditions. Significant mixing may reduce the reaction rates of equilibrium-limited reactions and change selectivity of reactions in series. A fluidized bed can be designed to enhance the desired product by changing the flow conditions within fluidized beds through vessel and distributor shapes and locations plus flow and particle properties. It is also common to introduce internals that form the flow as preferred. From both and engineering and optimization view it is important to properly simulate the reactions in a gasification fluid bed. Traditionally two classes of models have been proposed to describe the performance of fluidized bed reactors, the pseudo homogeneous approach and the two-phase approach [1]. Several articles have been published on the CPFD and CFD modeling and model validation of combined fluidized bed reactor hydrodynamics and kinetics [2-5]. This presentation compares a pseudo homogeneous and CFPD method to model a coal gasification fluid bed reactor. The first method uses time average flow patterns from a two-dimensional CPFD simulation to estimate a steady-state combination of Plug Flow and CSTR regions with mass transfer. The second method uses dynamic 3D CPFD. Time averaged dynamic results provide pseudo-steady state results to compare with the first method.

2 Background

2.1 Models

The pseudo homogeneous model divides the fluid bed in two parallel flow patterns corresponding to bubbles and emulsion phase. Bubble flow is assumed plug flow almost void of solids and emulsion is assumed CSTR at minimum fluidization void fraction and interstitial flow. In addition there is mass exchange between the two phases. Bubble fraction, minimum fluidization conditions and mass exchange coefficient can be correlated from literature, simulated or determined experimentally. Distributor and freeboard region may be modeled separately.

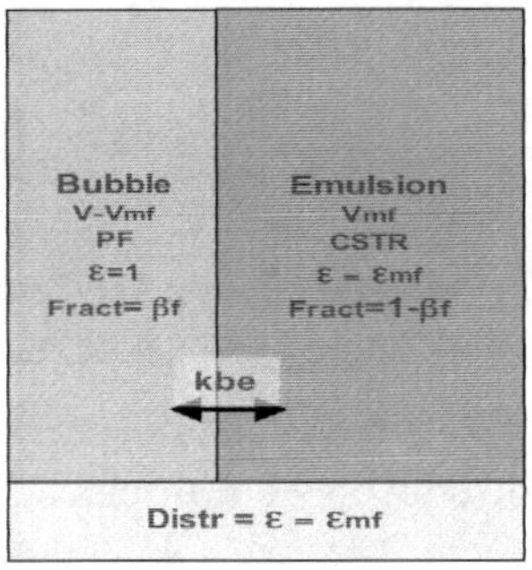

Figure 1 Pseudo Homogeneous Model

The CPFD model uses a Euler-Lagrange approach where the fluid phase is described as a continuum by solving the Navier- Stokes equations while the dispersed (solid) phase is treated by tracking a large number of particles through the calculated flow field. The dispersed phase has coupling with the continuum phase and can exchange momentum, mass and energy. Barracuda™ was chosen as the CPFD software for our simulations. In the Computational Particle Fluid Dynamics (CPFD) scheme of Barracuda™ each particle has three-dimensional forces from fluid drag, gravity, static–dynamic friction, particle collision and other forces. In this method, the collision force on each particle is modeled as a spatial gradient [5]. A numerical particle in CPFD method is identified as the particles where are grouped with the same properties such as species, size, and density. The numerical particle is an approximation like the numerical control volume where the property of the fluid is considered even.

2.2 Reaction Kinetics

The gasifier reactions are very complex. In this work we consider the coal gasification reaction model from BARRACUDA-CPFD package, the equations are listed in Table 1.

Table 1: List of reactions considered in the model.

Chemical reactions	Reaction rate (mol/cm³.s)	Reference
$C + H_2O \leftrightarrow CO + H_2$	$R_1 = K_1\left(\frac{\varepsilon_s \rho_s X_{sl}}{12}\right)\left(P_{H_2O} - \frac{P_{H_2}P_{CO}}{\exp(17.29 - 16326/T_g)}\right)$	Syamlal and Bisset
$C + CO_2 \leftrightarrow 2CO$	$R_2 = K_2\left(\frac{\varepsilon_s \rho_s X_{sl}}{12}\right)\left(P_{CO_2} - \frac{P_{CO}^2}{\exp(20.92 - 20282/T_g)}\right)$	Syamlal and Bisset
$C + 2H_2 \leftrightarrow CH_4$	$R_3 = K_3\left(\frac{\varepsilon_s \rho_s X_{sl}}{12}\right)\left(P_{H_2} - \left(\frac{P_{CH_4}}{\exp(-13.43 + 10999/T_g)}\right)^{0.5}\right)$	Syamlal and Bisset
$2C + O_2 \rightarrow 2CO$	$R_4 = K_4\theta_s \exp\left(-\frac{13590}{T}\right)P_{O_2}$	Yoon *et al.*
$CO + H_2O \leftrightarrow CO_2 + H_2$	$R_5 = K_5[CO]^{0.5}[H_2O] - K_6[H_2]^{0.5}[CO_2]$	Bustamante *et al.*

2.3 System Description

The reaction system considered is illustrated in Figure 2. It is a 2D bench scale fluid bed with 15cm width, 1 cm thickness and 120 cm total height. Air and steam mix is fed uniformly to the bottom of the reactor. Recycled ash and fresh coal are injected through two different side ports of the reactor. In these simulations the ash recycle is independent of the ash effluent although the input and output are tuned to close the time average mass balance.

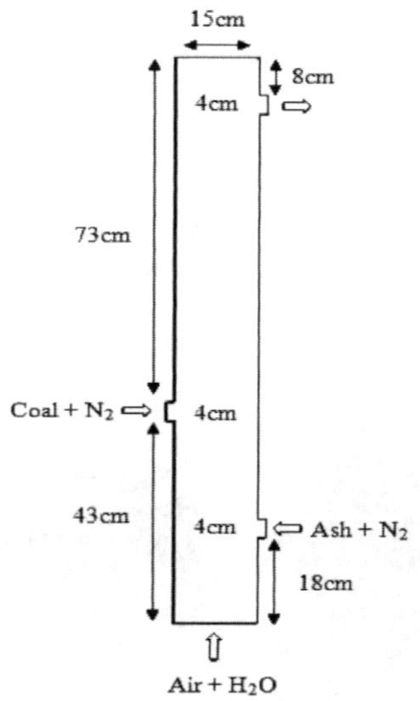

Figure 2: Schematic of gasification reactor

The system was solved using BARRACUDA-CPFD package The CPFD coal gasification model was established in a computational domain of 15cm×120cm×1cm using a grid of 18×120×1 to mesh the domain.

The pseudo homogeneous model initially assumed two reactors in series, The main (dense) bed was modelled as Plug Flow and CSTR in parallel while the freeboard (more dilute) region is modelled as a series of CSTR in series above after a dense Plug Flow/CSTR dense bed. Distributor region was considered zero in these simulations.

3 Results and Conclusion

A typical CPFD simulation result is shown in Figure 3. The system consists of a dense bed of bubble and emulsion with a more dilute freeboard region. The simulations indicate a fairly high concentration of solids in the freeboard as well so it is important to include this region in reaction calculations. The solid size distribution in freeboard is not too different from the dense bed but there is a larger portion of larger particles in the dense bed. Thus the Freeboard region is an important part of the reaction region so it was in the simulations. Since the bed looks fairly homogeneous a single-phase model with CSTR in Series for the whole reactor was also evaluated.

Effluent concentration results from one of the Barracuda simulations are shown in Figure 4. The simulation appears to reach a pseudo steady state after approximately 2 seconds. Thus an average of the remaining 8 seconds appears to provide reasonable steady state results to compare with the pseudo homogenous model.

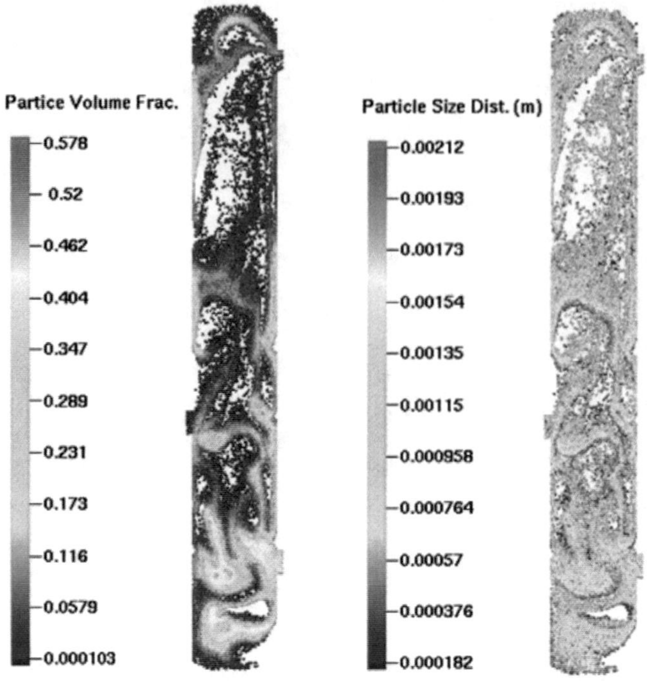

Figure 3: The contour of particle volume fraction and particle size distribution at 10 s simulation.

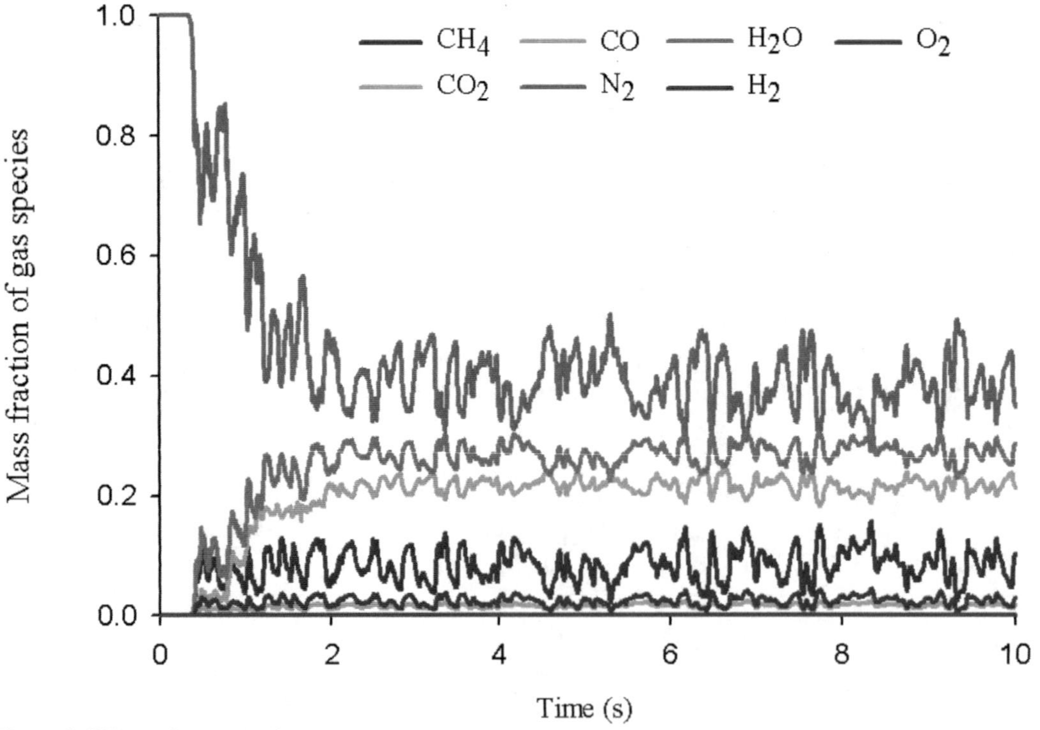

Figure 4: Effluent Concentrations

The comparison between CPFD results and pseudo homogeneous results provides useful insight about the details expected of either method. The results are reasonable with the pseudo homogenous method for similar flow conditions and it allows a more in-depth study of kinetics and reaction scenarios (shrinking core vs. shrinking particle) for a variety of process conditions. This approach is very useful to highlight design or process changes and optimize production.

CPFD combined with reaction is more time consuming so does not allow the same flexibility of conditions within short time frames. However it is superior to the pseudo homogeneous approach in evaluating and highlighting the dynamic and spatial effects. In addition the Barracuda™ results also allows the direct visual illustration of particle flow and entrainment as well as direct evaluation of design changes.

In conclusion the two methods are complementary. The pseudo homogeneous approach will allow engineers to quickly study and get a get a better understanding of a variety of process conditions. The CPFD simulation can then be targeted to the most promising conditions and is very useful to investigate large-scale systems as well as scale-up and design changes prior to implementing them in practice.

References

1- Wen-Ching Yang, N Handbook of Fluidization and Fluid-Particle Systems (Marcel Dekker, Inc., New York, 2003

2- Sebastian Zimmermann and Fariborz Taghipour, CFD Modeling of the Hydrodynamics and Reaction Kinetics of FCC Fluidized-Bed Reactors, Ind. Eng. Chem. Res. 2005, 44, 9818-9827

3- Jonghwun Jung , Isaac K. Gamwo, Multiphase CFD-based models for chemical looping combustion process: Fuel reactor modeling, Powder Technology 183 (2008) 401–409

4- Binlin Dou, Valerie Dupont, and Paul T. Williams, Computational Fluid Dynamics Simulation of Gas-Solid Flow during Steam Reforming of Glycerol in a Fluidized Bed Reactor, Energy & Fuels 2008, 22, 4102–4108

5- Dale M.Snider, Three fundamental granular flow experiments and CPFD predictions, Powder Technology 176 (2007) 36–46

6- M. Syamlal, L.A. Bisset, METC Gasifier Advanced Simulation (MGAS) Model, DOE/METC--92/4108, DE92 001111. 1992

7- H. Yoon, J. Wei, and M. Denn, A Model for Moving-Bed Coal Gasification Reactors, AIChE Journal, 24 (1978)

8- F. Bustamante, R. M. Enick, A. Cugini, R. Killmeyer, B.H. Howard, K. S. Rothenberger, M. Ciocco, B. Morreale, Uncatalyzed and Wall-Catalyzed Forward Water-Gas Shift Reaction Kinetics, AIChE Journal, 51 (2005) 1440–1454

(280f) The Dry Gasification Oxy-Combustion (DGOC) Power Production Process

Michael E. Walker[1], Donald J. Chmielewski[1], Javad Abbasian[2] and Marco J. Castaldi[3], (1)Chemical and Biological Engineering, Illinois Institute of Technology, Chicago, IL, (2)Chemical and Environmental Engineering, Illinois Institute of Technology, Chicago, IL, (3)Earth & Environmental Engineering (HKSM), Columbia Univeristy, New York, NY

Proposed is a novel coal conversion process, dubbed the Dry Gasification Oxy Combustion (DGOC) Cycle, because of its unique two stage conversion process. Feed coal will be partially oxidized at very high pressures (60 bar) in an oxygen-blown gasification unit, using recycled flue gas as a gasification agent (~65% CO2, 35% H2O b/v). This is in contrast to typical gasification systems that use steam as a gasification agent. In addition, the reducing environment of the gasifier will provide an opportunity to perform in situ sulfur removal through sorbent based capture.

The second stage oxy-combustion unit will also utilize recycled flue gas; for flame temperature moderation. The oxy-combustion stage will provide energy to raise steam for power generation, while maintaining a process effluent that is very concentrated in CO2. This effluent stream, already at high pressure, will utilize ambient cooling to 1) flash out the water from the effluent stream and 2) condense remaining CO2 into a high purity liquid stream, ready for further compression and sequestration.

Analysis of a preliminary design was carried out using process simulation models developed with Aspen Plus and Matlab. Results suggest that DGOC can achieve carbon capture and sequestration (CCS) goals with only a 3.4% detriment to plant efficiency (compared to the estimated 10% for current CCS technologies). Results also suggest that DGOC will remain competitive against the IGCC process, in terms of fresh water consumption.

See more of this Session: Gasification Process Development
See more of this Group/Topical: Process Development Division

(304a) Experimentally Validated Model Based Optimization of Filter Drying Process

Ekneet Sahni, Department of Pharmaceutical Sciences, University of Connecticut, Storrs, CT, John Strong, Abbott Laboratories, North Chicago, IL, Billie Buzon, Pfizer Inc., Groton, CT and Bodhisattwa Chaudhuri, Department of pharmaceutical sciences, University of Connecticut, Storrs, CT

Drying is traditionally defined as the unit operation which converts liquid, solid or semi-solid feed material into a solid product of significantly lower moisture content. It is one of the most commonly used unit operations in pharmaceutical, catalyst, polymer, paper, food and mineral industries in the preparation of dry granules by thermally removing volatile solvent from the wet solid. The knowledge of granular flow, mixing and heat transfer patterns in a filter dryer is critical to optimize the design and operation of the equipment which is mainly employed in drying of API(s) in pharmaceutical manufacturing. Our study focuses on quantitative investigation of mixing and heat transfer in a filter dryer in the quest to determine the optimum drying conditions. We use experiments and discrete element method (DEM) based numerical model, to improve the understanding of agitated vacuum drying process, and hence develop quantitative approach for optimization. We did not study the filtration operation done by this filter dryer and have only studied the drying operation. Most of the DEM-based heat transfer work has been either two-dimensional or in static granular beds. To the best of our knowledge no previous work has used three-dimensional DEM to study heat transfer in granular materials in agitated filter dryer. Once the dryer vessel is charged with the slurry (1L), nitrogen is purged through the drying chamber. Low pressures are generally used to keep the granular cake from becoming so compressed that the crystals fuse together. The impeller of the dryer suspended at the center is rotated by a set of computer-controlled motor drives. The impeller can rotate in both directions to better dig into the cake to help break it up, remove it from the filter media and finally to dry. In our experimental study, glass beads (2.4 -2.9 mm) or lactose is used with appropriate amount of ethanol as the solvent. Before the main experimental run, the impeller is rotated so as to ensure uniform wetting of the granular system. The drying chamber is purged with pure nitrogen flow at 0.1 L/min. The granular bed is then dried by applying heat to the inner wall of the vessel. Silicone oil/propylene glycol is used to heat the jacket sorrounding the external wall of the vessel. During each batch run, at repeated intervals, the impeller is stopped and a thermocouple is placed through one of the ports in the dryer to measure the temperature of the bed as a function of time. At the same time, the samples of wet glass beads are withdrawn using a cup sampler and stored in tared vials. The vials are then dried under house vacuum (26½ in.) at 323 K overnight and analyzed for estimating the loss in weight in the drying experiments. The change is weight data leads to the measurement of loss of drying and the drying rate as a function of material properties and operational conditions of the filter dryer. Series of experiments are performed to check the effect of various system variables: impeller speed, fill ratio of the dryer and the wall temperature of the dryer on the drying performance of the two granular systems (glassbeads or lactose with ethanol). Discrete Element Method (DEM) is employed to simulate granular flow, mixing and heat transport in the filter dryer. DEM explicitly considers inter-particle and particle-boundary interactions, providing an effective tool to solve the transient heat transfer equations. Typical system with glass beads and lactose powder are numerically simulated using appropriate material properties and validated by the experimental findings. A parametric study for simulations is also being performed to check the effect of wall temperature (318 K, 338 K, 353 K), fill ratio (25%, 50%, 75%), impeller speed (5rpm, 12rpm, 25rpm) and particle size on drying performance. The solvent concentration and temperature profile from simulation are in consensus with our experimental results. In both experiments and simulations, higher wall temperature showed an increase in the drying rate and much higher rise in average bed temperature, for both glass beads and lactose systems, thereby decreasing the total time for drying operation. Also at given wall temperature and air velocity, the increase in fill volume (bed depth) resulted in a decline in the drying rate. Furthermore, as the rotational speed of the impeller decreases, the heat transfer in the granular bed increases. The heat transfer coefficient are also calculated for different parameters under study. Hence the optimal drying condition is determined be operating at an impeller speed of 5 rpm, wall temperature of 353 K and with 50% fill load.

See more of this Session: Solids Handling Considerations and Challenges in Pilot and Demonstration Plants
See more of this Group/Topical: Process Development Division

(304b) Modeling the Caking Behavior of Corn Distillers Dried Grains with Solubles (DDGS)

Rose Prabin Kingsly Ambrose and Klein Ileleji, Agricultural and Biological Engineering, Purdue University, West Lafayette, IN

Distillers dried grains with solubles (DDGS) is a co-product of the corn grain ethanol manufacturing process. DDGS is primarily used as feed, so the quality and shelf-life of DDGS are important for the ethanol industry to sustain the profitability of corn ethanol production. The quality and shelf life depends on physical and chemical stability of DDGS which are mainly influenced by equilibrium moisture content. The hygroscopic nature of this granular solid allows it to absorb or lose moisture when exposed to ambient conditions. The adsorption/desorption can lead to bridging between particles, thus causing the product to cake, which is currently a major hurdle in marketing DDGS. Some of the factors which influence the caking processes are chemical composition, moisture content, environmental relative humidity and temperature. The cyclical changes in temperature and humidity, during storage and transportation, influence the moisture diffusion, moisture migration due to temperature gradient and sintering and lead to a caked bulk product. In this study the process of moisture migration and temperature changes in DDGS bulk is modeled to predict the caking process in DDGS. The model will be tested by comparing the results with flow properties of DDGS conditioned at different environmental conditions. The results from this study will help in finding the optimum moisture content and environmental conditions for storage and transportation of DDGS.

See more of this Session: Solids Handling Considerations and Challenges in Pilot and Demonstration Plants
See more of this Group/Topical: Process Development Division

(304c) Understanding Granular Behavior in Designing Solids Processing Equipments

Suman K. Hajra, Thomas O'Brien and Chris Guenther, National Energy Technology Laboratory, U.S. Department of Energy, Morgantown, WV

Granular materials are widely used in industries such as power plant, cement, iron & steel, fertilizers, pharmaceutical, construction, mining and agricultural. Examples of granular materials are coal, sand, soil, gravel, sugar, salt, rice and tablets. Incline chute, conveyor belt and pneumatic systems are often used to transport the granular solids in industries. In industry, rotating cylinders, conical hoppers, silos and V-blenders are often used as solid mixing devices. Solid processing operations such as calcinations of limestone, reduction of iron oxide, clinkering of cementitious materials, cereal drying, calcinations of petroleum coke, and waste incineration are carried in the rotating cylinders. Rotating cylinders have certain advantages such as their ability to handle large variety of feed materials, provide long residence times and ease of cleaning. In spite of their versatile applications, the design of the solid processing equipments is ad hoc due to lack of understanding of internal processes. Moreover, one of the most interesting features of heterogeneous granular materials is their ability to de-mix or segregate under external agitation rather than to mix further. Hence, segregation causes problems in maintaining the uniformity of the bed materials and thereby affects the quality of final product. Though segregation is a common phenomenon in shear flow, vibration, shaking, and chute flow, yet understanding is incomplete. The scales - up is difficult due to the poor understanding of flow, mixing and segregation of granular materials in the solid processing equipments. Even the innovative way to improve the design of solid processing equipments is rare. In this study, experimental, numerical simulation and mathematical models are presented to understand the flow, mixing and segregation in granular materials and consequently to arrive at means to improve the design of the solid processing equipments. Scale -up rules are also discussed.

See more of this Session: Solids Handling Considerations and Challenges in Pilot and Demonstration Plants
See more of this Group/Topical: Process Development Division

How to Avoid Caking and Moisture Migration in Process Equipment

Kerry Johanson, Material Flow Solutions Inc.

Caking is the significant increase in bulk unconfined yield strength with storage time. It is caused by the formation of crystal bonds between particles. In general, the steps to caking are: first, the formation of zone or regions where the local temperature and humidity are high enough for the bulk material to pick up significant surface moisture. Second, coalescence of surface moisture on the particle forms pendules between particle surfaces. Third, dissolution of soluble components. Fourth, a change in local equilibrium condition (temperature or humidity) results in moisture migration away from the particle surface. A thermal-gravimetric method can be used to measure equilibrium moisture isotherms for the components in a mixture. This information is used to determine the locations in the mixture that attract moisture when the humidity is high. Sorption isotherms determine the potential amount of moisture at these particle junctions. The rate at which moisture will adsorb onto the particle surface from ambient air is measured. Rate constants allow computation of moisture transfer rates on particle surfaces as a function of the local temperature and humidity. A differential mass and temperature balance over the volume occupied by the material in the process equipment allows calculation of the temperature and relative humidity driving force causing moisture migration in the bulk material. Finite element calculation techniques can be employed to compute the local humidity, local temperature, and local solids moisture content in process equipment. These calculations are combined with structure calculations of the bulk to estimate the zone and areas subject to caking. Finally, caking can be estimated from contact force models. The net result is a model that can be used to determine zones and regions where caking is problematic. This mode is used to optimize process equipment design to avoid caking problems. This paper presents the general methodology to accomplish this task and relates it to real world situations.

Moisture Sorption Tests

Bulk materials absorb moisture due to the size of pores in the particle structure, the hygroscopic nature of the material, and the ability to form crystalline water at ambient temperatures [1],[2],[3]. The moisture sorption for a powder material was measured using moisture sorption test equipment from VTI. Approximately 50 mg of material was placed on a small weigh bin and suspended on a Cahn balance in a controlled humidity and temperature environment. The temperature was controlled at a constant 34 deg C. The relative humidity was varied between 20% RH and 98% RH and the weight was monitored as a function of time. The weight measurements were recorded each minute until the change in weight approached a prescribed low minimum set point value. No change in weight suggests that the solid has reached equilibrium at the prescribed relative humidity. However, moisture sorption tests often require long periods of time to fully equilibrate. It is possible that the minimum change set point value be met between two consecutive weight measurements without the solid fully reaching the equilibrium moisture content. Moisture sorption is a first order process which is governed by a first order rate law given in equation 1.

$$\frac{dq}{dt} = K_g \cdot (q^* - q) \qquad [1]$$

Where

q^* is the equilibrium moisture content for the prescribed relative humidity.
q is the current solid moisture content at the current time (Wt_{H2O} / Wt_{Solid}).
K_g is the Sorption Rate Constant (1/sec).

The rate of moisture loss or moisture gain is directly proportional to the difference between the current moisture content and the equilibrium value. This equation can be integrated for constant humidity conditions to yield an equation relating the solid moisture content as a function of time (equation 2).

$$q = q^* \cdot \left(1 - exp(-K_g \cdot t)\right) + q_o \cdot exp(-K_g \cdot t) \quad [2]$$

Where:

q^* is the equilibrium moisture content for the prescribed relative humidity.
q_o is the initial solid moisture content (Wt_{H2O} / Wt_{Solid}).
q is the current solid moisture content at the current time (Wt_{H2O} / Wt_{Solid}).
K_g is the Sorption Rate Constant (1/sec).
T is time (sec).

This equation is depicted graphically in Figure 1. With moisture data as a function of time, equation 2 can be fitted to this data to obtain the true equilibrium moisture concentration as well as the Sorption Rate Constant. This information is then used to estimate the solids moisture content in any piece of process equipment.

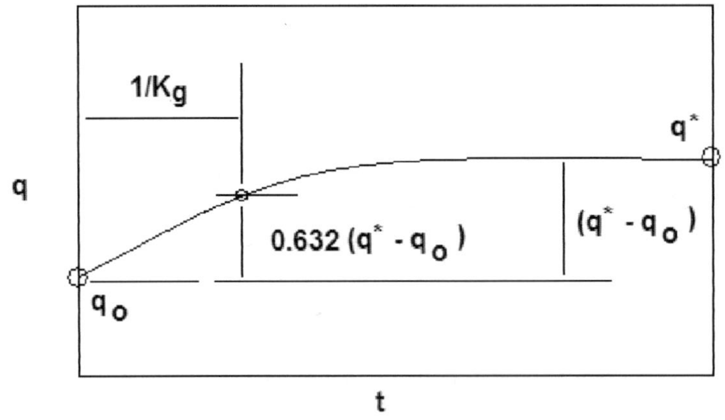

Figure 1. Schematic of typical moisture sorption curve

Table I – Moisture sorption parameters

Relative Humidity (%)	Equilibrium Moisture Content q* (%)	Sorption Rate Constant Kg (1/sec)
30.02	1.03%	0.000897
40.07	2.35%	0.000726
49.94	4.09%	0.000556
59.91	5.91%	0.000381
69.84	9.68%	0.000195
74.51	13.16%	0.000106
79.58	17.44%	0.000075
84.96	24.26%	0.000040
89.57	36.14%	0.000040

Plotting the equilibrium moisture content versus relative humidity gives the standard sorption isotherms used in moisture uptake analysis [4]. It is important to point out that this data is for conditions of, first, drying the material at 34 deg C and 20% relative humidity. This condition was considered the baseline moisture condition, and all data presented in this paper is relative to this condition. This data is also for a constant temperature of 34 deg C. Another temperature would yield another moisture sorption curve.

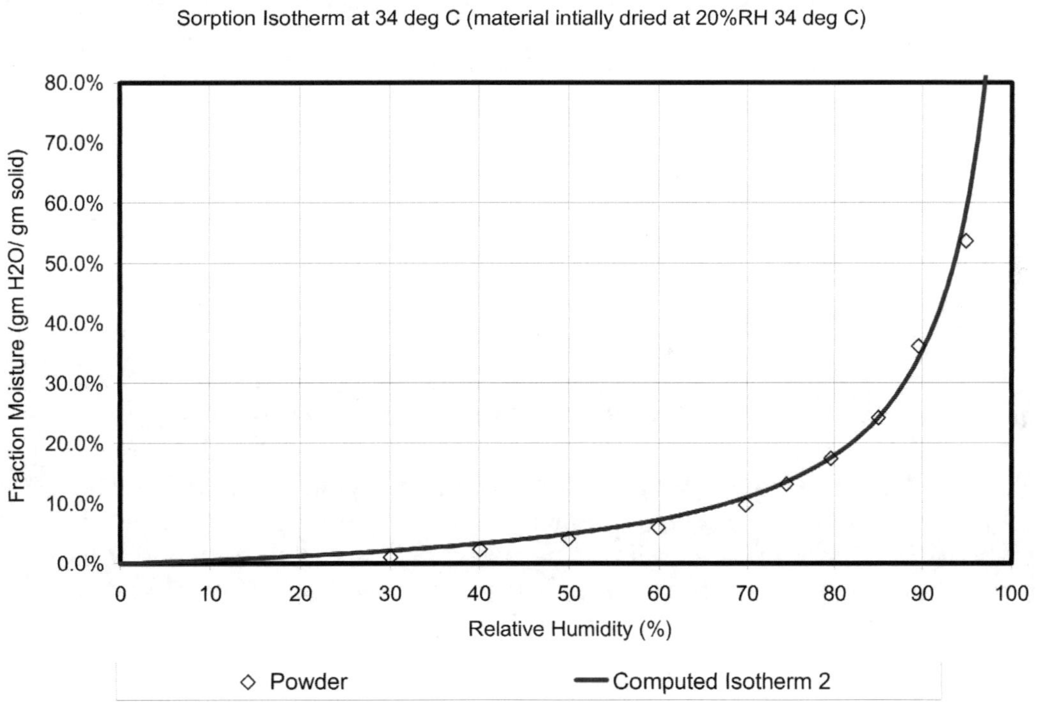

Figure 2. Sorption Isotherm from moisture test data

Figure 2 indicates a very steep moisture sorption curve at high relative humidities. However, there is significant moisture in equilibrium with lower relative humidity. The normal moisture content (q) of the powder tested is 19%. This would be in equilibrium with relative humidity values of about 80%. If this material were placed in a bin or process vessel and gas passed through it, then a simple equilibrium analysis suggests that drying would occur if relative humidity of injected air was below 80%. However, this is a gross over simplification. We also measured the moisture sorption rate constant for this material as a function of relative humidity (Figure 3).

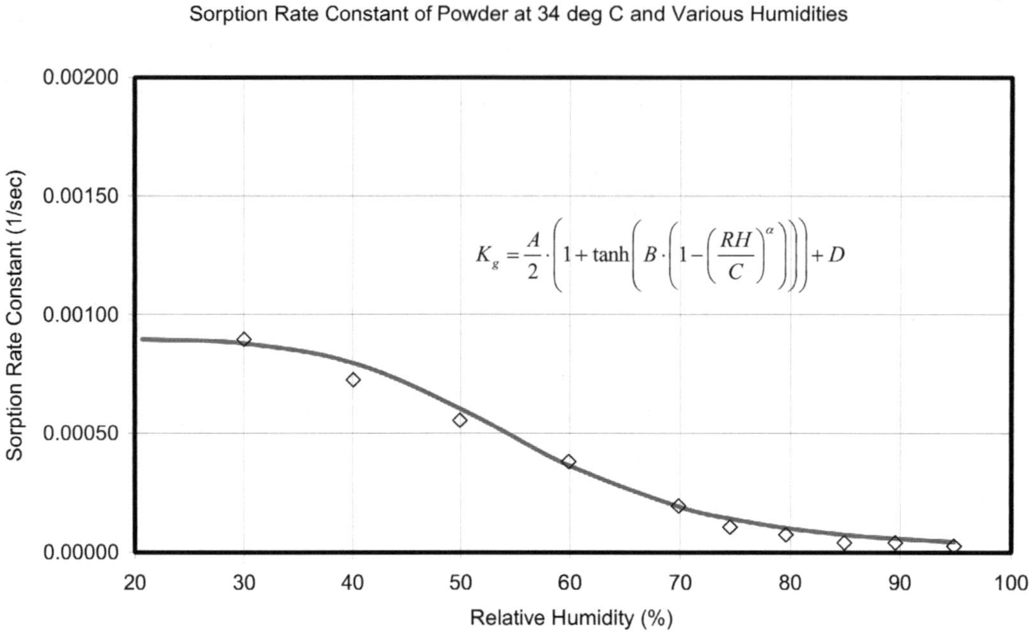

A	0.000887 1/sec
B	50
C	56
D	0.0000121 1/sec
α	0.06147

Figure 3. Sorption Rate Constant as a function of relative humidity

We plotted the Sorption Rate Constant as a function of relative humidity. The initial rate constant remains unchanged for relative humidity values below 35%. When the relative humidity rises above 35%, there is a significant decrease in the moisture pickup rate constant. This indicates that the rate of moisture pickup or loss for relative humidities near 100% is slow. This further suggests that, even if drying does occur under moist air conditions, it will likely occur at very slow rates.

Bin Moisture Estimates

We used the measured data to estimate the solids moisture content for a bin subject to gas counter flow. The bin is 2.8 meters in diameter and necks down to a conical hopper opening of 0.4 meters. The overall height of the bin is about 8 meters. A finite element analysis was performed on the silo using Darcy's law to describe the pressure drop across the permeable solid. To simplify calculations, axial-symmetry was assumed and gas was injected at three positions in the hopper section of the bin. The calculations assume uniform injection around the perimeter. Darcy's law provided us with gas velocity calculations in the silo. It is important to note that, due to the moderately large permeability of the material, the pressure profile and gas velocities came to equilibrium very quickly in the simulation compared to the time constant of other processes such as heat transfer and solid's drying effects. The small changes in temperature and water vapor concentration were assumed to have no effect on the computed velocity profiles. Thus, we computed the velocity profiles independently of other variables in the calculations. This velocity profile remained essentially constant for all times. The following gas velocity profile (Figure 4) was generated during the course of this simulation and assumes the material properties used in Table II.

Table II – Material properties used in the calculation of silo gas velocity and temperature

Material Property	Value	Material Property	Value
Ambient pressure	101352 Pa	Porosity	0.5
Solid's permeability	1.169e-9 m^2	Air temperature	312 °K to 302 °K
Bulk density	1270 kg/m^3	Air viscosity	1.85e-5 Pa s
Particle density	2400 kg/m^3	Heat capacity of solid	1500 J/kg °K
Bulk temperature	307 °K	Heat capacity of air	1005 J/kg °K

Notice the steep pressure gradient around the injection location. Gas entering at this location is at high velocity, but decreases significantly as the gas travels up the hopper. Much of this reduction in velocity is due to the changing cross-sectional area. However, some of this velocity difference is due to pressure drop losses. The total pressure drop across the bin is about 840 Pa for injection rates of 3.85 m^3/min (136 cfm). It is important to note that the velocity profile and gas pressure gradients seem to level out, producing a mostly uniform profile across the bin diameter.

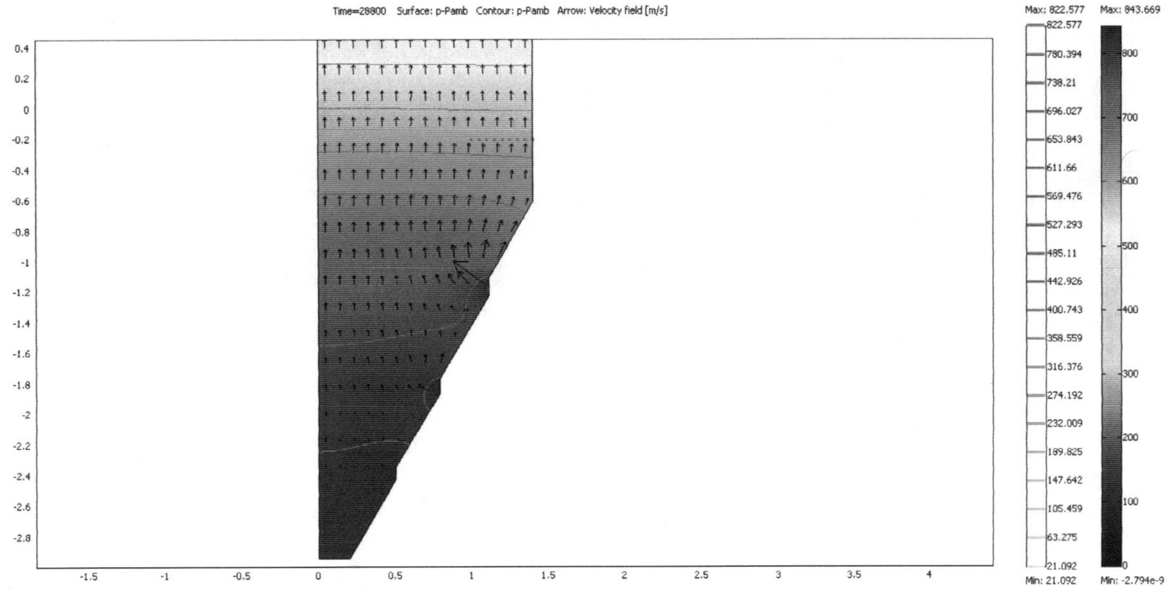

Figure 4. Gas pressure profile for a total injection rate of 3.85 m^3/min (136 cfm). Arrows show gas velocity field through hopper

We modeled the typical temperature fluctuation that would occur during a day-night temperature cycle. This was modeled by changing the air temperature as a function of time over a 24-hour period and using convective heat transfer coefficients to determine the expected transient temperature in the bin during that 24 hour cycle. Conduction was assumed to be the main means of heat transfer. However, the outside walls of the bin and hopper section were exposed to a convective heat transfer boundary condition where the convective heat transfer coefficient was assumed to be 4 W/m^{2o}K. This corresponds to a mild breeze around the bin. The heat transfer coefficient around the top of the bin was assumed to be about 2 W/m^{2o}K, which is more indicative of free surface convection. The temperature of material entering the silo was assumed to be at the temperature of the surrounding air (307°K or 34°C). We also imposed a sinusoidal variation in air temperature of $\pm$ 5 deg C during a 24 hour period. We used a conductive finite element model to describe the temperature profiles in the bin. Figures 8 and 9 show the expected temperature profile at the hottest part of the temperature cucle and at the coolest part of the cycle. Notice that the cool cycle shows a hot region (yellow) between the bin core material and the cool bin wall.

The temperature profiles were assumed independently of the concentration effects. That is to say, concentration differences did not cause significant changes in temperature of the bulk material. The reverse is not true. Temperature does cause significant changes to the predicted concentration profiles. Just like the gas velocities, these temperature profiles were computed independent of and prior to simulation of concentration profiles.

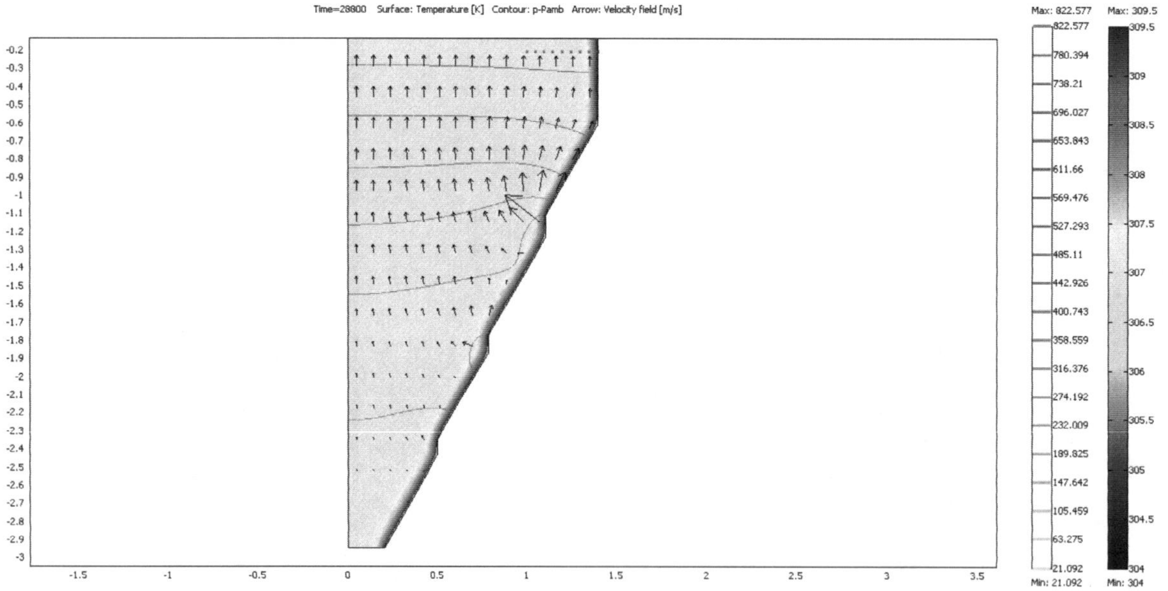

Figure 5. Expected temperature profile at 8 hours storage in bin

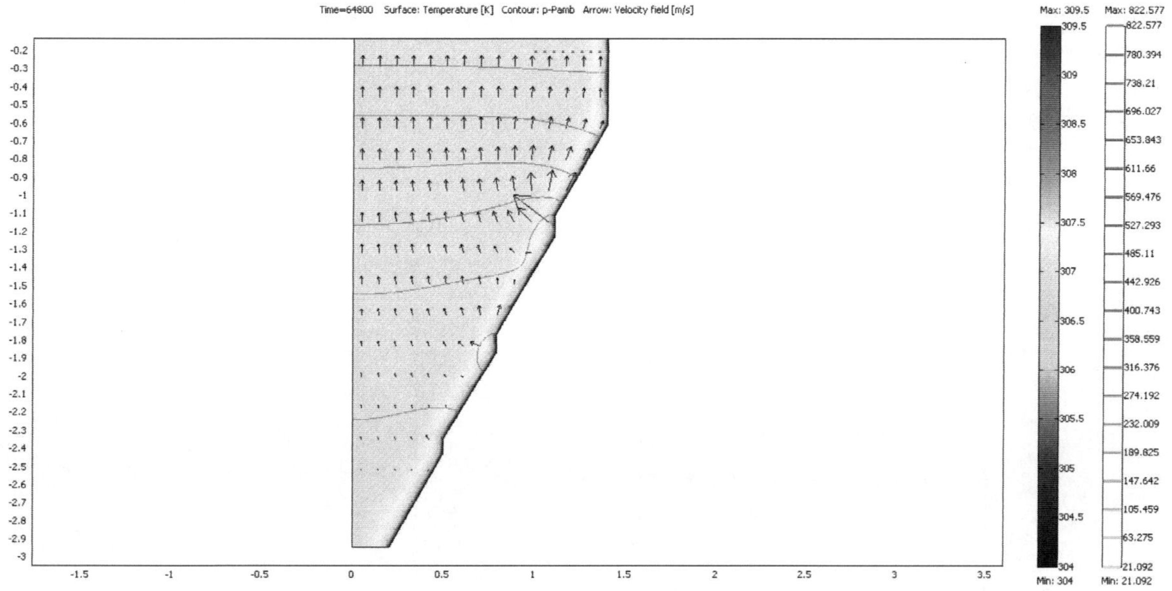

Figure 6. Expected temperature profile at 18 hours storage in bin

It is obvious from the above figures that the ingress of the temperature profile occurs within 0.2 meters of the silo surface. The maximum temperature difference for these daily temperature swings is about 5°C. This small temperature difference is not usually important, except that it changes the concentration of water vapor in material near the bin walls. Since the target equilibrium moisture content is a very strong function of relative humidity near saturated condition, small changes in local water vapor concentration can have a large effect in local moisture content of the bulk material.

We used a finite element analysis to predict the concentration of water vapor in the silo using a mass transfer convection diffusion equation coupled with the moisture sorption isotherm data generated above. The differential equations used in this analysis are found below.

$$\frac{\partial c}{\partial t} - \nabla \cdot D\nabla c = -\frac{(1-\epsilon)}{\epsilon} \cdot \frac{\rho_s}{M_{H2O}} K_g \cdot (q^* - q) \qquad [3]$$

$$\frac{\partial q}{\partial t} = K_g \cdot (q^* - q) \qquad [4]$$

Where:

K_g is the Sorption Rate Constant

q is the moisture content of the solid gm_{H2O} / gm_{Solid}

q* is the equilibrium moisture content from the sorption isotherm.

C is the concentration of water vapor in the air surrounding the solid.

D is the diffusion of water vapor in the air (0.0000042 m^2/s)
ρ_s is the solid particle density.
M_{H2O} is the molecular weight of water.
ε is the porosity.

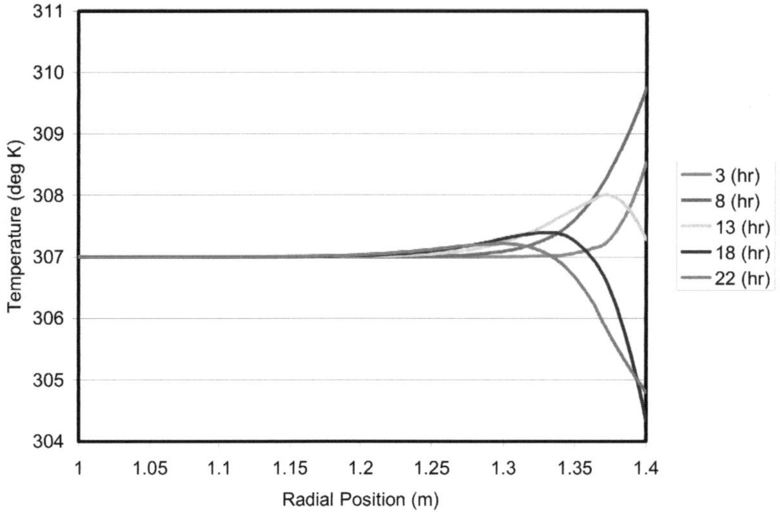

Figure 7. Temperature profiles at selected time increments

We assumed that the moisture concentration leaving the top free surface is equal to the moisture concentration of the ambient air. We also assumed a convective transfer of moisture entering the bulk material through the injection ports at 50% relative humidity. The initial concentration of water in the solids was assumed to be constant (19%), and that moisture content is in gm_{H2O}/ gm_{solid}. The initial water vapor concentration in the air was assumed to be from a saturated condition based on the local temperature of the bulk material.

Water vapor concentrations at storage times of 8 hours show concentrations 20% above those that the saturated air can hold (see Figure 8). This implies a zone of condensation which is generated within 2-hours of initiating the cooling cycle right after the maximum temperature was established in the bin. This is a critical condition since it can produce material wet to the touch. If subsequent drying occurs in this region of the bin, then crystal bonds may form between adjacent particles resulting in caking. Figure 9 shows that the subsequent injection and storage for 12 additional hours would result in less water vapor in the region that was initially high in water vapor. This suggests that the transition from the cone to the vertical section in the bin is very sensitive to caking and lump formation. Comparing the solids water concentration near the top of the conical hopper at 8 hours storage and 18 hours storage shows significant drying which will produce caking and lumps in this bin (Figures 10 and 11). This analysis has identified a critical flaw in the injection system and storage bin configuration. This methodology then provides a means of generating robust designs.

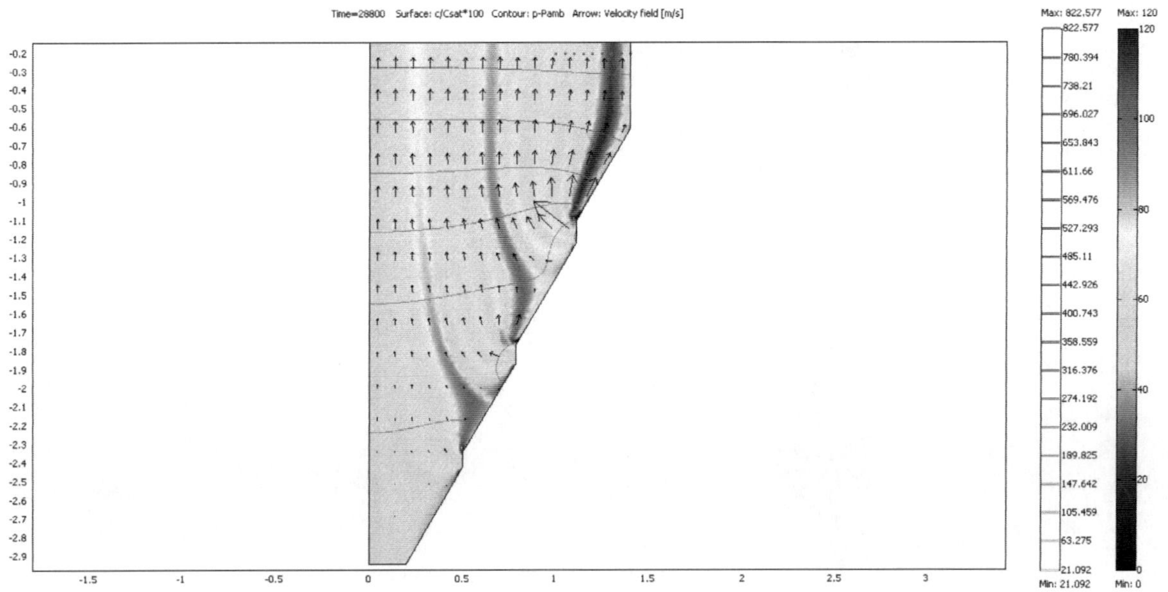

Figure 8. Expected water vapor concentration profile at 8 hours storage in bin

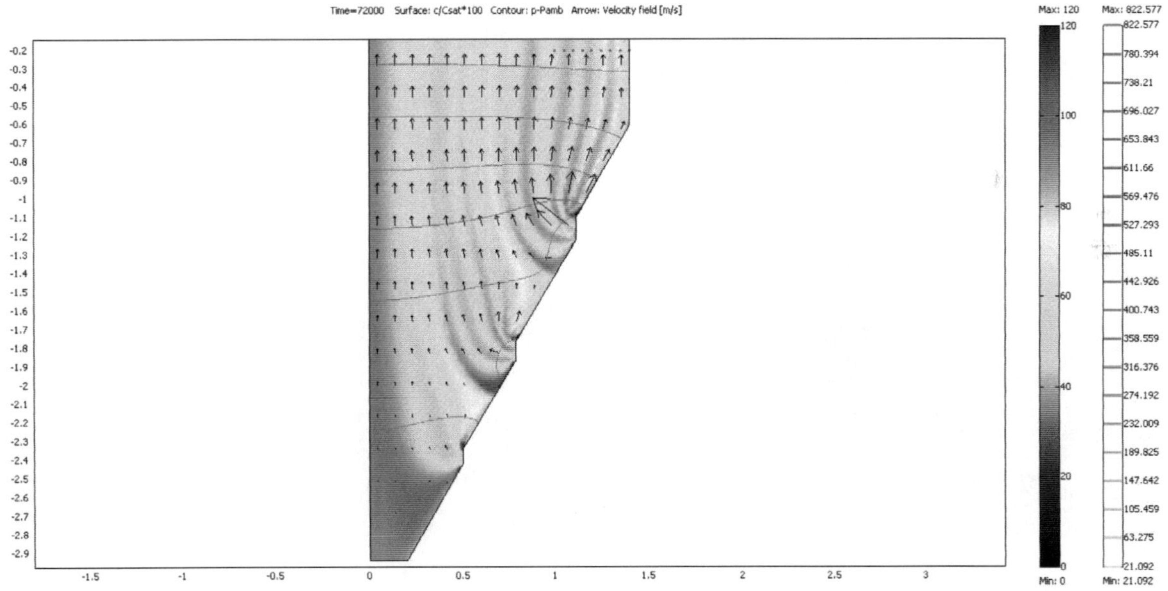

Figure 9. Expected water vapor concentration profile at 20 hours storage in bin

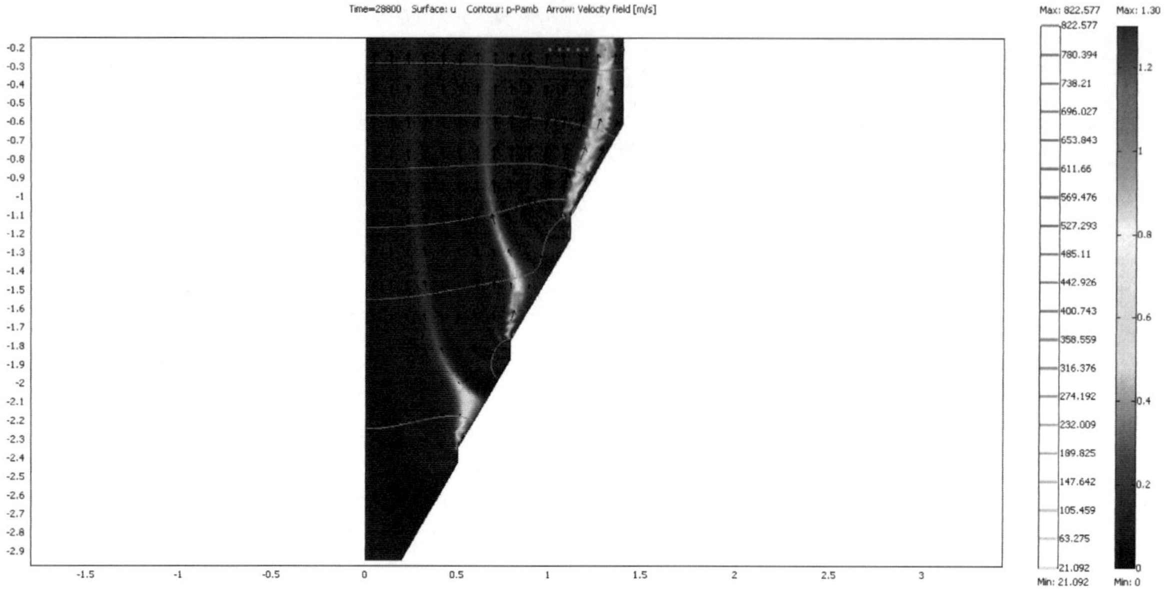

Figure 10. Expected solids water concentration profile at 8 hours storage in bin

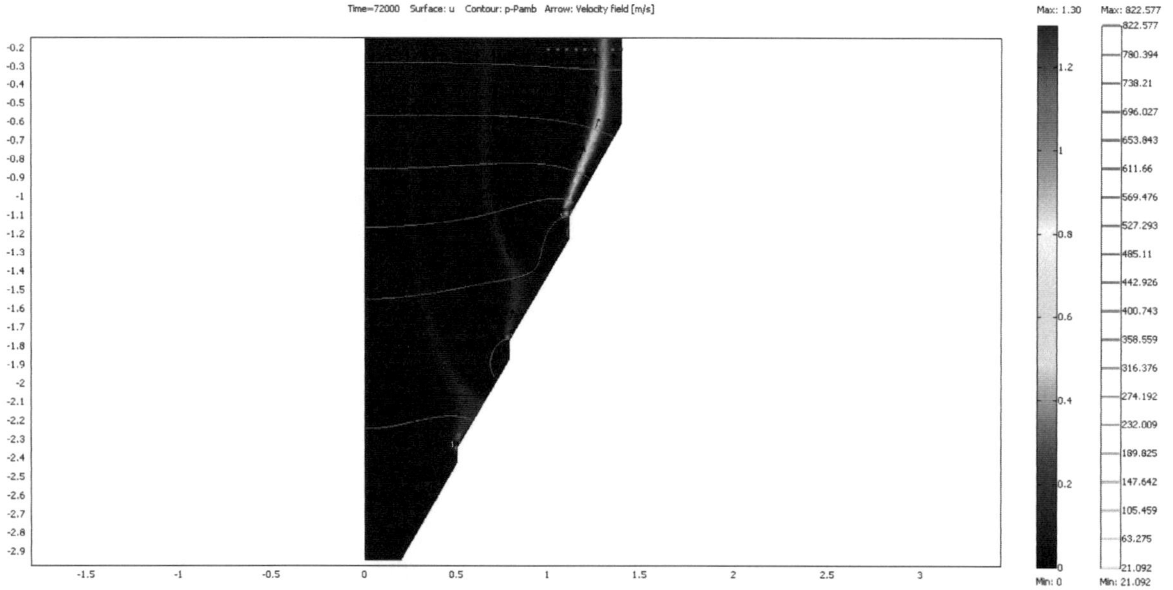

Figure 11. Expected solids water concentration profile at 20 hours storage in bin

4.0 Conclusions

Moisture sorption testing can help identify regions in process equipment that are sensitive to caking or significant lump formation. It is important to note that the relative humidity of the injection gas was 50%, well below the equilibrium concentration for the initial moisture content of the material. A simple analysis suggests that drying would occur with this material without significant problems. However, local gas flow rates and temperature fluctuations create zones where condensation can take place, resulting in caking problems during storage.

5.0 References

1. Vidya Swaminathan and Dane O. Kildsig, An Examination of the Moisture Sorption Characteristics of Commercial Magnesium Stearate, *AAPS PharmSciTech* 2001, 2 (4) article 28, *December 3, 2001*

2. M. HOWES, REFINED SUGAR CONDITIONING AND STORAGE, *Proceedings of The South African Sugar Techlologists,* March *1966*

3. Purutyan, Herman, Pittenger, Brian H, Tardos, Gabriel I, Prevent Caking During Solids Handling, Chemical Engineering Progress, May 2005

4. Tardos, G.I, Nicolaescu, I.V and Ahtchi-Ali, B., Imgress of Atmospheric Moisture into Packed Bulk Powders, Powder Handling & Processing, Vol. 8 No. 1

(304e) Segregation During Pneumatic Conveying of Blended Plastic Pellets

Nestor A. Vasquez, R&D, The Dow Chemical Company, Midland, MI, Karl Jacob, The Dow Chemical Company, Midland, MI, Parvinder Walia, Dow Automotive, The Dow Chemical and Alex Voets, RWTH Aachen University

The transport of plastic pellets in dilute phase pneumatic conveying is largely used in the chemical industry and it has been the subject of study in numerous publications. Many of these studies have focused on the analysis of particle dynamics and collision models, showing how the pellets travel along the conveying line. Thus it has been shown that some materials convey significantly slower than others due to intense bouncing. However, the transport of blended materials, in which the component pellets travel at very different velocity when conveying individually, has rarely been analyzed. The following work focuses on the potential segregation that can occur during the pneumatic conveying of two or three components blends during the transport along the pipeline and at the receiver.

In order to analyze whether segregation developed during pneumatic conveying, a series of conveying trials were performed with blends of two and three components as well as with the individual materials. High speed video analysis was used to measure individual particle velocities. The results of the study showed that:

- Harder materials show very little contact with the pipe wall. - Soft (elastomeric) materials show intense bouncing and slow velocity when they are conveyed individually. - When conveyed as a blend, all components (hard and soft) move with a very similar conveying velocity. The mixture reduces the bouncing of the soft pellets, allowing them to travel faster than in pure state. - No detectable segregation was found during conveying in either the two component or the three component blends. - Once the system is in steady state, segregation will not be an issue unless long conveying distances are used. In this case transient effects could become important and should be studied in more detail.

The segregation at the reception hopper was analyzed by testing two different geometries for the receiver inlet and three types of conveying line materials. The key conclusions were:

- Segregation is highly affected by the development of static charge. - The conveying line material has a strong effect on the development of static charges. - In terms of receiver design, the center inlet with pellet distributor showed the best performance in preventing segregation. - The tangential inlet promotes segregation due to the formation of an internal vortex that intensifies the bouncing of the particles and produces an inclined level of solids. - Reducing the air velocity, without reaching the saltation condition at the pickup point, reduces the segregation at the receiver due to less static charge build up and less bouncing of the pellets.

(304f) Nrel's Integrated Biorefinery Research Facility - Pilot Scale Solids Handling Risk Mitigation

Jane Fisher[1], Dan Schell[2], Rick Elander[2] and Dave Sievers[2], (1)National Bioenergy Center, National Renewable Energy Laboratory, Golden, CO, (2)National Renewable Energy Laboratory, Golden, CO

To achieve commercial-scale production of cellulosic ethanol at a cost that is competitive with gasoline, it is crucial to understand the entire integrated biorefining process and how one stage of the process can impact performance of the others. With the addition of the Integrated Biorefinery Research Facility (IBRF) at the National Renewable Energy Laboratory (NREL), the cellulosic ethanol industry will have access to a significantly expanded pilot plant and biochemical conversion process research facility. At NREL, the goal is to improve the cost effectiveness of cellulosic ethanol production processes to accelerate commercial scale deployment of these technologies. A number of aggressive government policies are guiding NREL's approach to cellulosic ethanol research and development, including the 2007 Energy and Independence and Security Act requiring 36 billion gallons of renewable fuels by 2022, and President Obama's New Energy for America Plan calling for 60 billion gallons of advanced biofuels by 2030. The IBRF's new integrated research and development and flexible process piloting capabilities will enable a wide variety of biofuels technology developers to reduce risks associated with scaling up to demonstration and full-commercial scales. Many of the risks to be mitigated are related to solids handling, which presents a variety of issues and challenges that the cellulosic ethanol industry must resolve to successfully transition the conversion technologies from pilot to demonstration scale and ultimately to commercial scale. Design considerations as well as lessons learned during facility commissioning and startup will be discussed.

See more of this Session: Solids Handling Considerations and Challenges in Pilot and Demonstration Plants
See more of this Group/Topical: Process Development Division

(319a) Generation of Hydrogen-Rich Syngas through Gasification of Fuel Cane Bagasse in a Downdraft Gasifier

C. Andrea Jordan, Ming Tham and Galip Akay, Process Intensification and Miniaturisation Centre, School of Chemical Engineering and Advanced Materials, Newcastle University, Newcastle upon Tyne, United Kingdom

Gasification of biomass for the production of electricity using small scale downdraft gasifier plants is an important route in the development of distributed generation systems matching the distributed nature of biomass availability. By coupling these reactors with high temperature fuel cells, this offers the best opportunity for optimising energy production by high efficiency conversion of low calorific value feedstock. Although high temperature fuel cells are thermally compatible with downdraft gasifiers and are relatively insensitive to micro-contaminants higher efficiencies of conversion of syngas to electricity can be obtained with increased hydrogen content in the syngas. Syngas composition is dependent on a wide range of factors including feedstock composition, gasifying agent, biomass moisture content and gasifier operating conditions. In this study the potential for the production of hydrogen-rich syngas from gasification of fuel cane bagasse in a 50 kWe capacity downdraft gasifier was investigated, increased concentration of hydrogen was observed by varying several parameters. This gasifier has been scaled up to 1 MWe and 1.7 MWe units and the operating data has shown that the performance of both systems is very similar to that of the original small scale (50 kWe) gasifier. The performance of the 50 kWe biomass gasifier was evaluated in terms of equivalence ratio, syngas composition and cold gas efficiency.

See more of this Session: Challenges in Small Scale Syngas Pilot Plants
See more of this Group/Topical: Process Development Division

188384

Scale-up of Membranes for Separation of Hydrogen from Syngas for Carbon Capture

D. H. Anderson, C. R. Evenson, J. D. Faull, and D. S. Jack
Eltron Research and Development Inc., Boulder, CO, U.S.A.

Abstract

In partnership with the U.S. Department of Energy (DOE), Eltron Research & Development Inc. is developing a process technology that separates hydrogen from carbon dioxide in syngas, enabling hydrogen production and clean power generation with carbon capture in coal-fired Integrated Gasification Combined Cycle (IGCC) power plants. The process is based on membranes that permeate pure hydrogen and retain the CO_2 at high pressure to aid its eventual sequestration. Techno-economic analysis has shown a significant improvement in the cost of electricity with carbon capture using the membrane technology in comparison with conventional technology. In addition, membrane separation can achieve nearly complete carbon capture depending on process configuration. Membranes are currently tested at the bench scale in reactors designed for lifetime stability and impurity testing under simulated syngas conditions. Eltron has partnered with Eastman Chemical Company in scaling up this membrane technology for pilot plant testing on coal-derived syngas.

IGCC Plant with CO_2 Capture Using Hydrogen Transport Membrane

Figure 1 shows an example of a process flow diagram of a coal-fired IGCC power plant with CO_2 capture that uses Eltron's hydrogen transport membrane (HTM) for separation of hydrogen from the synthesis gas stream. Coal is gasified at high pressure (>500 psig) and temperature (~1000℃) using steam and nearly pure o xygen from an air separation unit (ASU). The syngas product contains primarily H_2, H_2O, CO, CO_2 plus impurities such as H_2S. The impurities are cleaned from the syngas, which is then sent to a water-gas shift reactor system to convert most of the CO to CO_2 by the reaction:

$$CO + H_2O \rightarrow CO_2 + H_2$$

Hydrogen is separated from the syngas with the HTM to provide hydrogen fuel to produce electricity with a gas turbine in a combined cycle power plant. The remaining syngas contains primarily CO_2 and water at near gasifier pressure, where the CO_2 is sent to compression and sequestration after water removal and purification.

Advantages of the HTM are:

- permeation of pure hydrogen
- CO_2 is retained at high pressure for sequestration
- potential for high (99%+) carbon capture

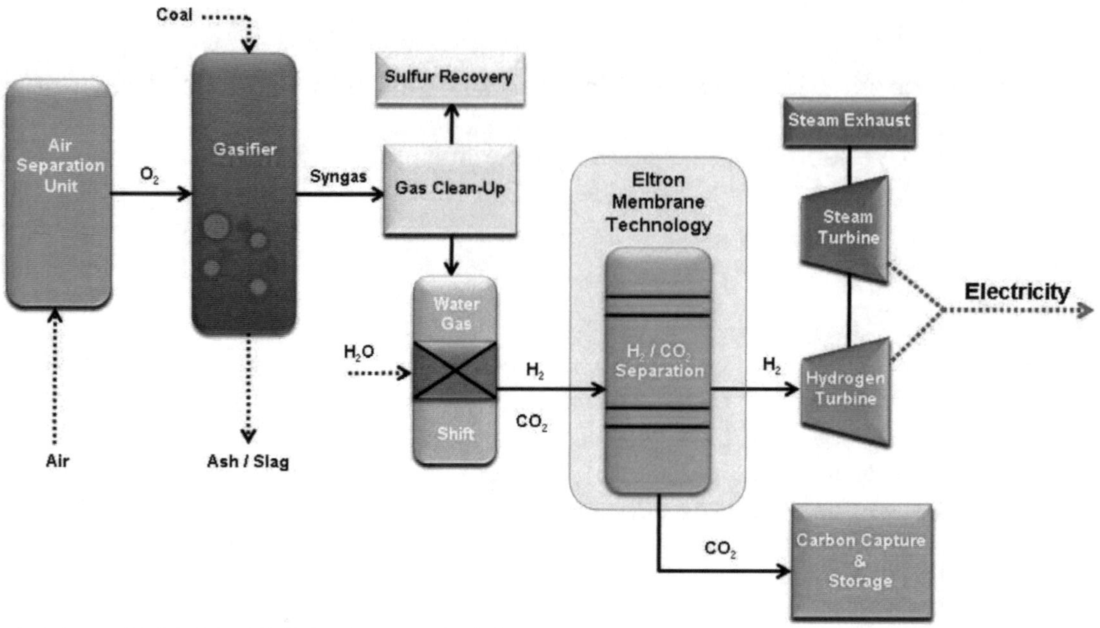

Figure 1. Use of HTM in IGCC plant with CO_2 capture.

Membrane Technology

Figure 2 shows the general concept for Eltron's Membrane System. A dense, multilayer membrane is used to separate hydrogen from a high pressure water-gas shift feed stream. The bulk membrane is a dense metal alloy with high hydrogen permeability. On each side of the bulk membrane is a catalytic layer. On the high pressure (or feed side) of the membrane, a hydrogen-dissociation catalyst is employed. On the low pressure (or permeate side) of the membrane, a catalyst is used to reduce dissociated hydrogen back to molecular H_2. In the presence of a hydrogen partial pressure driving force, hydrogen is transported across the membrane in dissociated form. The system can be operated with a sweep gas for direct injection into a hydrogen turbine or without a sweep gas for pure hydrogen production. The system results in pure hydrogen on one side of the membrane and a high pressure stream of CO_2 and steam with a small amount of CO on the other side of the membrane.

There are two key parts to Eltron's dense metal alloy membranes. The first is the bulk hydrogen separation metal. This metal alloy must have a high flux, possess the mechanical strength to withstand high total differential pressures and high hydrogen partial pressures, and must not alloy with the catalyst layers on either side of the metal alloy. The second key part to Eltron's membrane is the catalyst layers deposited on either side of the bulk metal alloy membrane. On the feed side of the membrane, the catalyst layer must effectively dissociate hydrogen, be resistant to a reasonable amount of hydrogen sulfide, and must not react with the bulk membrane.

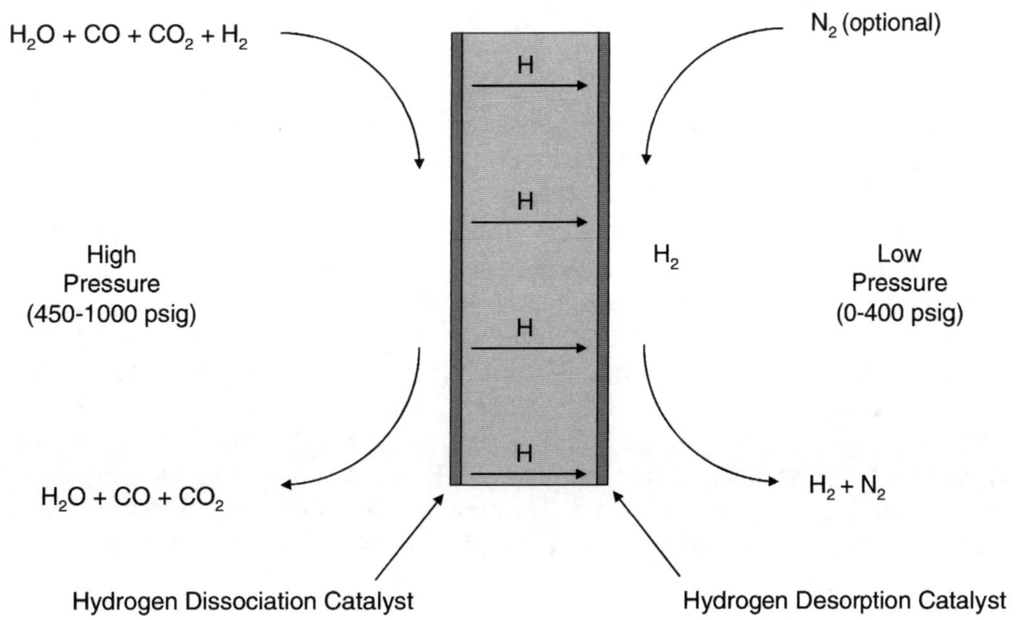

Figure 2. Hydrogen transport membrane schematic.

Figure 3 shows an example of hydrogen flux through a 230 micron thick metal alloy membrane tested in one of our laboratory test modules at 340℃ with a 40% hydrogen feed stream at 450 psig. A sweep gas was used on the permeate side of the membrane at 50 psig. Figure 3 shows that, under these conditions, the membrane had a stable permeation or hydrogen flux rate of 180 mL/min/cm^2 (354 SCFH/sq.ft.).

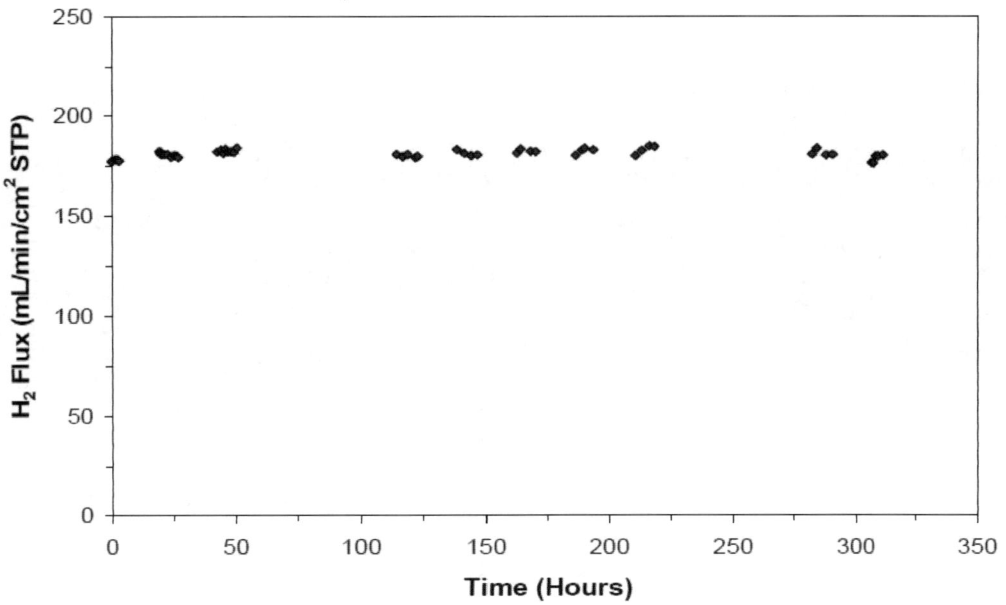

Figure 3. H_2 flux vs. time at conditions for an alloy membrane tested at 340℃ and 400 psig differential pressure across the membrane.

Economic Benefits of Hydrogen Transport Membrane

The economics of using Eltron's hydrogen transport membrane system for CO_2 capture in IGCC plants were outlined in a techno-economic analysis. Three separate cases are shown in Table 1. The base case (Case 1) uses conventional cold gas (Selexol) technology for sulfur removal and CO_2 separation downstream of a high pressure General Electric (GE) gasifier, quench cooling and water-gas shift. Case 2 uses the GE gasifier, followed by conventional cold gas amine (MDEA) treating for sulfur removal, water-gas shift, and Eltron's membrane system for H_2/CO_2 separation. Radiant-convective cooling of syngas from the gasifier is used in Case 2 to avoid high moisture content in the feed to the amine absorber before the shift reactors. Thermal efficiency is improved in this case (partly due to the gasifier cooling method), while cost of electricity decreases slightly. A similar configuration is used in Case 3 (see Figure 4) as in Case 2, substituting warm gas cleaning for the desulfurization step. The cost of electricity in Case 3 decreases by 9.5 $/MWh over the base case, with a 6% increase in thermal efficiency (see Table 1). Another benefit of the combination of warm gas cleaning and the hydrogen membrane is a significant increase in CO_2 capture from 90% to 99%.

Table 1. Summary of Economic Results for the Techno-Economic Analysis of CO_2 Capture in IGCC Plants.

Pre-combustion Gas Cleaning & CO_2 Capture Method	Case 1: 2-Stage Selexol	Case 2: Cold Gas Cleaning & Eltron Membrane	Case 3: Warm Gas Cleaning & Eltron Membrane	Improvement of Case 3 over Base Case
Thermal Efficiency	27.4%	32.0%	33.4%	6.0%
% CO_2 Captured	90%	90%	99%	9%
Cost of Electricity ($/MWh)	115.5	114.5	106.0	9.5

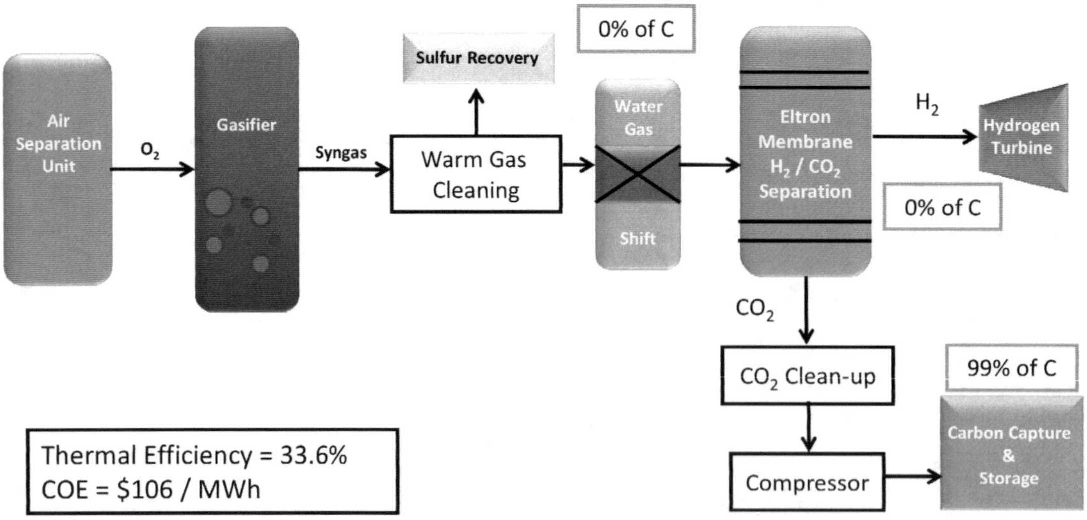

Figure 4. Case 3: CO_2 capture with warm gas cleaning and Eltron's H_2 separation membrane.

Scale-up

Eltron has partnered with Eastman Chemical Company in a U.S. Department of Energy project to scale-up this technology. Eltron and Eastman are currently constructing a 12 lb/day module for operation on a coal-based syngas slipstream in Kingsport, Tennessee by the end of 2010. This step will be followed by design, fabrication, and operation of a 250 lb/day module.

In addition, the U.S. Department of Energy has recently awarded Eltron R&D a project to accelerate development of a 4–10 ton/day module. Design of this module will be based on the results of the smaller scale modules. Eltron is currently down-selecting an appropriate site for this pre-commercial scale-up step.

2010 AIChE Annual Meeting, Salt Lake City, UT

Extrapolating Syngas Pilot Plant Data through CFD Modeling

Umesh Shah[1], Kerry McKenna[1], Alex Gorodetsky[2], Yakov Gordon[1]

[1]Hatch Limited, 2800 Speakman drive, Mississauga, Ontario L5K 2R7, Canada (www.hatch.ca)

[2]Alter NRG Corp., 700, 910-7 Avenue SW, Calgary, Alberta T2P 3N8, Canada (www.alternrg.com)

Introduction

2010 AIChE Annual Meeting, Salt Lake City, UT

Alter NRG Plasma Gasifier (APG) is well suited for gasification of the following feedstock:

- Biomass
- Municipal Solid Waste
- Hazardous Waste

Commercialization of Plasma Gasification

- Process simulations to predict viability of the commercial scale gasifier.
- Pilot Plant test work to prove the process.
- Scale-up from pilot scale to commercial scale.

WorkingTogether SAFELY

Design Approach and Tools

Process Simulators

- Predict the process efficiency based on theory
- Process simulators provide overall heat and mass balance
- Tuned based on the reaction mechanism/kinetics of existing commercial scale precedents
- Limited ability to factor economies of scale

Gasification pilot plants

- Technology demonstration and testing of a particular feedstock
- Provide data using which full scale plant design can be optimized
- Expensive to make design changes
- Difficulty in extrapolating pilot plant data to full scale design
- Provide data only at discrete locations. Many times difficult to interpret.

Computational fluid dynamics (CFD) models

- CFD models provide detailed temperature/velocity/composition distribution
- Easily adaptable to study many what if scenarios at a minimal cost
- Simulate full scale units and optimize its design.
- Require simplifying assumptions due to the complexity of the process
- Necessary to validate modeling assumptions using pilot plant data.

WorkingTogether SAFELY ALTERNRG HATCH™

2010 AIChE Annual Meeting, Salt Lake City, UT

Alter NRG Plasma Gasification (APG) Demonstration Facility

- Testing, development and evaluation using high temperature plasma heating
- 30 years and $100 million in research and development
- Well instrumented to measure temperature and gas compositions
- Serves to
  - Test a particular feedstock and determine optimal design conditions
  - Validate computer model assumptions

Westinghouse Plasma's world-class facility
WPC's 48 tonne-per-day commercial demonstration facility located in Madison, PA

WorkingTogether SAFELY   ALTER NRG   HATCH™

2010 AIChE Annual Meeting, Salt Lake City, UT

Pilot Plant Challenges

Expensive to make design changes for specific process requirements.

Provides limited distribution of flow and temperature.

Difficult to interpret all measurements due to dynamics and complexity of the process.

Full scale units often differ in design/shape/feeding system from the pilot scale units.

Difficult to ensure similar design conditions (flow/temperature/pressure) in pilot units as is planned for full scale.

WorkingTogether SAFELY ALTERNRG HATCH™

Computer Models

Process simulators such as Hysys, Aspen, VMGSim, etc. to model heat and material balance of the process.

Computational fluid dynamics (CFD) tools, e.g. Ansys Fluent/CFX, StarCD, etc. to model detailed flow, temperature, pressure and composition distribution through out the reactor volume

Alter NRG s plasma gasification technology may be defined as wet bottom counter current moving bed where the oxidants are injected into the packed bed of fuel/coke.

- Complex physical and chemical processes
- Computer models require simplifying assumptions
- Assumptions require validation based on pilot plant data or support of research in literature/elsewhere

Working Together SAFELY ALTER NRG HATCH™

APG Process - CFD Model Challenges

- Moving packed bed of irregular shape/size solids. Not possible with existing CFD modeling capabilities.
- Heterogeneous and homogeneous chemical reactions
  - A number of reactions, however, global reaction scheme can be defined based on literature.
  - Homogeneous gas phase reactions Fast under high temperature and approach equilibrium quickly. Mostly mixing limited
  - Heterogeneous gas-solid reactions dependent on solid fuel properties Difficult to model. Simplification/tuning and validation are required
- Jet penetration and high voidage in the proximity of tuyere gas entry
  - How to model increased voidage for the given tuyere gas entry momentum and bed properties?

To produce a reliable CFD model, validation with pilot plant data in addition to a number of systematic simplifying assumptions are required.

WorkingTogether SAFELY ALTERNRG HATCH™

CFD Model Definition and Validation Requirements

2010 AIChE Annual Meeting, Salt Lake City, UT

- Solid bed   Porous media approach
- Bed porosity   Variable porosity as a function of gas momentum and distance from wall
- Gasification reactions   global reaction scheme
- Validation requirements
  - Bed resistance   Pressure drop as a function of velocity
  - Tuyere gas penetration depth   Tuning porosity vs. gas momentum relationship based on slag melting cupola/blast furnace experience and experimental data/correlations
  - Temperature distribution   Tune the global reaction rates to reproduce the APG pilot plant temperature measurements

WorkingTogether SAFELY   ALTERNRG   HATCH™

Tuyere Gas Jet Penetration Depth Correlation Based Prediction

2010 AIChE Annual Meeting, Salt Lake City, UT

- Gas penetration depth defines the oxidant distribution its contact with the solid bed.
- Experimental slag melting cupola
- Solid bed granules 20-40mm
- Air tuyere injecting gas at about 45m/s
- CFD modeling with previously stated assumptions to determine bed velocity and density profile

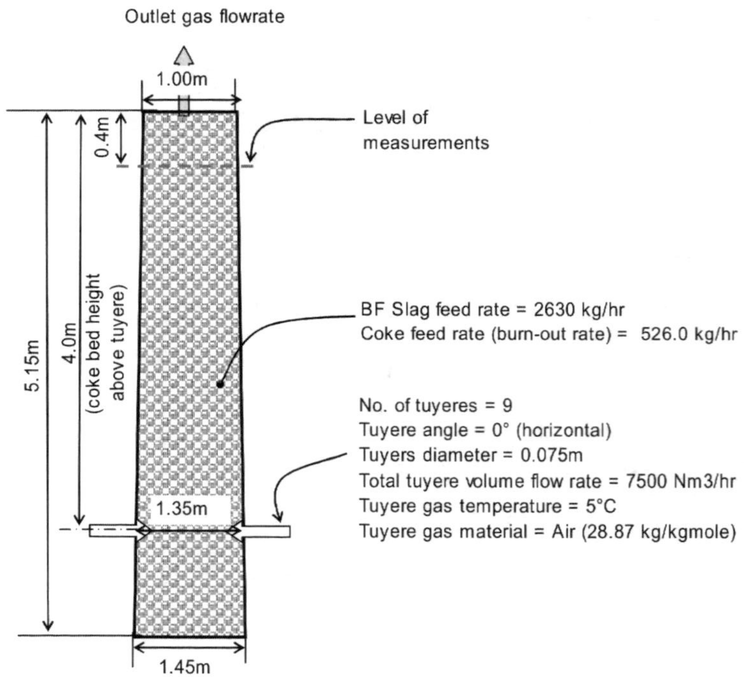

WorkingTogether SAFELY ALTERNRG HATCH™

Tuyere Gas Jet Penetration Depth - Semi-empirical correlation

2010 AIChE Annual Meeting, Salt Lake City, UT

- Semi-empirical correlation predicting raceway diameter for blast furnace pellet bed or any packed bed with similar conditions
- Function of bed properties, tuyere design, and gas properties.
- Applicable to APG reactor condition with validation.
- Hatch s years of experience in the area of blast furnace and cupola designing to validate/interpret the model predictions.

Jet Penetration Prediction for Experimental Cupola

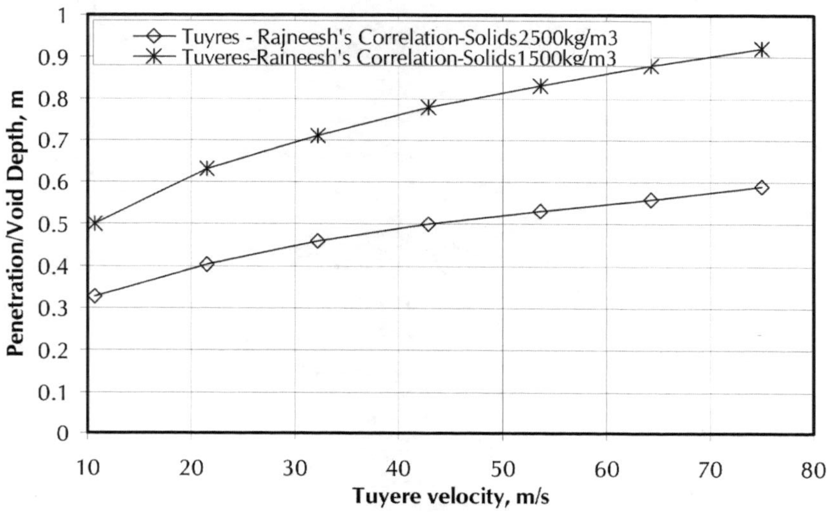

WorkingTogether SAFELY   ALTERNRG   HATCH™

Packed Bed Modeling   Porosity Definitions

2010 AIChE Annual Meeting, Salt Lake City, UT

It is known that the porosity of a packed bed is higher at the walls due to contact between solids and flat walls.

Porosity vs. Distance from reactor wall

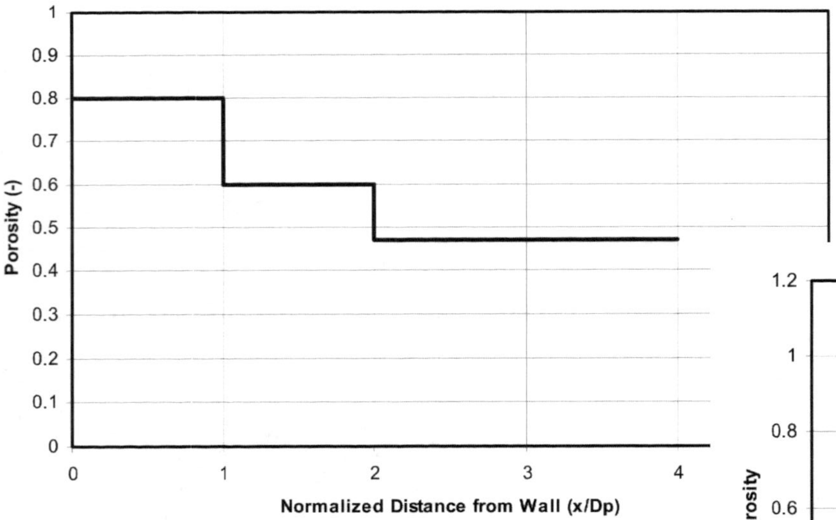

Porosity vs. Gas momentum

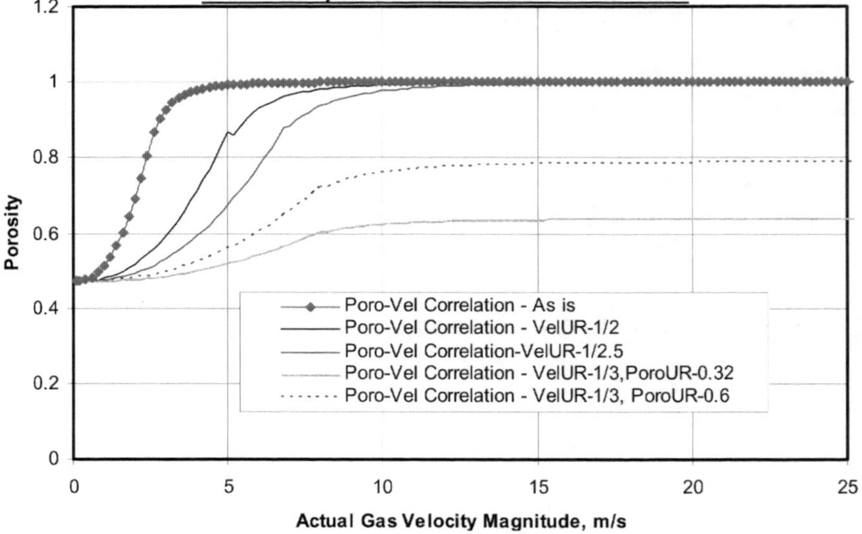

WorkingTogether SAFELY   ALTERNRG   HATCH™

Tuyere Gas Penetration Depth Correlation Implementation in CFD Model

2010 AIChE Annual Meeting, Salt Lake City, UT

- Porosity increase near the tuyere due to high gas momentum
- Literature based correlation implementation in CFD model
- Jet penetration or void diameter prediction
- Significant influence on tuyere gas distribution and potentially the process efficiency.
- Made it possible to simulate the flow through the packed bed of APG reactor.
- Validation required for final design of full scale reactor.

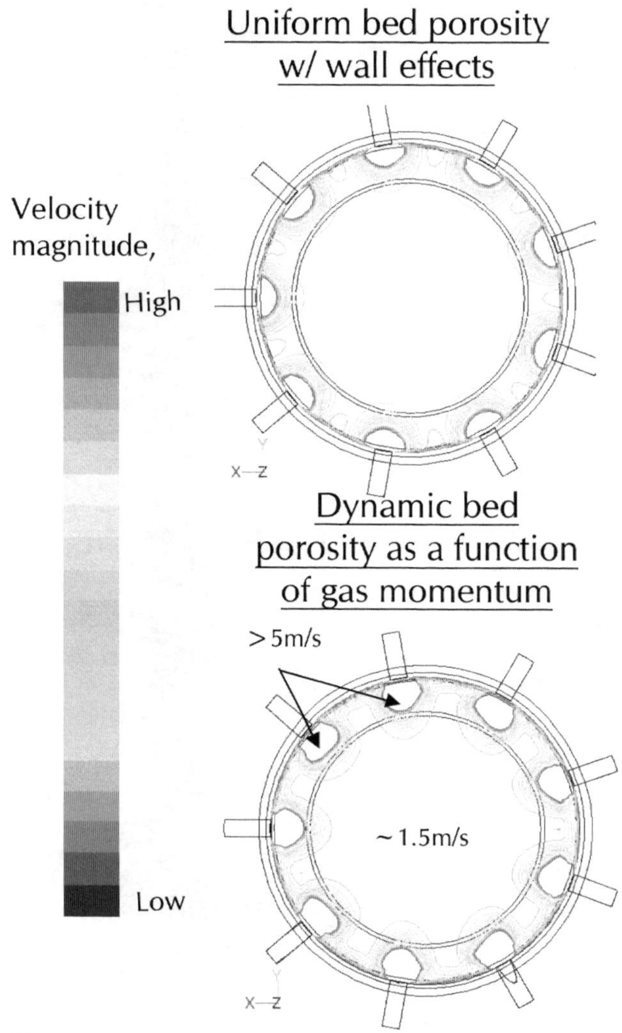

WorkingTogether SAFELY ALTERNRG HATCH™

Tuyere Gas Penetration Depth Correlation Implementation in CFD Model

2010 AIChE Annual Meeting, Salt Lake City, UT

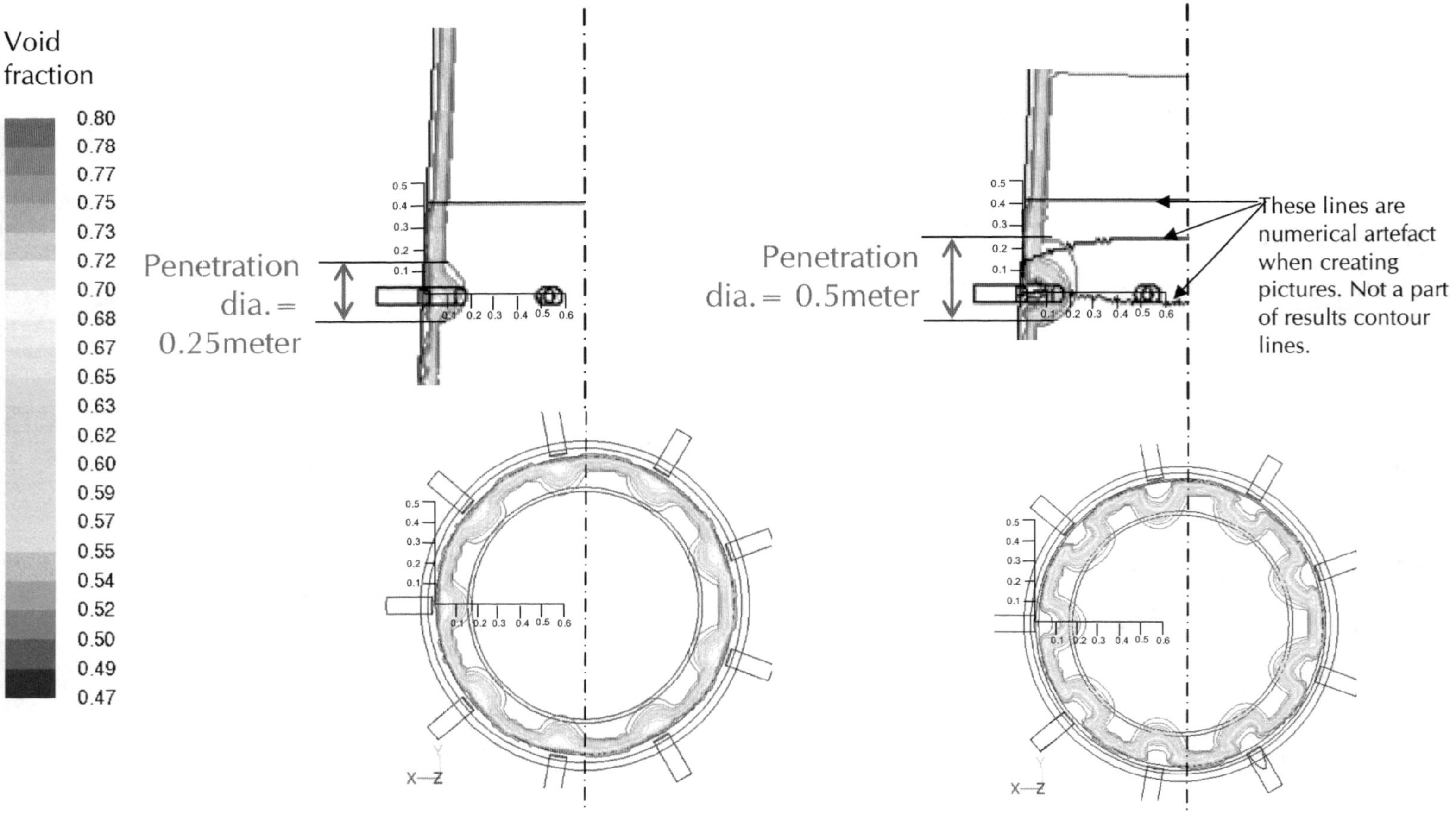

WorkingTogether SAFELY · ALTERNRG · HATCH

2010 AIChE Annual Meeting, Salt Lake City, UT

Model Results vs. Measurements

CFD model captures the bed resistance effect well in qualitative terms.

The data not sufficient to validate the tuyere gas penetration depth or any near tuyere high voidage formation.

Validation data required.

Hatch s experience in designing blast furnace/cupola added additional confidence interpreting/validating the predictions.

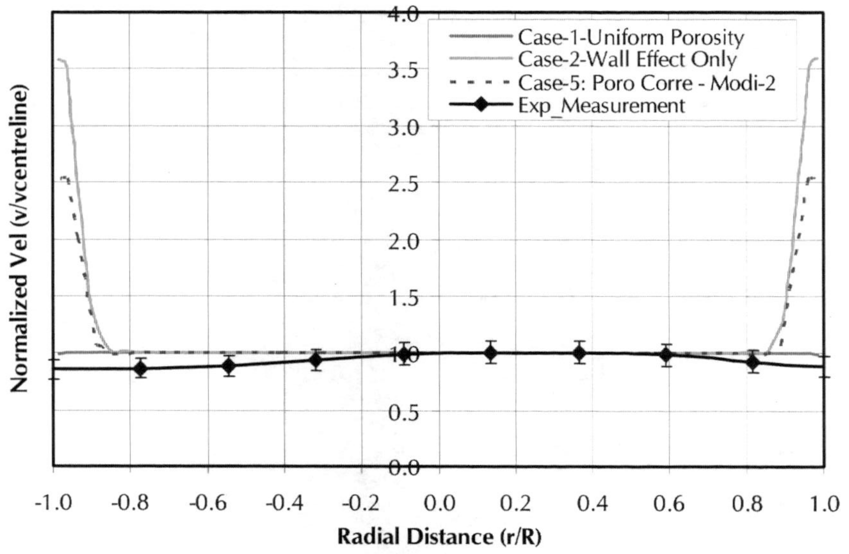

WorkingTogether SAFELY ALTERNRG HATCH™

2010 AIChE Annual Meeting, Salt Lake City, UT

Temperature Distribution APG Pilot Plant

- APG pilot plant cupola shaft data to calibrate the CFD model heat/mass transfer
- CFD model purpose
  - Extrapolate the measured temperature data
  - Tune reaction rates to match the temperature data
  - Identify the flow distribution that resulted into the measured product gas distribution
  - Apply the calibrated CFD model to the optimization of full scale design

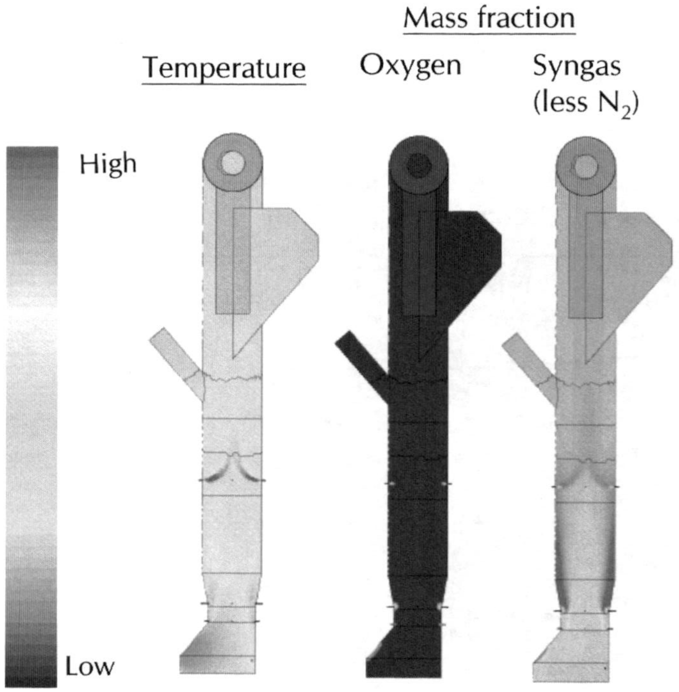

WorkingTogether SAFELY ALTERNRG HATCH™

Application of CFD Model to Full Scale Design

- Bridge pilot plant data and full scale design
- Ensure that the overall process criteria are met
- Tune the design for a specific feedstock/condition
- Provide opportunity to further optimize the design
- Study sensitivity to various parameters
- Produce visual images that stimulate design engineers to ask why? and what if? → Better design

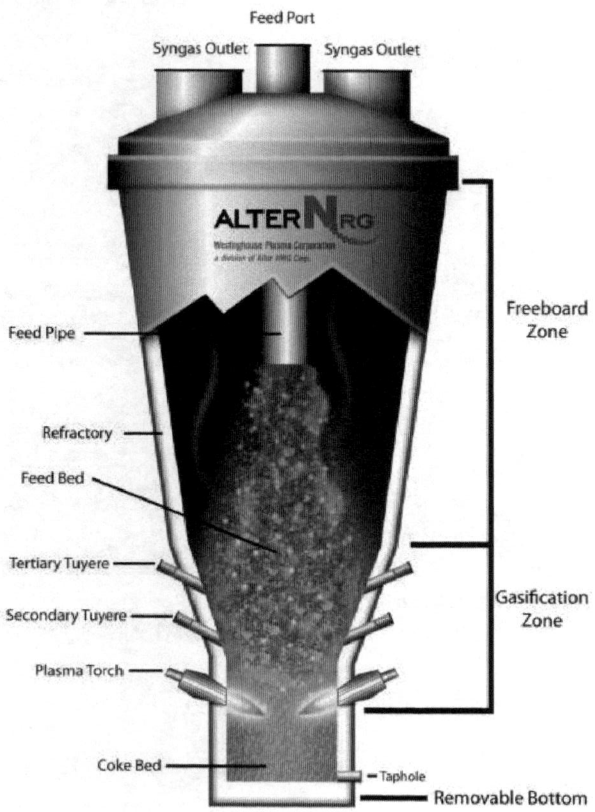

Working Together SAFELY ALTER NRG HATCH™

Application of CFD Model to Full Scale Design

Purpose: Improve the oxidant distribution through the fuel bed

Porosity-Velocity Correlation Tuned/validated ,

- Defined fuel bed height and its porosity/permeability
- Variable porosity
- Bed shape was defined to material angle of repose : Flatter the bed, better is the gas distribution
- Optimized tuyere size and shape combining model predictions and years of experience designing blast furnace/cupola.

Tuyere size:Base case

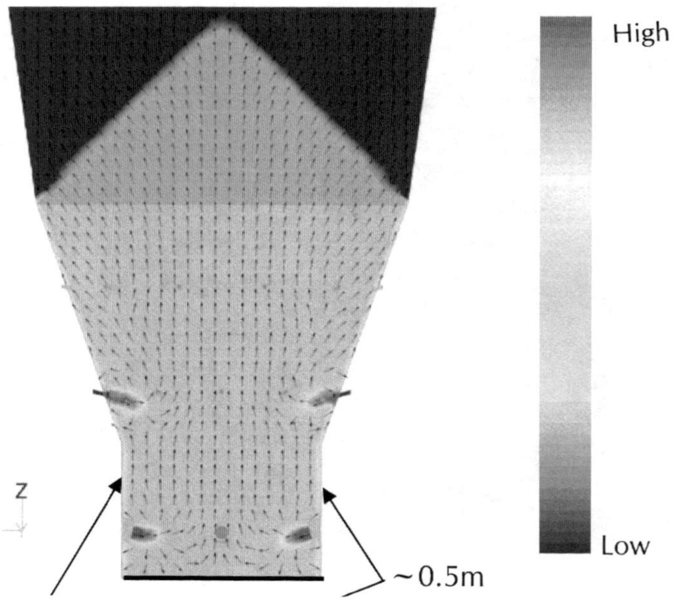

CFD implementation of porosity-velocity correlation was modified to match the penetration depth (depth of high voidage region) based on Rajneesh s semi-empirical correlation

WorkingTogether SAFELY   ALTERNRG   HATCH™

2010 AIChE Annual Meeting, Salt Lake City, UT

Application of CFD model to Full Scale Design

Purpose: Improve the oxidant distribution through the fuel bed

Smaller sized tuyere improved oxidant distribution by 10%.

Secondary tuyere gas mass fraction

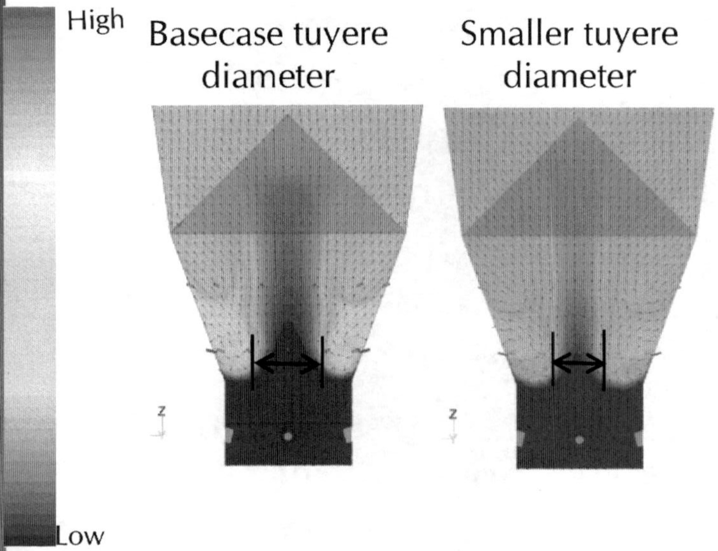

Tuyere gas distribution through the charge bed (Volumetric coverage)

Design case	Total tuyere gas distribution
Basecase tuyere diameter	80% of charge bed volume
Smaller tuyere diameter	88% of charge bed volume
% improvement	~10%

WorkingTogether SAFELY   ALTER NRG   HATCH™

Summary

It was demonstrated that a combined use of pilot plant testing/data and computer models can serve as a valuable tool for commercial gasifier design.

APG process involves moving packed bed with interphase heat/mass transfer and chemical reactions. It poses significant challenges to the current state of the art CFD techniques. CFD model simulating the APG process was developed using systematic simplifying assumptions.

Pilot plant data, previous research in the field, and extensive experience in the design/optimization of blast furnace/cupola were utilized to interpret, tune and validate the CFD model predictions.

Such calibrated models are valuable to process/design engineers to study many what if scenarios and assist in the optimization of a full scale reactor.

2010 AIChE Annual Meeting, Salt Lake City, UT

References

1. Gordon, Y., Mirkova, R. and Shvidkiy, V. Regularities of gas flow formation in shaft furnaces with tuyere gas entry. Investia of Vusov, Chernaya Metallurgia, Vol. 2, 1990, pp. 89-92.
2. Rovnushin, V., et al. Cokeless treatment of titania-magnetite ores. Moscow, Metallurgy, 1988, p. 63.
3. Goldshtik, M.A. Transfer Processes in Granular alyers. 1984, pp. 8-18.
4. Augier, Laroche, and Brehon. CFD-Based Adsorption Modelling. 6th International Conference on CFD in Oil and Gas, Metallurgical and Process Industries, 10-12 June 2008

(319d) Integration and Operation of Pilot Plants in a Commercial Gasification Facility

Glenn Shoaf[1], Paul Fanning[1], Michael Brannon[1] and Jerry Schlather[2], (1)Gasification Process Development, Eastman Chemical Company, Kingsport, TN, (2)Eastman Chemical Company, Kingsport, TN

Eastman Chemical Company has operated a coal gasification facility in Kingsport, TN, for over 27 years. However, until recently, Eastman did not have the ability to use its commercially derived syngas to test alternate processes, sorbents or catalysts. In partnership with the Department of Energy (DOE) and others, Eastman has installed a number of pilot plants that can use commercial syngas as their feedstock. This paper will discuss the challenges faced with the integration and operation of these R&D units in a commercial gasification facility.

See more of this Session: Challenges in Small Scale Syngas Pilot Plants
See more of this Group/Topical: Process Development Division

PRESSURIZED OXY-BLOWN FLUIDIZED BED BIOMASS GASIFICATION SYSTEM

Dr. Ravi Randhava & Dr. Richard Kao
Unitel Technologies, Inc., Mt. Prospect, IL

Abstract

SynGest Inc. intends to build and operate multiple mini-plants to make anhydrous ammonia fertilizer from biomass, and engaged Unitel to design its front-end gasification system to produce hydrogen-rich syngas from field dried corn cobs.

The essential requirement stipulated by the customer had to do with minimizing the formation of methane, volatiles such as BTX, and tars. The HarvestGas process developed by Unitel for SynGest includes a pressurized three-stage progressively expanding bed gasifier that is fluidized with oxygen and steam. Any methane, volatiles and tars exiting the disengagement section of the gasifier are reformed by means of a downstream oxy-blown autothermal reactor. A cryogenic air separation unit provides the oxygen for the gasifier and the autothermal reformer, plus pure nitrogen for the ammonia synthesis section.

Background Notes

According to the U.S. Department of Energy (USDOE), by the end of 2010, the worldwide gasification industry will exceed 70,000 MWth of syngas output from 155 plants and 451 gasifiers. In terms of feedstock choices, coal continues to maintain its lead, followed by petroleum (including fuel oil, refinery residue, naphtha, etc.) and natural gas. Biomass accounts for approximately 2-3% of the syngas that is currently produced, but this number is projected to grow to 4-5% by 2015.

Industrial level gasification of biomass is a relatively new practice. Most of this activity involves the use of air-blown gasifiers to make low BTU gas for steam boiler applications. A limited amount of biomass conversion is being carried with oxy-blown gasifiers to make a syngas fuel for turbines. In both cases, the objective is to maximize the methane content in the end product, thereby achieving a proportionately higher heating value.

SynGest's goal in biomass gasification is the exact opposite of the above – to minimize the amount of methane in the syngas, thereby creating a more desirable starting point for manufacturing chemicals such as ammonia, dimethylether (DME) and methanol. In these cases, methane is a bothersome inert that generates inefficiencies in the conversion process.

After reviewing a variety of gasifiers, it became readily apparent that shoehorning existing designs would not meet SynGest's requirements. Instead of spending more time and money at trial-and-error retrofits, Unitel elected to design a new gasifier from the ground up, one that is specifically tailored to optimize the conversion of biomass into a syngas that contains minimum amounts of methane and tars.

For the record, under the auspices of the Xytel Group, Xytel-Bechtel and now Unitel, both authors have been involved in designing and constructing hundreds of pilot- and mini-plants, including a variety of gasification systems.

The SynGest Project

SynGest's first bio-ammonia mini-plant will be located in Menlo, Iowa, 45 miles west of Des Moines. The company has purchased 75 acres of land and intends to start construction in 2011.

The SynGest Menlo plant will use 150,000 tons of locally supplied corn cobs per year to manufacture 50,000 tons of bio-ammonia annually, enough to fertilize 500,000 acres of nearby Iowa farmland under corn. Figure 1 graphically shows the "circles of influence" associated with a 150 tons per day (TPD) mini-plant.

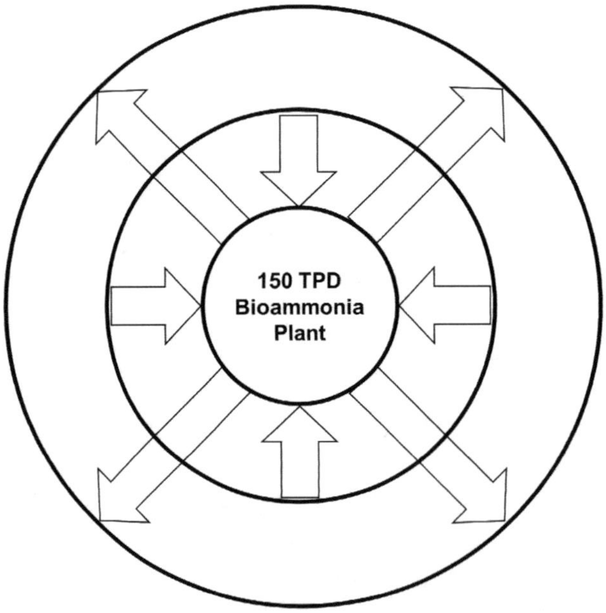

Corn cob feed supply radius: 30 miles
Anhydrous ammonia product distribution radius: 50 miles

Figure 1. Project Logistics

A block schematic of the SynGest process and summary material balance data are presented in Figure 2 and Figure 3 respectively.

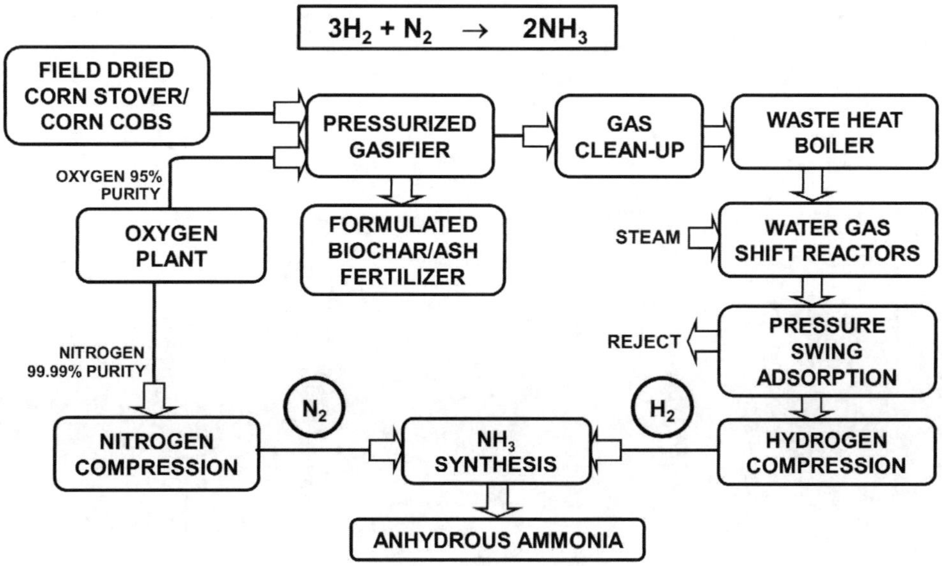

Figure 2. SynGest's Bio-Ammonia Process

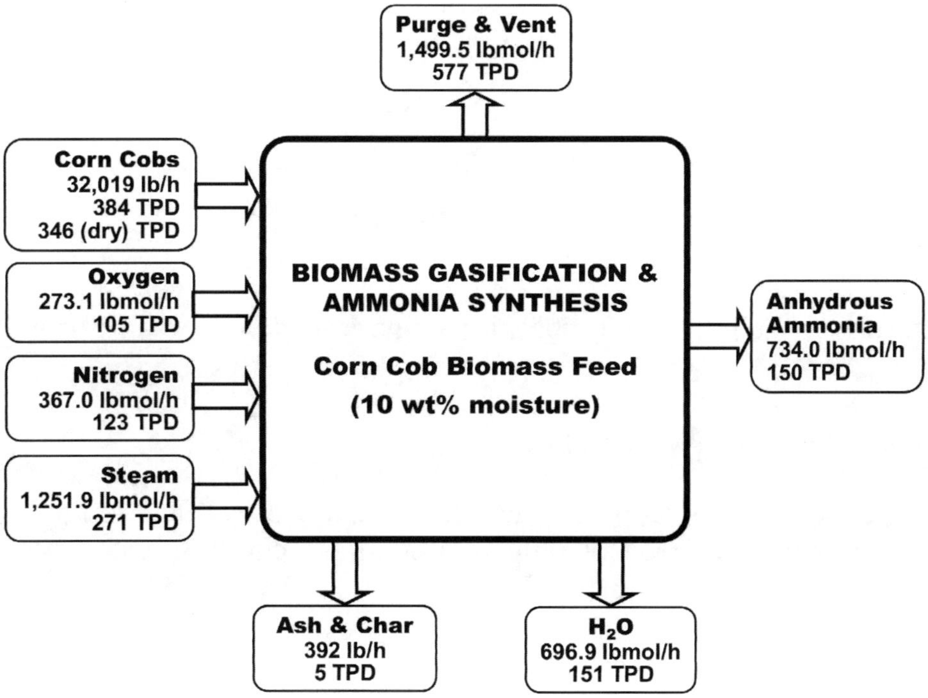

150 TPD Mini-Plant

Figure 3. HarvestGas Process Overview

A block schematic of the biomass gasification process designed by Unitel for the SynGest application is shown in Figure 4.

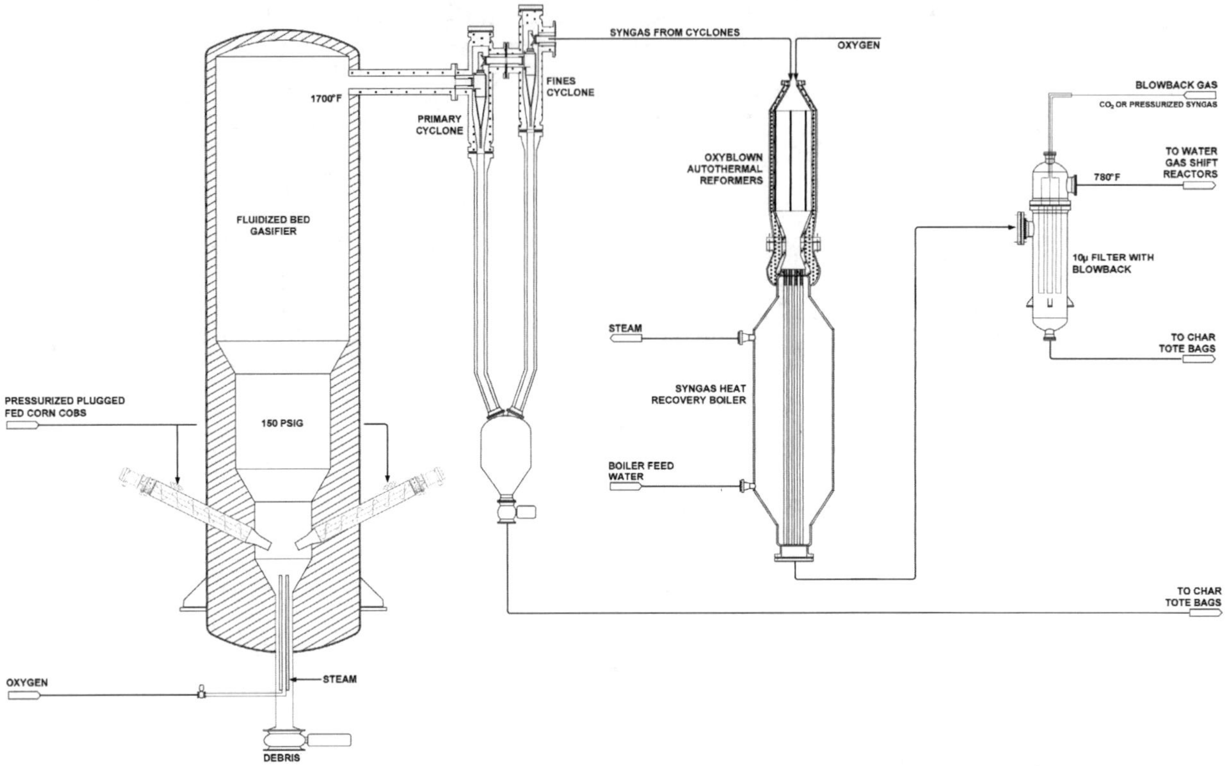

Figure 4. Biomass Gasification System

Following the schematic from left to right, the primary sections of the HarvestGas system include:

- A pressurized plug biomass feed unit.
- A pressurized multi-stage progressively expanding oxy-blown fluidized bed gasifier. Pressure: 15-300 psig. Depending upon location, temperatures within the gasifier can range from 1,450F to 2,000F.
- Cyclones for disengaging the ash and fines.
- An oxy-blown autothermal reformer for converting residual tars, benzene and other volatiles into additional syngas.
- A syngas heat recovery boiler followed by a final filter with blowback.

Gasifier

The geometry of the gasifier for generating syngas to make 150 TPD of bio-ammonia is presented in Figure 5.

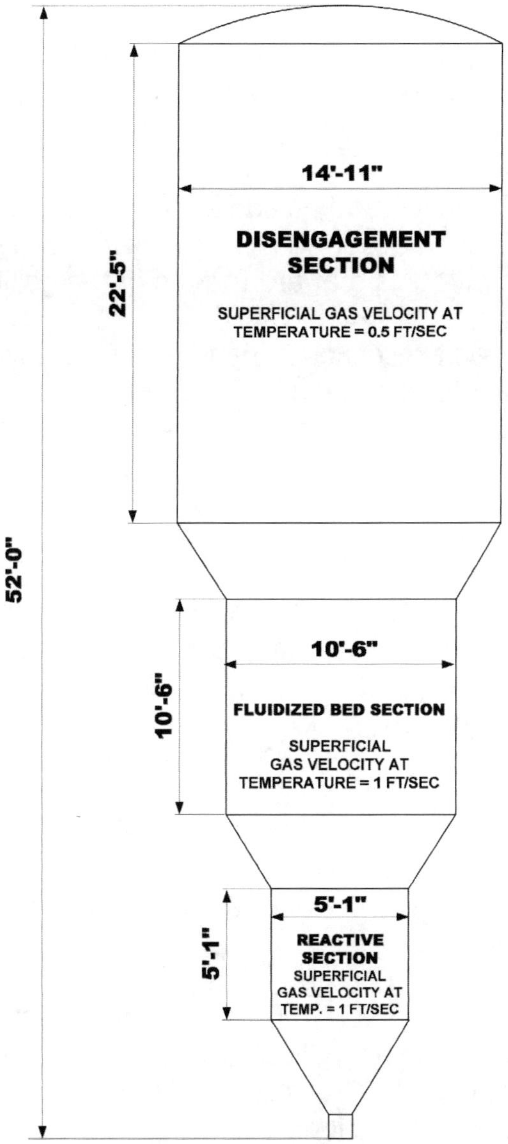

Figure 5. Three Stage Fluidized Bed Gasifier

Biomass is continuously introduced into the reactive section. The location and angle of entry of the feed depend upon the physical properties of the feedstock. Ignition and initial heat-up actions are facilitated by the addition of natural gas through the oxygen feed line. The reactive action of the oxygen is modulated and contained by the steam that is also injected at the bottom of the gasifier. The oxygen and steam entry locations are shown in Figure 4.

The steam shroud helps to stabilize the oxygen lance. Past experience with oxy-blown gasifiers has shown that, unless properly managed, the oxygen flame can create some serious problems. This is more than ever true when the flame is surrounded by biomass that is continuously and simultaneously being consumed and replenished. If the flame is not adequately constrained, it may almost appear to have a temperamental mind of its own that can cause the flame front to change its angle and "dance."

A measured amount of olivine sand is added and maintained inside the gasifier. In addition to facilitating the fluidization action, the olivine helps to crack a significant amount of the tars and other hydrocarbon products that are formed in the gasifier into syngas

The products from the reactive section expand and flow upwards into the fluidized bed section where they are subjected to intense mixing. The expansion is modulated by the taper of the transition zone between these two sections. Based upon our calculations, we are looking at a residence time of 15-30 minutes in the fluidized bed section. As the reactants fill up this section, they will continue to expand and flow upwards into the disengagement section.

A certain amount of fluidized bed activity will continue to take place in the lower part of the disengager, characterized by a dense phase that is 2-3 feet in depth. The freeboard height of this section must be sufficient to enable adequate separation of the particulate matter (char and ash) from the syngas product.

The gasifier shell is made of high strength carbon steel, wherein the thickness of the steel depends upon the exact operating pressure. In the reactive section, the carbon steel is lined with castable refractory to handle a working temperature of 2,000F, with its thickness calculated so that the external temperature is about 300F, thereby allowing a controlled amount of heat to flow out radially.

The reactive section with its castable refractory is connected to the rest of the gasifier with flanges so that it can be occasionally pulled apart for maintenance and upkeep. The two upper sections are lined with spray-on refractory that is easy to set up and is not subject to the hard duty required by the reactive section.

Oxy-Blown Autothermal Reformer

The reformer serves to convert any residual tars, benzene, toluene and xylenes in the gasifier effluent into additional syngas. Most of the methane in this gas mixture is also reformed into syngas.

Autothermal reforming combines the heat effects of partial oxidation and steam reforming reactions by feeding the humid syngas and oxygen into the reformer. This process is carried out in the presence of a catalyst that controls the reaction pathways, and thereby determines the relative extents of the oxidation and steam reforming reactions. The presence of steam, oxygen and the use of an appropriate catalyst enables lower temperature operation and greater product selectivity to favor the formation of H_2 and CO, while inhibiting the formation of coke (solid carbon).

Using the SynGest gasification system as a reference, Table 1 shows a) the composition of the syngas leaving the gasifier versus b) the composition of the syngas after the oxy-blown autothermal reforming step.

Table 1
Syngas Composition

CORN COBS	
COMP.	WT%
C	46.58
H	5.87
O	45.67
N	0.47
S	0.05
ASH	1.36
TOTAL	100.00
Moisture	10.00

	SYNGAS FROM GASIFIER	AFTER OXYBLOWN AUTOTHERMAL REFORMER
COMP.	VOL%, WET	VOL%, WET
H_2	26.65	34.23
CO	29.17	30.47
CO_2	16.15	15.18
CH_4	6.03	2.00
N_2	0.22	0.20
H_2O	21.76	17.90
H_2S	0.02	0.02
TOTAL	100.00	100.00
TEMP.	1750 F	1540 F
PRESSURE	150 PSIG	142 PSIG
FLOW	5,234 ACFM	5,365 ACFM
FLOW	13,726 SCFM	14,791 SCFM

HarvestGas Applications

The syngas generated by the HarvestGas system has a H_2/CO molar ratio that is inherently close to 1. Some significant applications of this biomass derived gas product are cited below, along with the approximate H_2/CO molar ratio that is preferable in each instance:

- Dimethylether (DME), H_2/CO = 1, no need for ratio adjustment.
- Fischer-Tropsch mixed alcohols, H_2/CO = 1.25, ratio can be adjusted by water gas shift (WGS).
- Fischer-Tropsch diesel, H_2/CO > 2, with WGS.
- Methanol, H_2/CO = 2.25, with WGS.
- Hydrogen for fuel cells, H_2/CO > 50, with WGS.
- Ammonia, H_2/CO > 50, with WGS.
- Ethanol via catalysis, H_2/CO = 1.25, with WGS.
- Ethanol via fermentation of syngas, H_2/CO < 1, with controlled addition/injection of CO_2 into the biomass gasifier.

(319f) Experience with a Pressurized Fluidized Bed Biomass/Black Liquor Gasifier

Kevin J. Whitty[1], Daniel Sweeney[2] and Brett Christensen[2], (1)Dept. of Chemical Engineering; Institute for Clean & Secure Energy, The University of Utah, Salt Lake City, UT, (2)Institute for Clean & Secure Energy, The University of Utah, Salt Lake City, UT

In 2005 the University of Utah started up a small-scale (200 kW) steam-blown, pressurized fluidized bed black liquor gasifier. During the first year of operation, many operational challenges were encountered and overcome. By the second year, the system was able to operate very reliably and several successful campaigns of 100+ continuous hours were carried out. Recently the system has been modified to process more conventional solid biomass. Feeding solid biomass into a pressurized system is always a challenge. This paper describes the experience with this gasifier over the past five years, including startup, shutdown, operation, control, and sampling of sygnas and tars.

See more of this Session: Challenges in Small Scale Syngas Pilot Plants
See more of this Group/Topical: Process Development Division

(319g) Commissioning of a 500 KW Pressurized Entrained-Flow Coal Gasifier

Kevin J. Whitty[1], Travis Waind[2] and David Ray Wagner[2], (1)Institute for Clean & Secure Energy, Dept. of Chemical Engineering, The University of Utah, Salt Lake City, UT, (2)Institute for Clean & Secure Energy, The University of Utah, Salt Lake City, UT

The University of Utah recently commissioned a small pilot-scale, slurry-fed, oxygen-blown, pressurized entrained-flow coal gasifier. The system is designed for operation at pressures to 450 psi (31 bar) and is rated for approximately 70 kg/hr of coal at maximum pressure. Generally speaking, startup of the system has gone well. But, there have been challenges along the way. This paper describes the experience with this system to date, including challenges with feeding, operation and control of the system.

See more of this Session: Challenges in Small Scale Syngas Pilot Plants
See more of this Group/Topical: Process Development Division

(335a) Membranes Technologies for Chemical Processes to Produce Clean Water and Energy, Biofuel and Pharmaceutics

T.S. Chung, Chemical and Biomolecular Engineering, National University of Singapore, Singapore, Singapore

Clean energy and water shortage, affordable healthcare and global warming are four major concerns globally resulting from resource depletion, record high oil prices, clean water shortages, high costs of pharmaceuticals, and changing climate conditions. Among many potential solutions, advance in membrane technology is one of the most direct, effective and feasible approaches to solve these sophisticated issues for chemical industries. Membrane technology is a fully integrated science and engineering which consists of materials science and engineering, chemistry and chemical engineering, separation and purification phenomena, statistical mechanics-based molecular simulation, process and product design. We will cover several areas such as:

1. The basic principle of membrane formation 2. Membranes in chemical processes for energy development (CH_4, H_2 and biofuel) and CO_2 capture 3. Membranes in chemical processes for emerging desalination technologies: forward osmosis and membrane distillation 4. Membranes for pharmaceutical separation and life science

Email: chencts@nus.edu.sg Websites: http://www.chee.nus.edu.sg/people/faculty_chungneal.html http://www.chbe.nus.edu.sg/membrane

See more of this Session: High-Throughput and Compact Chemical Processing Technologies
See more of this Group/Topical: Topical G: Innovations of Green Process Engineering for Sustainable Energy and Environment

(335b) Microreactor for Industrial Applcation: Technical and Economical Benefits

Sergio Pissavini, Corning SAS, Avon, France

A glass based high throughput microreactor technology is presented illustrating the basic design window, technical data and configuration. An overview of some class of reactions performance is provided as compared to batch process. For three typical reaction a comparison between typical plant design batch versus micro-reactor is presented with an economical and footprint analysis comparison. The econmical and technical advantage are then projected versus their "green" impact.

See more of this Session: High-Throughput and Compact Chemical Processing Technologies
See more of this Group/Topical: Topical G: Innovations of Green Process Engineering for Sustainable Energy and Environment

DESIGN ENHANCEMENTS OF AN HIGH TEMPERATURE, CERAMIC μ-CHANNEL HEAT EXCHANGER

M. A. Wilson, J. N. Cutts and J. R. Fellows

Ceramatec, Inc.

Salt Lake City, UT

ABSTRACT

The recent developments in the energy and process industries have lead to more efficient, higher-temperature processes and compact equipment. By using small features such as micro-channels, pin fins or plate fins the length scale for which transport occurs is minimized and thus increases the heat transfer. By integrating high-thermal conductivity ceramic foams into these compact heat exchanger designs, even greater benefits of reduced hydraulic diameters and increased surface areas can be realized in environmentally durable, structurally strong materials. In this paper we will discuss several design configurations coupled with the inherent properties of preferred ceramic materials to assess the effects of foam pore size and overall porosity. These foams have been laminated into micro-channel assemblies and co-fired into monolithic devices for operational testing. The interfacial thermal-contact resistances between the porous foams and the dense films were eliminated by a co-firing process. Within these studies hydrodynamic and heat transfer models were used to predict the performance and viability of next generation high temperature heat exchangers.

INTRODUCTION

Open-cell macro-porous silicon carbide (SiC) foams, which provide high surface area to mass ratio, are well suited for heat exchange (HX) designs requiring good corrosion resistance and strengths, at elevated temperatures[1], while also providing excellent thermal conductivity. Strength and permeability, however, are among several considerations that are required in the proper selection of SiC foams used in any particular HX design. The characterization of strength or permeability is often a function of the morphology of a given foam structure, with differing porosities, strut configurations and associated pore per inch (ppi) values. To make proper choices in the selection of a foam used within an HX device, it is important to understand the correlation between the structural morphology and the attendant effects on strength and flow characteristics. Several SiC foams, processed using a replica technique, and designated initially by ppi values, have been processed and prepared to study how the morphology of the sintered foam impacts the strength of the foam body under compressive loads, as well as how the flow rates vary as a function of pore distribution and porosity values.

Figure 1 illustrates how these porous structures can enhance the heat transfer performance in a gas to liquid heat exchanger. For example, fluid channels are small for the relative high-density, high-heat transfer characteristics and low pumping power required for the fluid. The foam side (or gas side) of the heat exchanger provides high surface area and a small length-scale to enhance the heat transfer. Heat is rejected to the micro-channel walls where it is conducted through the porous matrix into the gas-side foam matrix. The cross-section of the gas-side channels is large due to its low-density, low-heat transfer characteristics and low pressure drop constraints. The gas-side integrates "extended fins" throughout the channel increasing the surface area from which heat is convected and reduces the length-scale between "cool gas" and "hot walls". This foam increases this heat transfer area from the traditional x-y planar surface to the surfaces of fibrous struts and membranes reaching into the 3rd dimension that make up the porous body.

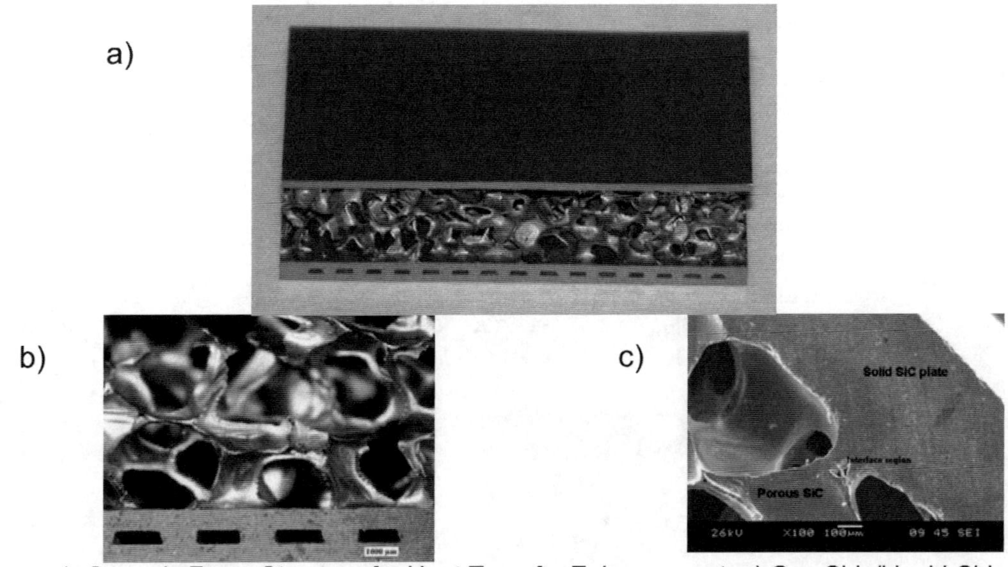

Figure 1. Ceramic Foam Structure for Heat Transfer Enhancement, a) Gas-Side/Liquid-Side Repeat Unit, b) Optical Micrograph Showing Detail of Porous Structure, c) SEM Micrograph of Porous/Dense Interface

The scope of this research was to develop porous ceramic structures and to evaluate their functionality in heat exchanger applications. In its entirety, this research will include the evaluation of the following characteristics of these foams and provide engineering correlations enabling the design and optimization of foam structures for specific applications. However, since this research is incomplete the characteristics to be discussed in this paper are indicated below:

1) Mechanical Strength – is characterized by the compressive strength of porous foam.
2) Permeability – is characterized by the pressure drop using near ambient air.
3) Net Heat Transfer Performance – Preliminary evaluations were completed wherein the heat transfer and pressure drop in foam structures were measured and are reported.

EXPERIMENTAL PROCEDURE

Various foam samples were machined into heat transfer evaluation coupons (1" x 1" x 0.25"). These foam samples were made by multiple vendors (Ceramatec, Ultramet, Duocell, Poco and Vesuvius) and represented multiple materials (silicon carbide, carbon, and aluminum).

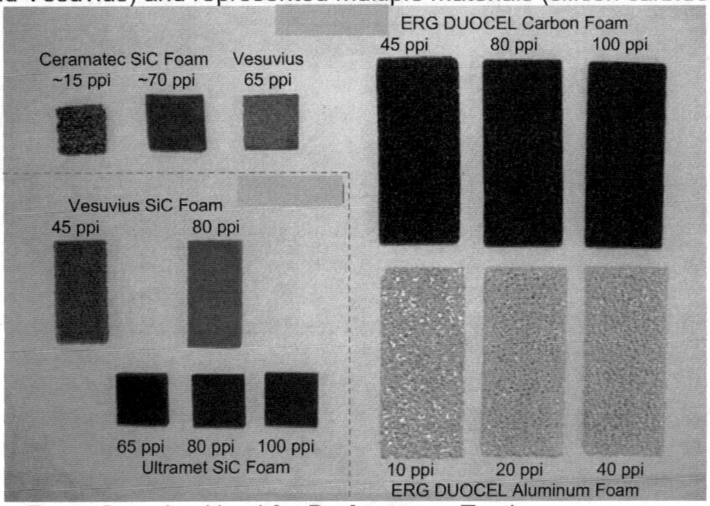

Figure 2. Foam Samples Used for Performance Testing

The heat transfer evaluation tests were configured such that the largest surfaces (1" x 1") were in thermal and mechanical contact with copper blocks to maintain isothermal surfaces, and air flowed through the porous structure in a cross-wise manner as seen in Figure 3. Using this apparatus, the pressure drop and the gas stream temperature change could be measured. Additionally, by measuring the temperature gradient in the copper block the heat flux and the temperature at the copper/foam interface could be interpolated. Given these temperatures and flow rates, an energy balance could be used to assess the net heat transfer characteristics of the porous foam structures.

Subsequently, other foam samples were fabricated to better assess the mechanistic behavior of the foam structures with respect to its mechanical strength and gas permeability. These foams were characterized in terms of their geometric and physical parameters; these included: porosity and pore size.

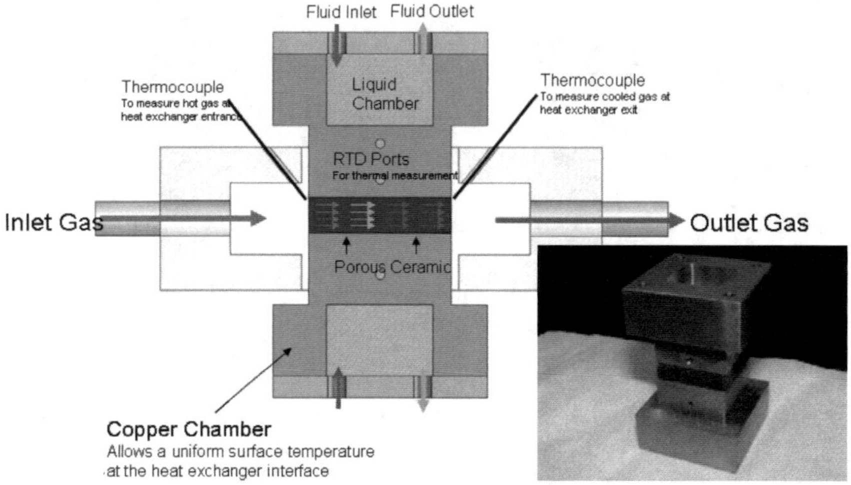

Figure 3. Heat Transfer Performance Testing of Foam Structures

Processing of ceramic foams, as described by Studart *et al.*[1] was accomplished by utilizing the replica technique, with synthetic templates. Upon dip-coating synthetic templates, samples underwent organic burnout, with subsequently sintered. Using commercial foams designated by pores per inch (ppi) designation, ten columns of 19 mm diameter by 57 mm length of sintered SiC foam were fabricated for each ppi designation (20, 30, 60, 80, and 100 ppi). Figure 4 shows optical photographs of typical sintered foam from each of the group.

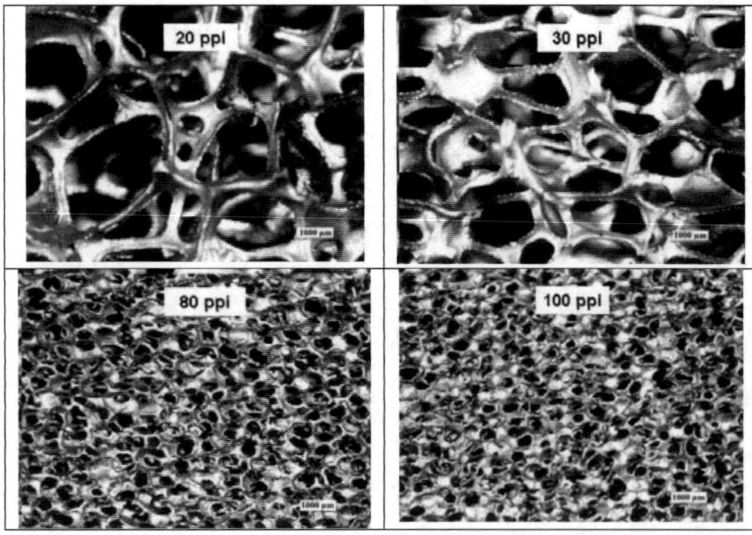

Figure 4. Optical photographs of various sintered SiC foams used in this study.

Within each group of ten, five specimens were tested in compression for strength measurements, while the other five where subjected to flow tests to determine permeability characteristics. All compression specimens were loaded in compression with a compliant foam layer was added to the top and bottom of each column to minimize uneven point loads. A typical compression test set-up is seen in Figure 5, with 60 ppi SiC foam being tested.

Figure 5. Compression test set-up.

Flow tests were conducted, with air at room temperature, by sealing individual specimens within a heat shrink wrap and flowing air at various flow rates from 5 to 20 standard liters per minute (slpm) through these columns as seen in Figure 6.

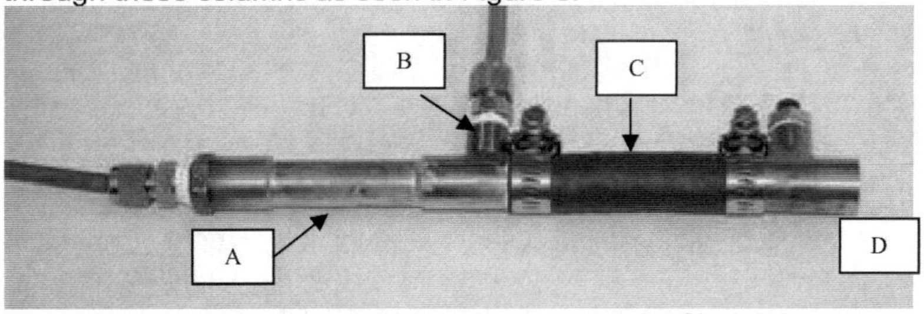

Figure 6. Flow test device. A) gas inlet, B) inlet pressure port, C) shrink wrapped SiC foam, D) gas outlet

RESULTS AND DISCUSSION

HEAT TRANSFER EVALUATIONS

These preliminary tests measured the heat transfer and pressure drop performance data of foam structures. The heat transfer performance is computed as the heat transferred (mass flow rate * Cp *ΔT of air) divided by the log mean temperature difference between the gas and constant wall interfaces. The pressure drop was measured from the inlet stream to the outlet. It was seen in the heat transfer and pressure drop data, different pore structures (manufacturer's products) even with similar PPI values have very different performance characteristics (Figure 7). It was apparent that the traditional pores-per-inch classification of these foams is insufficient for predicting their permeability. In fact, the 80 ppi foam manufactured by Vesuvius had a porosity of 58% while the Ultramet and Ceramatec 80 ppi foams had porosities between 80% and 90%. Optically, the major difference between the Ultramet and Ceramatec foams was the characteristics of the "windows" connecting adjoining cells whereby gas must pass through to permeate from upstream to downstream. Further characterization of these windows may include diameter or area and percent blockage.

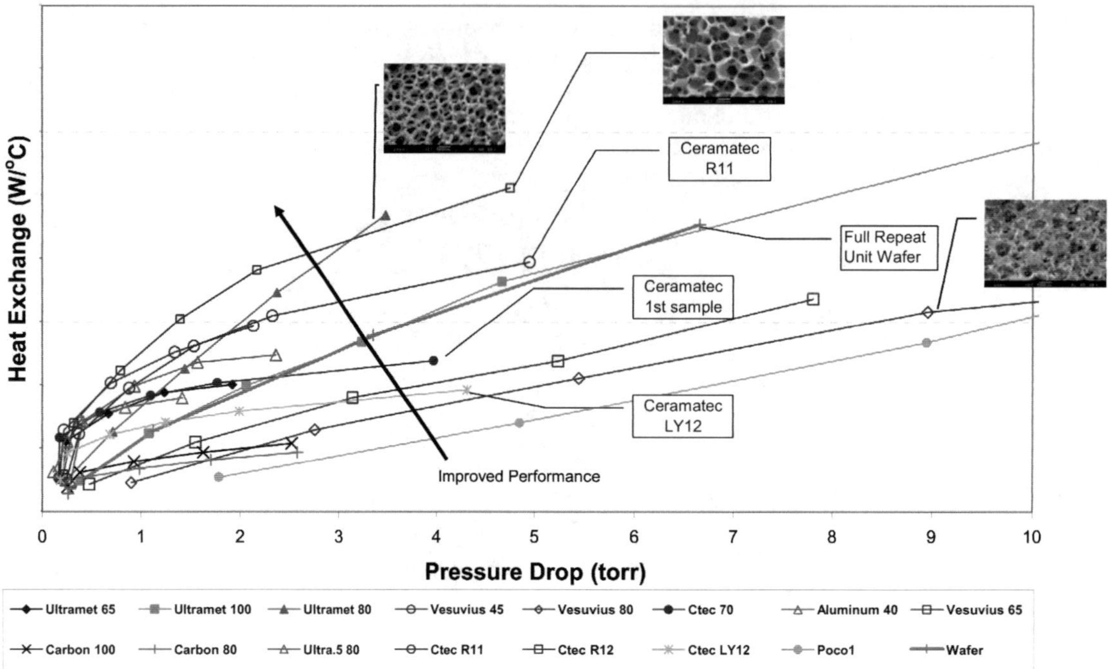

Figure 7. Relative Performance of Foam Structures

Ultimately, the objective of this research is to develop heat transfer correlations and pressure drop correlations such that foam-based heat exchangers can be optimized and compared to other compact heat exchanger designs for design selection. Kays and London[2] have compiled the performance of many compact heat exchangers for design comparisons. Appropriately, these comparisons are made using traditional non-dimensional variables (scaling parameters) such as the Reynolds number, Stanton number, Prandtl number and the friction factor. In order to define these scaling parameters, local geometric dimensions must be characterized. This lead to the second round of samples to better characterize the mechanisms influencing the performance of these foam structures as related to heat exchangers.

COMPRESSION RESULTS

The geometrical and morphological characteristics as well as the strengths for each group of sintered foam are averaged and tabulated in Table I. Upon crushing the foams under a compressive load, data was tabulated to determine the stress (σ^*) at which the cylindrical foam structure failed.

Table I. Morphological Parameters Characterizing Porous Foam Samples with Associated Average Strengths and Strength Constants.

Group PPI	Sintered PPI	Porosity (%)	Stress σ^* (MPa)
20	19.3	93.5	0.85
30	28.0	93.0	0.91
60	66.1	77.1	1.58
80	83.9	89.6	1.34
100	95.6	88.0	1.17

Data for all specimens tested in compression was tabulated, with the relationships of strength (σ^*) to porosity and ppi shown in Figure 8. Here it can be seen that strength would not correlate well with pore spacing. However when comparing strength as a function of porosity, we find, that strength increases nonlinearly as porosity decreases, as would be expected[3]. Hence, porosity is a more predictive parameter in determining strength of the foam than the pore spacing.

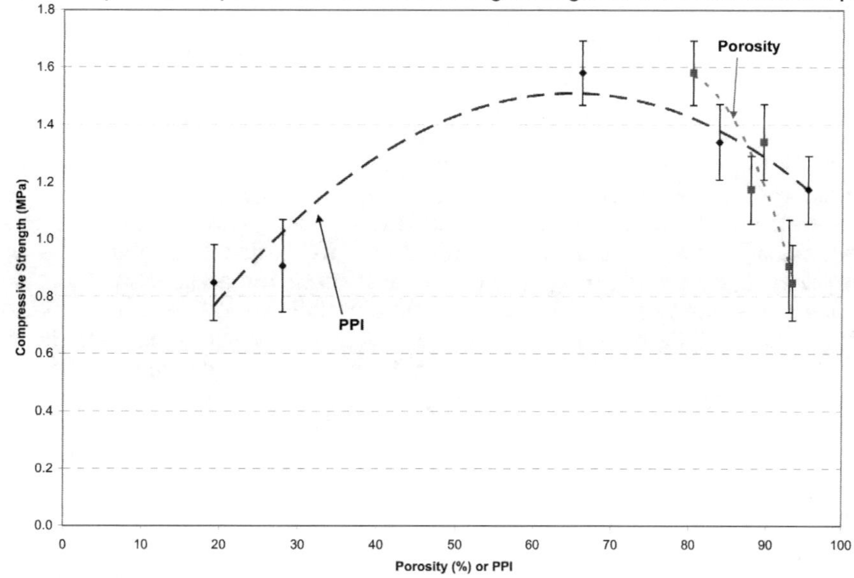

Figure 8. Strength as a function of porosity (%) and PPI.

PERMEABILITY RESULTS

Figure 9 shows the general trend of permeability for all samples tested, with pressure drop (normalized) increasing as porosity decreases. As also indicated in Figure 9, the pressure drop seems to align with inverse porosity and foam ppi designation.

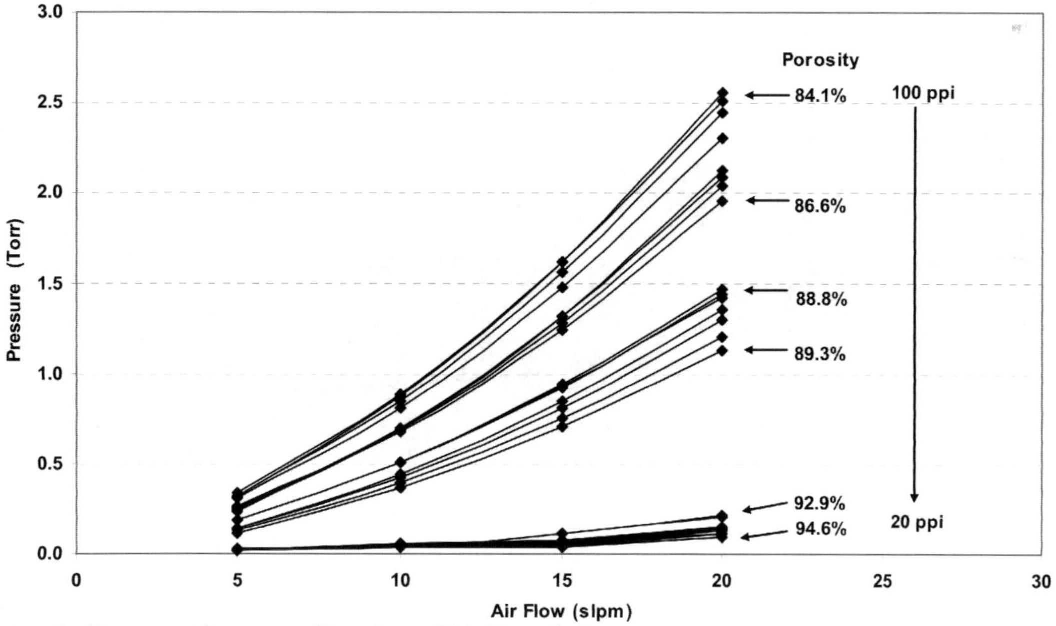

Figure 9. Pressure Drop as a Function of Air Flow Rate.

The change in pressure, or pressure drop ΔP, is a function of several variables given by[4]

$$\Delta P = f\left(\frac{l}{D}\right)\rho\frac{V^2}{2} \tag{1}$$

where f is the friction factor, l is the tube length, D is the tube diameter, ρ is density of the gas that flows through the tube, and V is the free stream velocity of the gas.

As discussed previously, the window area between cells may limit and determine the flow characteristics of the foam. It is well know that for laminar internal flows the friction factor (f) is inversely proportional to the Reynolds number (Re)[7]. For example in circular, unblocked pipes the relationship becomes f=64/Re; for high aspect ratio rectangular, unblocked ducts this is f=96/Re[5]. By assembling and reducing this pressure drop data for flow through porous foams this relationship is similar, yielding a proportionality constant of about 156 as seen in Figure 10, based on the dimensions of the cell to cell windows. This last correlation shows that to best tailor foam to meet specified pressure drop performance, the window characteristic of diameter is of primary importance. It should be noted that the dimension of these windows will vary with pore spacing and porosity. In other words for a preferred window area, there is range of porosities that can be attained for a given pore spacing of the synthetic template, and vice versa. Therefore, once the target flow and pressure drop have been determined, the necessary window diameter for the foam can be engineered and constructed using a specified range of pore openings per unit length and/or porosity characteristics.

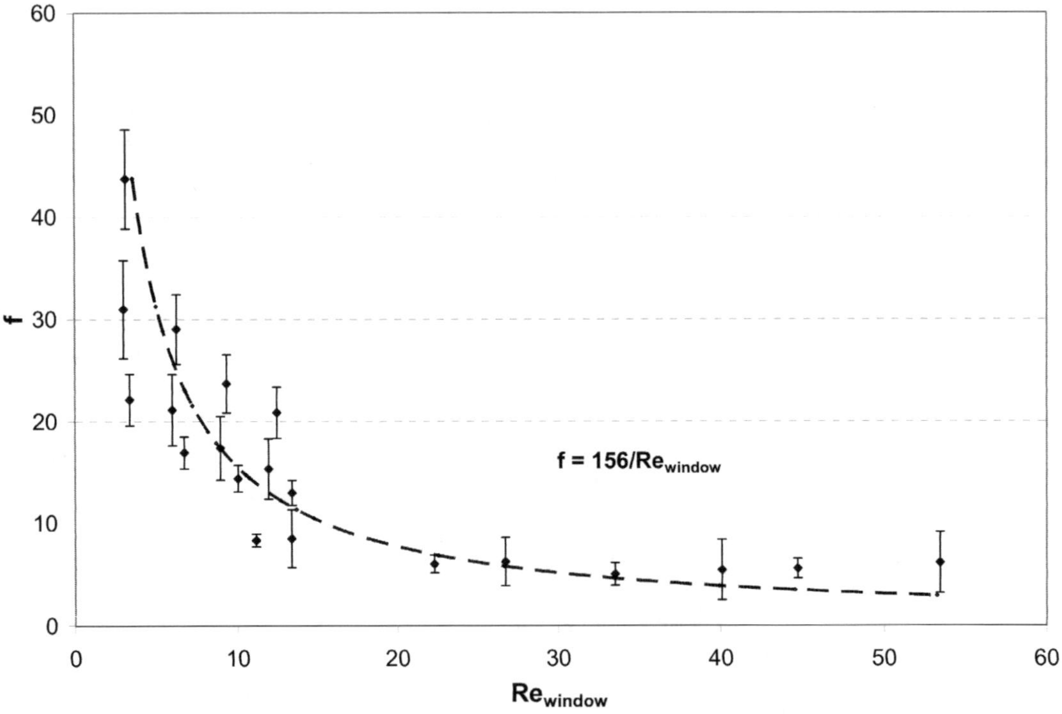

Figure 10. Friction factor as a function of Reynolds number through cell window.

SUMMARY AND CONCLUSIONS

SiC foam is an ideal candidate material for HX systems due to its thermal properties and strength. This work shows how to better quantify SiC foams produced using the replica method and synthetic templates.

The strength of the SiC foams tested are strongly dependent on porosity rather than pore spacing. As porosity decreases there is an increase in strength under compressive loading, with a maximum load of 1.6 MPa, correlating to a porosity of 77%.

Further, flow tests of these foams have shown that pressure drop and friction factor are functions of both porosity and ppi. These data and correlations have shown that the relationship between the friction factor and the Reynolds number of the window between cells are of primary importance in determining flow characteristics and subsequent tailoring of foams.

By examining several foams structures and materials it was found that improved heat transfer performance can be attained by modifying the morphology of the porous structure. One metric in characterizing foams is the pore size (typically PPI – pores per inch) another is the porosity (% theoretical density). Typically, moderately high porosity foams perform best. Foams that have porosities in the 90^{+}% range have reduced surface areas and ligament cross-sectional areas, thus possibly degrading their capability to transfer heat from a thermal substrate to a flowing stream. Foams that have moderate porosities (less than about 60%) have excessive flow blockage and hence minimize the flow rates of the gas stream to be convectively heated or cooled. The optimum appears to be within the 70% to 90% porosities range. Further optimization will largely be determined by the micro-structure features of the foams (strut diameter, window blockage etc).

ACKNOWLEDGMENTS

This work was initiated under a Department of the Air Force SBIR contract FA8650-08-2821. Subsequent follow-on work was done under the internal sponsorship of Ceramatec, Inc.

REFERENCES

[1] A. R. Studart, U. T. Gonzenbach, E. Tervoort, and L. J. Gauckler, "Processing Routes to Macroporous Ceramics: A Review," *J. Am. Ceram. Soc*, 89[6] 1771-89 (2006).
[2] Kays, W. M. and London, A. L.; "Compact Heat Exchangers", Reprint Edition 1998, Krieger Publishing Company, Malabar, FL, 32950.
[3] R. W. Rice, "Porosity of Ceramics" (1998), CRC Press.
[4] Alan Mironer; "Engineering Fluid Mechanics", 1979, McGraw-Hill, Inc.
[5] J. P. Holman, "Heat Transfer", 5th Ed., (1981), McGraw Hill Book Co.

TOWARD LOW COST FABRICATION OF MICROCHANNEL PROCESS TECHNOLOGIES – COST MODELING FOR MANUFACTURING DEVELOPMENT

Steven D. Leith and Dale A. King, Pacific Northwest National Laboratory, Richland, WA
Brian K. Paul, Oregon State University, Corvallis, OR

Introduction

Chemical and energy conversion systems based on microchannel process technology (MPT) demonstrate high performance in applications in which rates are controlled by diffusive heat and mass transfer flux. MPT-based heat exchangers, absorbers/desorbers and chemical reactors all benefit from process intensification and have been used in a variety of mobile energy conversion systems including fuel reformers/converters, heat pumps and waste heat scavenging technologies. The service environments typical of MPTs often require the devices to be fabricated from metals such as aluminum, titanium, stainless steel or high temperature super alloys. Flow channels and associated critical dimensions in these devices can be as small as 50 um, but generally range from 100 to 1000 um in width and height with characteristic flow channel lengths normally in the mm to cm range. High surface area architectures (*e.g.* wicks or textured surfaces) are often included in the flow channels as well for enhanced mass transfer and/or catalytic functionality.

Heat activated cooling systems that utilize MPT devices are being developed at the Microproducts Breakthrough Institute (MBI) for applications in which reduced system size and weight are of high value and/or electricity generation costs are high (*e.g.* mobile space cooling units in forward deployed military field operations). A typical system is illustrated in Figure 1 and is based on an aqueous ammonia absorption/desorption cycle driven by an external heat source. In this particular system design, the desorber, absorber and heat exchanger represented schematically in (A) are MPT devices and are integrated with off-the-shelf components in the heat pump system (B). The absorber device (C) is the largest MPT component in the system and is used to regenerate concentrated ammonia solution by contacting ammonia vapor with a weak aqueous ammonia solution in a high surface area mixing architecture.

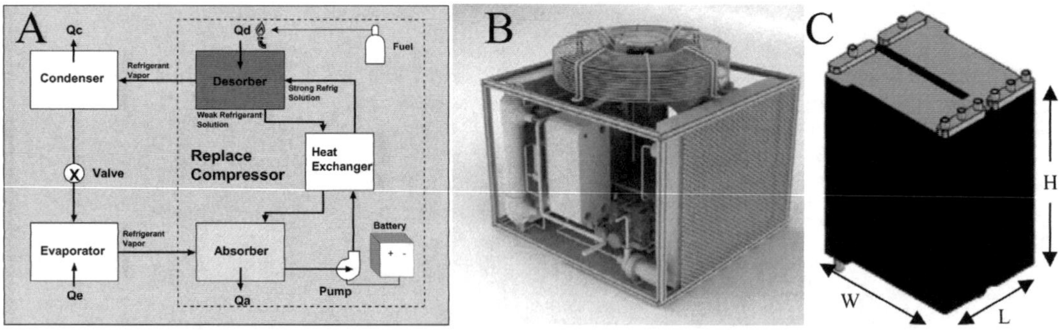

Figure 1. Ammonia absorption cycle heat pump system (A and B) utilizing MPT devices. The 5 kW system requires two absorbers (C) each with external dimensions of approximately 15 cm (W) x 15 cm (L) x 25 cm (H).

MPT enabled heat pumps demonstrate high COP for a given system size, are quiet in operation and can be driven by renewable energy, combustion of natural gas or, more notably, by waste heat from an internal combustion engine or fuel cell. This characteristic energy use flexibility, unit mobility and relatively small system size make MPT heat activated cooling systems performance- and cost-competitive with existing technologies in a number of military and commercial markets. To move from technology demonstration to market entry and commercialization, however, requires an better understanding of the manufacturing processes and associated cost structures necessary to fabricate components in production volumes. Recent work at the MBI has focused on developing volume-sensitive cost estimation models for predicting the manufacturing costs of MPT devices fabricated using a range of processing technologies. The process-based cost models are used to develop an understanding of the economic trade-offs between candidate processes and are utilized in a design for manufacturing approach to MPT device development. In this paper we present results and analysis of the cost modeling effort to date and apply the methodology in a case study of the heat pump absorber device designed, built and tested by Pacific Northwest National Laboratory.

Process-Based Cost Modeling

Process-based cost models have been used to estimate manufacturing costs for a range of materials and products including diamond films, carbon nanotubes, steel powder and injection molded plastics (1-4). Using this approach, the total manufacturing cost is broken down into a number of cost elements associated with each step in the production process. Each cost element is estimated separately and then all are summed to establish an overall manufacturing cost. Typical cost elements include capital equipment, labor, direct materials, indirect materials, energy/utilities, facilities and maintenance. Variation in each cost element is estimated through creation of functional relationships that quantify the impact of manufacturing variables (*e.g.* production volume, equipment through-put, labor and loading rates, electricity cost, etc.) on each cost element. The output from the process-based model is an estimate of cost of goods sold (COGS); typical overhead cost contributors such as sales and marketing, R&D, administration, management, profit and taxes are not considered. Application of the process-based cost modeling methodology to manufacturing of a device such as the absorber shown in Figure 1 can be used to illustrate the relative contribution of design, materials and fabrication process choice on final device cost. Identification of these cost drivers is a critical enabler in effectively prioritizing the fabrication processes that give the highest probability for commercial success. Cost model sensitivity analyses can be used to determine areas of technical and/or financial risk and are a useful tool in developing more sophisticated economic assessments; for example supporting a "make vs. buy" decision in manufacturing production. The process-based cost model is used here to identify primary cost drivers in the manufacture of the ammonia absorber device and to illustrate the interaction of design and COGS across a range of production volumes.

Case Study – Manufacturing Cost of MPT Absorber Device

The first step in the modeling methodology is to define the fabrication process and develop the functions and equations that relate manufacturing parameters to the individual cost elements. A typical fabrication process flow sheet for manufacturing of MPT devices is illustrated in Figure 2. The architecture of the absorber device is based on a stacked-shim fabrication process in which shims are first patterned (with flow channels and/or fluidic vias) and then bonded in a monolithic stack. Two specially machined end-plates are joined to either end of the device to allow addition of interconnects for fluidic communication to and from the device. The service environment of the MPT device has a great influence on the choice of shim and end-plate materials and, in turn, the material choice has a significant impact on available fabrication processes and processing conditions. In the present study, the service environment of the absorber requires mechanical strength at elevated temperature in corrosive fluids. To meet the material performance requirements of the prototype system, austenitic 316 stainless steel is used for both shim and end-plate architectures.

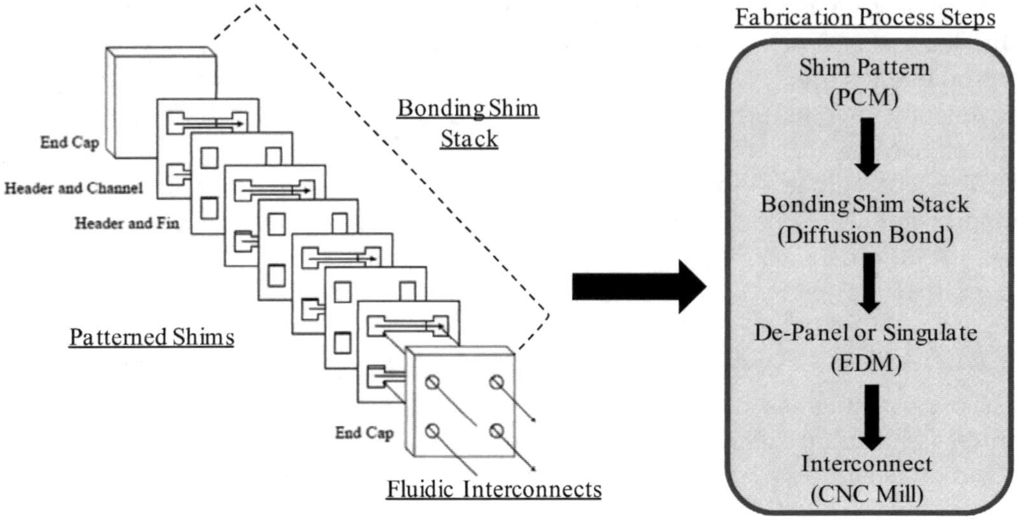

Figure 2. Typical construction scheme and fabrication process steps for stainless steel MPT device.

A number of fabrication processes have been used at MBI to build demonstration devices. The most-widely used process steps, however, appear in Figure 2 and are especially well-suited for fabricating metallic-based MPT devices at low (prototype) volumes. In the present study, this "default" manufacturing process flow is considered in which photochemical machining (PCM) is used to pattern shims, diffusion bonding for joining of the shim stack, electro-discharge machining (EDM) for final device perimeter shaping and CNC milling to define fluidic interconnects.

At demonstration and prototyping fabrication volumes, the cost associated with shim patterning and shim stack bonding has been a significant fraction of the total fab cost for MPT devices made at MBI. Typically, however, these fabrication processes have been performed by third party contract manufacturers and the cost of device fab is very likely not representative of that which could be achieved using a dedicated production line as is assumed with the cost model. Using the default process flow illustrated in Figure 2, the process-based cost model is used to predict manufacturing costs of the absorber over a range of production volumes representative of the anticipated market. Figure 3 illustrates the cost breakdown for manufacturing between 100 and 10,000 absorber units/year. Consistent with past prototyping efforts, at low volumes (100/yr) the majority (80%) of the manufacturing cost is contributed by the PCM shim patterning and diffusion bonding processes. At these volumes, the direct material cost of the stainless steel shims and end plates accounts for nearly all the remaining cost (15% of total) while the EDM and CNC milling processes contribute very little. As manufacturing volumes increase, however, substantial reductions in device cost are realized. In fact, at fairly modest production volumes of 1,000 to 10,000 absorber units annually, the unit COGS is estimated to be between $3,000 and $4,000. Consistent across all absorber production volumes is the cost contribution of the PCM process. Model simulations predict that PCM costs actually increase as a fraction of the total with increased volume production even while decreasing on a unit COGS basis.

The cost model is used further to explore the cost drivers inherent in the PCM process and to evaluate the financial viability of this option as a candidate fabrication option in an MPT production environment. Figure 4 illustrates the area-based production cost of using PCM to pattern stainless steel shims using an industry standard etching line designed for a maximum capacity of 40,000 m^2/yr. When the direct material cost (*i.e.* 316 SS shim stock) is removed from the analysis, the cost model predicts a decrease in manufacturing cost from approximately \$275/m^2 at low production volumes to a minimum of about \$50/m^2 at production volumes which approach maximum design capacity. For the absorber architecture, the \$50/m^2 COGS estimate is equivalent to approximately \$1.25 per shim. The rate of cost reduction with volume predicted for the PCM process is characteristic of a manufacturing process line designed for a specific production volume. At low production, the line is under-utilized and fixed costs associated with capital and labor are allocated to a small number of parts, driving the unit cost up. As production volumes increases, both capital and labor utilization improve and unit COGS decrease.

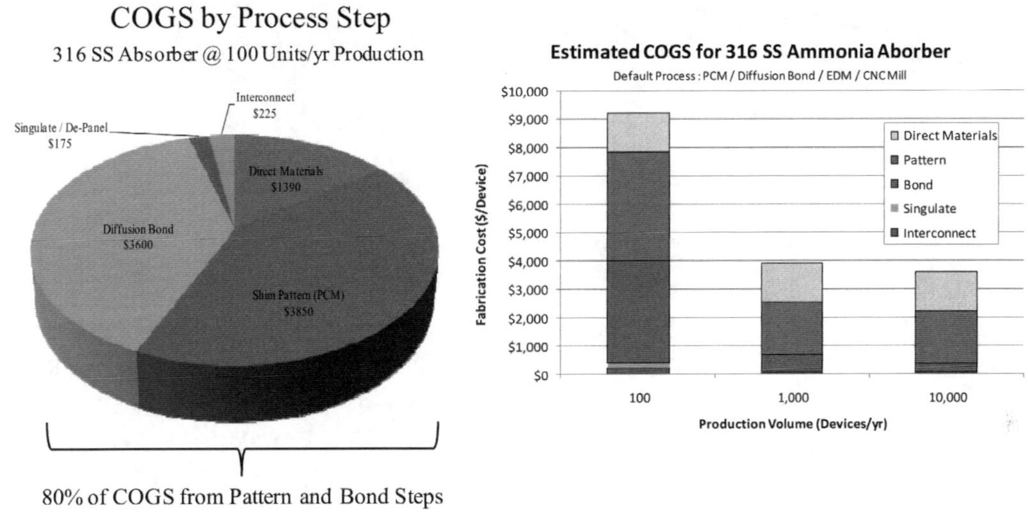

Figure 3. Manufacturing cost breakdown for fabrication of absorber device at 100, 1000 and 10,000 units/yr production.

Similar cost estimation results and breakdowns to those illustrated in Figure 4 can be obtained by applying the cost model methodology to other candidate processes of interest. For example, shim patterning can be performed using a number of other fabrication processes including electrochemical machining (ECM), stamping (piercing or coining), laser cutting and abrasive water jet cutting. Direct comparison of the cost associated with each of these processes for the patterning of specific shim architectures illustrates which candidate processes are likely to be the most economical in production. In this way, alternative shim designs can be considered that might enable a lower cost fabrication path.

Figure 5 summarizes the process costs for patterning shims with fluidic vias (A) and "blind" flow channels (B) representative of architectures required for the absorber (*i.e.* 500 um thick, 316 SS shims). Patterning of fluidic vias can be and has been performed using each of the patterning processes listed above. One design attribute that can have a significant impact on pattern cost is the density of vias on a given shim. For serial patterning processes (*i.e.* those that raster a point energy source across the shim surface) such as laser cutting and abrasive water jet, the via perimeter per unit area directly influences process through-put and, ultimately, cost. Alternatively, for parallel patterning processes (*i.e.* those which pattern the entire area at once) such as PCM and ECM, the intrinsic processing rate and final cost is independent of via pattern density. The piercing process, however, has characteristics similar to both parallel and serial processing and exhibits a modest sensitivity to via perimeter area. For consistency in comparison, the COGS estimates summarized in Figure 5 (A) are based on a shim architecture with a via perimeter of 180 m/m^2.

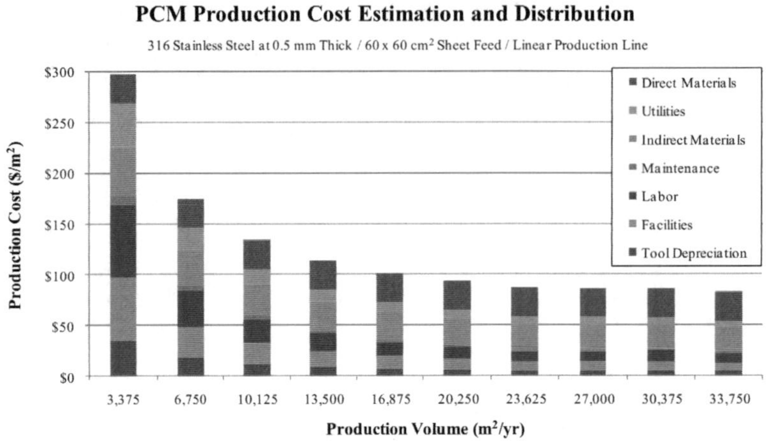

Figure 4. Estimated manufacturing cost for patterning shims using PCM

The summary of estimated costs for patterning fluidic vias in stainless steel sheet stock using the different candidate processes illustrates the volume sensitivity and relative costs associated with each option. At low production volumes, manufacturing costs increase for all process, but most sharply for PCM processing. At volumes of 24,000 m^2/yr and greater, however, the estimated COGS for all options except abrasive water jet fall between \$50 and \$75/m^2 (approximately \$1.25 and \$1.75 per shim for the absorber). This cost differential is modest, but can be significant for devices that require a high number of shims. The final choice of patterning process, therefore, will likely require consideration of process capability (e.g. minimum feature size, edge burr, sidewall profile, etc.) vs. pattern dimension tolerances.

Figure 5 (B) summarizes the cost of patterning flow channels, or blind features, in an MPT shim using the different candidate processes. Note that only three of the five patterning processes are included; the capability of both laser and abrasive water jet patterning of flow channels is questionable and not yet production ready. Coining of flow channels is also a challenging process and puts specific limitations on channel design. This approach, however, has been demonstrated for production of fuel cell components and given the similarity to coining MPT architectures is included in the analysis. Similar cost reduction with volume as with via patterning is predicted by the model; in fact the cost estimations for PCM and ECM are essentially identical for patterning both vias and flow channels. Both of these processes are capable of creating vias and flow channels in a single production run, a distinct advantage over alternative processes and a significant reason why PCM has been and looks to continue as a primary option for shim patterning in MPT manufacturing.

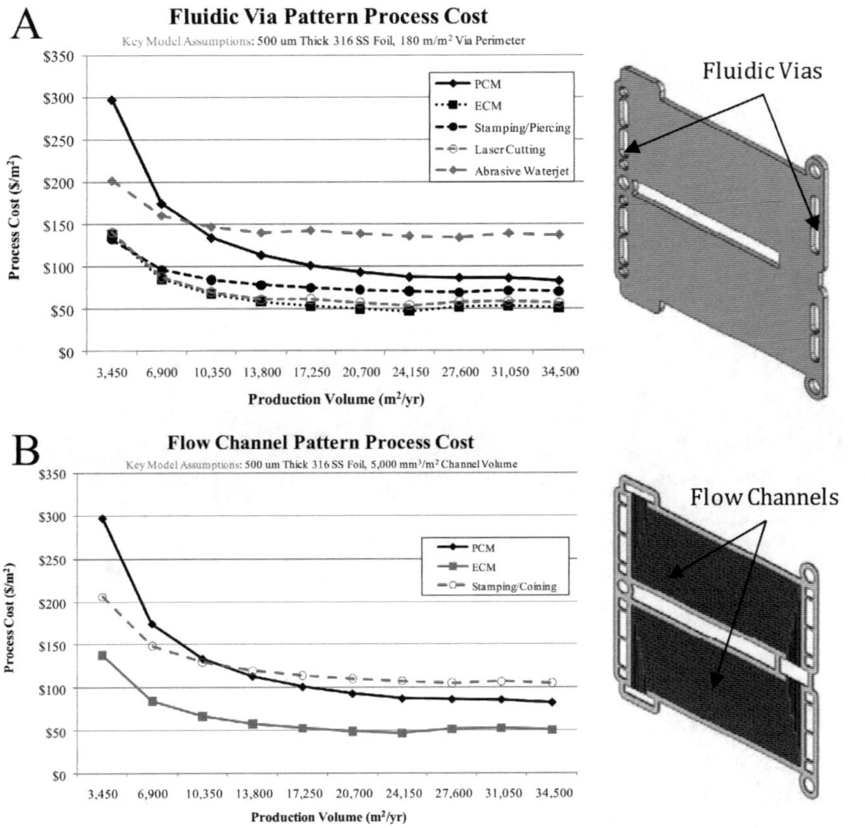

Figure 5. Cost comparisons for patterning shims using different candidate processes.

As illustrated in Figure 3, the bonding of shim stacks also contributes significantly to the overall manufacturing cost of MPT devices. In Figure 6, the cost of two candidate bonding processes are compared; diffusion bonding and laser welding. As with the serial patterning processes discussed above, laser welding cost is dependent on the total weld joint length in the device. As shown in Figure 6 (A), welding lengths between 6,000 and 120,000 cm per device are considered. Alternatively, diffusion bonding is not dependent on weld joint length, but rather on the number of devices bonded in a single process cycle. Two device through-puts are included in the cost estimation results for diffusion bonding summarized in Figure 6. The solid line simulates a bonding process in which four devices are bonded in each cycle while the dashed line represents cost for bonding 16 devices per cycle. Consistent with increased through-put, laser welding costs are minimized for devices which require the least number of joints while diffusion bonding costs are minimized as the number of devices per bond cycle increases. The simulation results summarized in Figure 6 (A) can be used to determine which joining process is most economical for a given device design. At low volumes, the estimated cost of laser welding is lower than that predicted for diffusion bonding, a direct result of the high capital cost associated with industrial scale diffusion bonding equipment. At higher volumes, however, the choice is not completely clear and will likely depend on process capability (*i.e.* yield) and an assessment of technical risk (*i.e.* the likelihood that bonding 16 devices simultaneously is feasible). The cost model can be used to explore the impact of yield through sensitivity analyses, but the assessment of technical risk and impact on capital investment choice is ultimately a business decision.

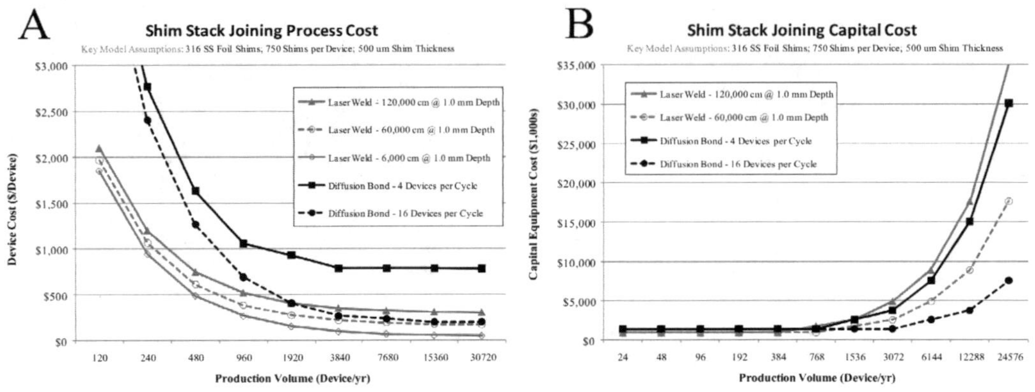

Figure 6. Shim stack joining process cost (A) and required capital equipment investment (B).

In addition to predicting manufacturing cost for the various candidate processes, the cost model simulation is also used to estimate the required capital investment to realize a given unit COGS. For example, consider the unit COGS and capital investment estimations for laser welding vs. diffusion bonding of a device that requires 60,000 cm of laser weld joining per device. From Figure 6 (A), the lowest unit COGS at high volume appears to be same for laser welding and diffusion bonding (16 devices per cycle). As discussed above, for such a situation in which two candidate processes are estimated to have identical COGS, a logical fabrication decision might take into account the capability attributes of one process vs. the other. Figure 6 (B), however, illustrates the difference in capital investment required between the two processes. From the standpoint of financial risk, one could reasonable argue that an investment in the diffusion bonding process at $4M is preferred over a $9M laser welding investment to achieve the same unit COGS. This example illustrates an important aspect in the interpretation and use of cost model results; reducing COGS is certainly necessary for competitive manufacturing, but the capital investment required to realize the unit COGS must be considered as well. Excessive investment in capital can increase the financial risk of manufacturing and ultimately stress the commercialization effort. While the process-based cost model discussed here can provide insight into the relationship between COGS and required investment, a more sophisticated economic analysis that includes market intelligence and cost targets is required to determine what investment level is appropriate or what manufacturing cost is too high.

Summary and Next Steps

1. Process based cost models have been developed at the MBI for estimating the manufacturing cost of MPT devices using a variety of candidate fabrication processes.

2. These cost model tools are being developed and used by the MBI to estimate production costs of MPT devices to assist technology commercialization partners evaluate MPT options for processing solutions in heat transfer and reaction engineering applications.

3. Costs for absorber and absorber-like architectures at relatively modest production volumes of 10,000 per year are in the $3000 to $4000/unit range. This is significantly lower than contract manufacturing cost and is likely an acceptable COGS target in one or more market segments. The cost model tools will be used to facilitate design for manufacture (DFM) activities aimed at further reducing costs for similar architectures while maintaining required device functionality. Near term efforts are focused on developing market intelligence in-house and in cooperation with commercialization partners to better match the MBI MPT value proposition with market needs and cost targets.

4. Capital investment for production of absorber and absorber-like devices at relatively modest production volumes of 100 to 10,000 annual units are on the order of $M to $10M dollars. The cost of a dedicated PCM line at these volumes is on the order of $1M and can be used to produce shims at a COGS as low as $1.50 each. In addition, PCM is capable of both via and flow channel patterning, suggesting that PCM will continue to be a process of interest in MPT manufacture for the near future.

Next Steps

1. Multi-variant sensitivity analyses need to be performed to better identify and quantify technical and financial risk to commercialization. Sensitivity analyses are underway at MBI to explore impacts of design, process and materials on manufacturing cost and capital investment.

2. Model extensions to include consideration of overhead costs and total cost of ownership parameters (*e.g.* device lifetime, disposal cost, operation cost, etc.) are being developed which will further strengthen the cost understanding in more market-specific applications.

References

1. Busch, J. (1994), "Cost Modeling as a Technical Management Tool", *Research Technology Management*, Nov/Dec, pp. 50-57.
2. Isaacs, J.A., Tanwani, A., Healy, M.L., Dahlben, L.J. (2010), "Economic Assessment of Single-Walled Carbon Nanotube Processes", *Journal of Nanoparticle Research*, **12**, pp. 551-562.
3. Kirchain, R. (2001), "Cost Modeling of Materials and Manufacturing Processes", in *Encyclopedia of Material Science and Engineering*, pp. 1718-1727.
4. Chiango, D.A., Isaacs, J.A., Miller, S.A. (2000), "Production of Steel Powder by Rotating Electrode Processes: Economic Analysis", *International Journal of Powder Metallurgy*, **36**, No. 4, pp. 49-56.

(335e) A Compact Integrated Warm Syngas Overall Cleanup System

Liyu Li[1], David L. King[2], Baowei Chen[3] and Chris Howard[1], (1)Hydrocarbon Processing Group, Pacific Northwest National Laboratory, Richland, WA, (2)Pacific Northwest National Laboratory, Richland, WA, (3)Biological Sciences, Pacific Northwest National Laboratory, Richland, WA

Gasification of coal or biomass to syngas followed by catalytic synthesis of ammonia, hydrocarbons or oxygenates provides a feasible strategy to meet the increasing demand for fertilizers and transportation fuels. This syngas has many impurities, including As, Be, Cd, Cl, Cr, Hg, K, N, Na, P, Pb, S, Sb, Se, V, Zn. Most of them need to be removed down to part per billion levels (ppb) due to their strong interactions with syngas reforming and synthesis catalysts. Although technical approaches exist for removal of these species, they are rather costly, employing solvents at ambient or lower temperature and backup sacrificial sorbents. The energy efficiency of the current process would be improved if all the contaminants can be removed at temperatures higher than the syngas reforming and chemical synthesis temperatures (greater than 200oC).

In this presentation, we will describe our recent progress in developing a compact, integrated system for overall impurity cleanup from warm syngases generated from water-quenched coal or biomass gasifiers. This compact system consists of four major sub-systems: 1) a particulate filter to remove solid impurities, 2) a high capacity sacrificial absorption bed to remove HCl, 3) two sequential absorbent beds to remove sulfur from more than 1000 ppm levels down to less than 60 ppb levels, and 4) a high capacity metal-based sacrificial absorbent to remove As, P, Se, and trace level S and HCl. Each sub-system was individually developed, and a compact integrated system was built up. This integrated system was demonstrated to be able to remove all the catalysts poisons in a simulated warm syngas stream and thus effectively prevented the downstream syngas reforming catalysts from being deactivated. This compact system can also be used in stationary hydrocarbon fuel processors for clean feed gas generation for fuel cells applications.

See more of this Session: High-Throughput and Compact Chemical Processing Technologies
See more of this Group/Topical: Topical G: Innovations of Green Process Engineering for Sustainable Energy and Environment

Pyrolysis and co-pyrolysis of biomass and coal in a free-fall reactor

Ligang Wei[1,2], Shaoping Xu[1*], Li Zhang[1]

(1. State Key Laboratory of Fine Chemicals, Department of Chemical Engineering, Dalian University of Technology, Dalian 116012, China; 2. School of Chemical Engineering and Material, Dalian Polytechnic University, Dalian 116034, China)

Pyrolysis and co-pyrolysis of solid fuels, like biomass and coal, is a promising process to produce liquid fuels and chemicals. The free-fall reactor has the advantages of a higher heating rate and a short but narrowly distributed residence time (in the order of seconds) of both solids and gas. It is a good approach to investigate the mechanisms of fast pyrolysis of solid fuel in a free fall reactor. In this work, pyrolysis and co-pyrolysis of biomass and coal were performed in a free fall reactor under atmospheric pressure. Effects of type and composition of feedstock, reactor temperature, steam, and catalyst on the characteristics of the process were investigated at 500~800°C. The biomass feedstocks are agriculture residues legume straw (LS) and tobacco stalk, forestry residues pine sawdust (SD) and fruit shells apricot stone, which have different chemical compositions. Two types of coal, low rank Dayan brown coal (DY) and high rank Tiefa bituminous coal (TF) were selected as coal samples.

The experimental results show that a higher temperature favors producing the hydrogen-rich gas. Biomass composed of higher holocellulose component produces a relatively higher hydrogen-rich gas yield than that composed of higher lignin. The characteristics of pyrolysis can be inferred according to the chemical compositions of biomass samples. The presence of steam increases the gas yields and reduces the tar and char yields. It is considered that an in situ steam reforming reaction of tar and water–gas shift reaction takes place even in a short gas residence time in the free-fall reactor. Dolomite reveals a comparably good performance in terms of catalytic activity of tar destruction and the consequential increase in the production of gases in thus a short vapor-catalyst contact time.

The results of co-pyrolysis of biomass and coal shows that the char yields decrease and consequently the liquid yields increase compared with the calculated values, even the compositions of the gaseous products from the blended samples are not in accordance with those of their parent fuels, It suggests that there exist synergetic effects in co-pyrolysis of biomass and coal even in a short gas residence time in the free-fall reactor. In the co-pyrolysis of LS/DY, the obvious synergy on the products yields was observed at 600°C and the biomass blending ratio of 73 wt.%. The type of feedstock influences greatly the product yields and the CO_2 reactivity of char from the co-pyrolysis. The synergy is more significant in the presence of LS with higher holocellulose and ash content than that in the presence of SD. The liquid yield from the co-pyrolysis of high rank TF is higher than that of low rank DY. The char reactivity from co-pyrolysis in the presence of LS is much higher than that in the presence of SD, while that in the presence of TF is more active than that in the presence of DY.

The liquid products from the pyrolysis and co-pyrolysis were separated by solvent extraction-column chromatography and then analyzed by GC/MS. The results show that, the yields of asphalts and long chain aliphatic hydrocarbons from co-pyrolysis decrease in comparison to the calculated values, while the yield of light molecular weight phenol derivatives increase. The contents of decalins, which cannot be found in the tar from pyrolysis of single fuel, were 43.37% in aromatic fraction from the tar in co-pyrolysis. All these results verify that the synergy exists in the co-pyrolysis of biomass and coal in a free fall reactor. It is a contribution of the reactions such as hydropyrolysis and hydrogenation in co-pyrolysis of coal and biomass under the hydrogen atmosphere supplied by biomass pyrolysis. Co-pyrolysis of biomass and coal favor producing low molecular weight compounds and improving the quality of the liquid product.

Acknowledgements. The authors sincerely acknowledge the financial support from the National Natural Science Foundation of China under Grant No.50776013, the National High-tech R&D Program of China (863 Program) under Contract No. 2008AA05Z407 and the Program for Liaoning Excellent Talents in University under Grant No. 2009R10.

*Corresponding author. Tel.: 0086-411-39893837 , Fax: 0086-411-83683467 , E-mail : huizixu@hotmail.com

(335g) Solar Thermochemical Production of Metal Nitrides and Ammonia Utilizing Transition Metal Reactants

Ronald Michalsky, Chemical engineering, Kansas State University, Manhattan, KS and Peter H. Pfromm, Department of Chemical Engineering, Kansas State University, Manhattan, KS

One of the major products of the chemical industry, ammonia is used widely as artificial fertilizer and has been proposed recently as sustainable transportation fuel. Using transition metal-based catalysts the Haber-Bosch process produces ammonia industrially from its elements. Globally, the Haber-Bosch process consumes up to 5% of all natural gas produced and 2% of the total energy production, with significant fossil-based carbon dioxide emissions. The process cannot easily be scaled down or performed in developing countries due to the required temperature, pressure, natural gas, and utilities. Based on previous research employing Al- or Cr-based reactants to produce ammonia by a solar thermochemical reaction cycle at near atmospheric pressure and 1000-2000°C and without the need for natural gas as reactant, the work presented here studies process and reactant modifications that might allow the solar-driven ammonia production from water and air to proceed at below 1500°C, i.e., at temperatures which would not require sophisticated furnace design and construction materials. Utilizing a laboratory scale Fresnel lens-based tubular solar thermochemical reactor, formation of metal nitrides from various transition metal oxides is studied at atmospheric pressure. Liberation of ammonia from the nitrides due to corrosion with steam will be focused. The effect of ammonia formation favored at low temperatures and metal nitride corrosion favored at high temperatures is addressed. Deviation from thermodynamic calculations for bulk materials is discussed.

See more of this Session: High-Throughput and Compact Chemical Processing Technologies
See more of this Group/Topical: Topical G: Innovations of Green Process Engineering for Sustainable Energy and Environment

(337a) Mitigation of Fouling in a Super-Decanter Centrifuge

Akiya Kuboyama[1], Hirofumi Inokuchi[1] and Christianto Wibowo[2], (1)SSOT Group, Mitsubishi Chemical Corporation, Okayama, Japan, (2)ClearWaterBay Technology, Pomona, CA

Steady plant operations without shut-downs or decreases in production due to operating problems are crucial for the bulk chemicals industry. In Mitsubishi Chemical Corporation (MCC), operating problems are responsible for an enormous loss. Fouling, plugging, corrosion, and product quality issues top the list of most common problems, accounting for over 25%. In order to eliminate or mitigate these problems, the Super Stable Operation Technology (SSOT) task force was created to develop technologies for their prediction and prevention. The first target was combating fouling and plugging. This talk illustrates an example of our approach in tackling the problem of fouling in the super-decanter centrifuge of one of MCC's aromatic compound plants. An engineering model is developed to gain a better understanding of the phenomena, such that possible solutions can be proposed.

See more of this Session: Industrial Applications of Process Synthesis
See more of this Group/Topical: Process Development Division

(337b) Challenges in Low Temperature Process Design

Danahe Marmolejo-Correa and **Truls Gundersen**, Energy and Process Engineering, Norwegian University of Science and Technology, Trondheim, Norway

The aim of this work is to show the challenges in Low Temperature Process Design and how thermodynamic tools make it possible to develop an efficient design. Low temperature processes need large quantities of energy to operate, hence any improvement in technology or design methodology have significant relevance in this industrial field. Large amounts of energy are needed to drive the compressors for the refrigeration cycles and energy consumption increases at lower temperature levels. Therefore, such processes must be designed as efficient as possible.

Pinch Analysis (PA), has become the primary thermodynamic tool for designing heat transfer networks [1]. PA concepts have been successfully used and extended over the years to fit various types of processes and design constraints i.e. [2-4]. However, PA has its limitations when it is applied for designing cryogenic processes since the only design variable used is temperature.

In low temperature process design, temperature is not the only important variable. The pressure level has an enormous influence on energy consumption, thus it must also be considered as a design variable. Often when temperature and pressure are design variables, stream phase may have a significant importance in the processes. In fact, many low temperature processes deal with the phase change as their main objective, i.e. LNG processes and air separation. Consequently, temperature, pressure and stream phase are the main design variables for low temperature processes [5]. For this reason, basic PA is not sufficient to deal with all the requirements for this type of designs.

Generally, process engineering uses the second law of thermodynamics to evaluate process performance since it indicates how efficient energy is used in the process [6, 7]. Exergy Analysis (EA) is not commonly used as a process design tool, more often to measure the quality of the resulting process, although it has a high potential to aid the design of efficient processes. The combined use of PA and EA as design tools for low temperature processes has given encouraging results in our research group. Aspelund et al. [8] proposed a design methodology (ExPAnD) combaining PA and EA and they used it in a Liquefied Energy Chain (LEC) [9] for utilizing stranded natural gas for power production with offshore liquefaction, ship transport, and onshore re-gasification.

The exergy content in a process stream can be divided into chemical exergy and thermo-mechanical exergy. The thermo-mechanical exergy can be further divided into temperature based exergy and pressure based exergy. Therefore, EA can handle the low temperature process design variables and it can also be used as the main process efficiency driver. EA definitions however, are still not homogeneous in the literature and within the scientific community. In particular, there are different definitions of exergy efficiency available, and this work suggests a general and standardized definition that is also suitable for sub-ambient processes.

References

[1] Linnhoff, B. and E. Hindmarsh, The pinch design method for heat exchanger networks. Chemical Engineering Science, 1983. 38(5): p. 745-763.

[2] Linnhoff, B. and V.R. Dhole, Shaftwork targets for low-temperature process design. Chemical Engineering Science, 1992. 47(8): p. 2081-2091.

[3] El-Halwagi, M., D. Harell, and H. Dennis Spriggs, Targeting cogeneration and waste utilization through process integration. Applied Energy, 2009. 86(6): p. 880-887. [4] Linnhoff, B., H. Dunford, and R. Smith, Heat integration of distillation columns into overall processes. Chemical Engineering Science, 1983. 38(8): p. 1175-1188.

[5] Gundersen, T., et al., An Overview of New Methodologies for the Design of Cryogenic Processes with an emphasis on LNG, in Proceedings of the 1st Annual Gas Processing Symposium. 2009, Elsevier: Amsterdam. p. 104-112.

[6] Feng, X. and X.X. Zhu, Combining pinch and exergy analysis for process modifications. Applied Thermal Engineering, 1997. 17(3): p. 249-261.

[7] Anantharaman, R., O.S. Abbas, and T. Gundersen, Energy Level Composite Curves--a new graphical methodology for the integration of energy intensive processes. Applied Thermal Engineering, 2006. 26 (13): p. 1378-1384.

[8] Aspelund, A., D.O. Berstad, and T. Gundersen, An Extended Pinch Analysis and Design procedure utilizing pressure based exergy for subambient cooling. Applied Thermal Engineering, 2007. 27(16): p. 2633-2649.

[9] Aspelund, A. and T. Gundersen, A liquefied energy chain for transport and utilization of natural gas for power production with CO2 capture and storage - Part 1. Applied Energy, 2009. 86(6): p. 781-792.

See more of this Session: Industrial Applications of Process Synthesis
See more of this Group/Topical: Process Development Division

(337c) Exergy-Analysis Based Refrigeration System Synthesis at Ethylene Plants

Jian Zhang[1], Yanqin Wen[2] and Qiang Xu[1], (1)Dan F. Smith Department of Chemical Engineering, Lamar University, Beaumont, TX, (2)Dan F. Smith Department of Chemical Engineering, Lamer University, Beaumont, TX

A refrigeration system generally works by transferring the heat from low-temperature sources to higher-temperature objects at the expense of electricity or mechanical work, such that the source temperature can be lowered or maintained at a certain value. It is very important for an ethylene plant operation because its performance is closely influence the product quality, energy use efficiency, and the plant profitability. Thus, the system should be studied at the very beginning of the conceptual design stage. However, the design of a chemical plant refrigeration system is a very challenging task. The complexity mainly relies on the following two aspects: i) the refrigeration system of an ethylene plant needs to cover refrigeration requests in a wide temperature range (23 ~ -100.8 Celsius degree). It is thermodynamically impossible for using only one refrigerant as the working media. Thus, multiple refrigerants have to be employed under this situation. ii) The refrigeration system needs to satisfy the cooling demands from many different units. Thus, each refrigerant will have multiple loops and cover different cooling tasks. None of the previous studies have been reported to simultaneously consider the optimal design of a refrigeration system with multiple refrigerants and multiple temperature levels. In this paper, a conceptual design model for ethylene plant refrigeration system synthesis has been developed. The refrigeration system considers multi-refrigerant and cascade multi-loop designs. It also includes the modeling of a multi-stage steam system for accounting the energy consumption to drive the refrigerant compressors. Meanwhile, the concept of exergy is employed to analyze the thermodynamic nature of the refrigeration system and thus to guide its conceptual design. The efficacy of the developed methodology has been demonstrated by designing a cascade refrigeration system for an ethylene plant.

See more of this Session: Industrial Applications of Process Synthesis
See more of this Group/Topical: Process Development Division

(337d) Fully Automatic Computer-Aided Design and Synthesis of Complex Separation Networks

Seon B. Kim, Chemical and Bioengineering, University of Illinois at Chicago, Chicago, IL and Andreas A. Linninger, chemical and Bioengineering, University of Illinois at Chicago, Chicago, IL

We propose a novel computer-aided design for complex heat-integrated column networks to identify optimal energy efficient configurations. Since various complex column configurations including Petlyuk, Kaibel, divided wall columns, side-stripper and rectifiers have better energy efficiency than simple separation networks, we target to design complex column networks to separate multicomponent mixtures. However, complex column network synthesis is a challenging problem because the rigorous mass, equilibrium, summation and heat (MESH) equations lead to non-smooth fragmented search spaces. Moreover, combinatorial complexity grows rapidly with the number of components to be separated. Our new computer-aided design methodology deploys an inverse design methods based on temperature collocation which substantially reduces the problem dimension. We will demonstrate new algorithmic approaches like the bubble point distance using polynomial algebra, as well as and a hybrid niche algorithm for the global search for feasible, effective and optimal energy efficient designs. This methodology allows for design and synthesis of the complex column network in a fully automatic fashion. The solutions obtained by our synthesis method agree closely with rigorous mass, equilibrium, summation and heat (MESH) computations verified with the commercial flowsheet simulator AspenPlus. Thus, our design method is essentially equivalent to rigorous flowsheet synthesis. In conclusion, the rigorous solutions to separation problems can be obtained in a fraction of valuable engineering design time by proposed algorithms. Furthermore, the automatic and rigorous flowsheet synthesis is apt to systematically address industrial-size process design problems such as the synthesis of energy-efficient separation networks, layout of biorefineries with novel feedstocks or sustainable process for reduction of greenhouse gases emissions.

See more of this Session: Industrial Applications of Process Synthesis
See more of this Group/Topical: Process Development Division

(337e) Comparison of Mixing in Rocking Reactors and a Laboratory Dyeing Reactors

Srinivas Hanumansetty and Edgar O Rear, Chemical,Biological and Materials Engineering, University of Oklahoma, Norman, OK

Common laboratory dyeing reactors used in the textile industry such as the Werner Mathis AG unit consist of stainless steel cylindrical canisters mounted on a carousel with a horizontal axis. Reactors are canted to the horizontal so that the reactant mixture in a canister circulates with each revolution. Mixing is driven by end-to-end flow as the canister revolves with movement of an air void from front to back to front. The mixing in a canister nearly mimics that in a rocking reactor. Intensity of mixing was determined experimentally by rate of dissolution of benzoic acid granules at 30 C as a function of fill volumes of canister, amount of solute and rotation speed of the carousel. Increasing rotational speed of the reactor system increases mixing but with increasing fill volumes the mixing in canister decreases. For example, only 35% of benzoic acid charged at 4g is dissolved in 900 ml of water with a rotational rate of 25 rpm but 50% of benzoic acid is dissolved in 250 ml of water. As the volume increases in canister, potential energy contribution to mixing decrease as the change in height of center of mass decreases for volume of liquid in canister. Optimum volume for highest mixing is found to be one third of the volume of reactor with higher rotation speed. Results obtained are compared to those found by Temkin et al. for a rocking reactor in which he determines specific mixing power. This included the rocking amplitude and frequency for different degrees of filling of vessel.

See more of this Session: Industrial Applications of Process Synthesis
See more of this Group/Topical: Process Development Division

(337f) Optimal Synthesis and Design of Three-Phase Separator for Upstream Oil and Gas Processes

Rajab Khalilpour, School of Chemical and Biomolecular Engineering, The University of Sydney, Sydney, NSW, Australia and Ali Abbas, School of Chemical and Biomolecular Engineering, The University of Sydney, Sydney, Australia

Three-phase separator (3PS) is the heart of wellhead processes. The extracted well stream is a multiphase flow containing sands, water, light and heavy hydrocarbons. The stream, after extraction is directed to the so called "topside" processes where it first enters the 3PS. It is then separated into three streams: produced water, crude oil and natural gas. Each of these streams afterwards undergoes various processes until reaching export/discharge qualities. Three-phase separator is therefore the most critical upstream process. If it is designed correctly, the three outlet streams will require less quality adjustment. Traditionally, a series of two or three 3PS have been used. Separators vary in internal design to handle different well stream qualities. For example, when water cut is low, boot type separator is used while for high cut, weir or gutter arrangement might be used. Coalescer might be arranged inside the separator to decrease water oil particles in produced water. When a network of a few separator is used then a combination of different internal designs might be selected. Synthesis of 3PS network followed by efficient sizing is therefore a complex problem which if solved properly can result in significant downstream operation cost savings. This paper outline the development of a mixed integer nonlinear program to handle this problem, and presents modelling work and case studies.

See more of this Session: Industrial Applications of Process Synthesis
See more of this Group/Topical: Process Development Division

(346a) Challenges in Developing A Bilayer Tablet

Admassu Abebe[1], Ilgaz Akseli[2], Kyle Martin[3], Omar Sprockel[3] and Alberto Cuitino[4], (1) Biopharmaceutics, Bristol-Myers Squibb Company, New Brunswick, NJ, (2)Mechanical and Aerospace Engineering, Rutgers University, Pisacataway, NJ, (3)Bristol-Myers Squibb Company, New Brunswick, NJ, (4)Mechanical and Aerospace Engineering, Rutgers University, Piscataway, NJ

In the last decade, interest in developing a combination of two or more Active Pharmaceutical Ingredients (API) in a single dosage form (bilayer tablet) has increased in the pharmaceutical industry, promoting patient convenience and compliance. Bilayer tablets can be a primary option to avoid chemical incompatibilities between API by physical separation, and to enable the development of different drug release profiles (immediate release with extended release). Despite their advantages, due to the use of different materials and complex geometric boundaries between the adjacent layers, the mechanical structures of this drug delivery system have become quite intricate, requiring complicated tablet architectures as well as patient-friendly administration which pose serious challenges to the pharmaceutical scientists/engineers. This oral presentation details the major challenges associated with bilayer compression and rational strategy to deliver the desired bilayer tablet performance.

One of the major challenges is lack of sufficient bonding and adhesion at the interface between the adjacent compacted layers which is often the result of an interfacial crack driven by residual stresses in the tablet propagating a finite distance within the tablet and leads to delamination (layer-separation) which may not always be apparent immediately after compaction (e.g., during storage, packaging, shipping). In addition, if the compacted layers are too soft or too hard, they will not bond securely with each other which can lead to compromised mechanical integrity. Other challenges during development include establishing the order of layer sequence, layer weight ratio, elastic mismatch of the adjacent layers, first layer tamping force, and cross contamination between layers. These factors, if not well controlled/optimized, in one way or another will impact the bilayer compression per se (inefficient or uncontrolled process) and the quality attributes of the bilayer tablets (sufficient mechanical strength to maintain its integrity and individual layer weight control). Therefore, it is critical to obtain an insight into the root causes to enable design of a robust product and process.

Since the adjacent compacted layers of a bilayer tablet are bonded together by mechanical means, understanding what influences the stress state, the mechanical properties of each layer and the resultant bilayer tablet, and compression parameters along with specialized techniques to predict failure as a function of layer properties and compression conditions are primordial to successfully developing bilayer tablets.

See more of this Session: Multi-Layer Units Manufactured by Compression
See more of this Group/Topical: Process Development Division

(346b) Fundamental Concepts and Practical Experimentation on Multi-Layer Tablets

Adam Procopio[1], Katherine Chin[1], Larrry Rosen[1] and Antonios Zavaliangos[2], (1)Pharmaceutical Sciences and Clinical Supplies, Merck & Co., Inc., West Point, PA, (2)Materials Science and Engineering, Drexel University, Philadelphia, PA

Multi-layer tablets present unique technical challenges in developing robust compositions and processes that meet the marketed drug product quality attributes. These challenges result from the increased engineering complexity and sensitivity in merging multiple formulations into a single dosage form such that the resulting mechanical performance is able to sustain subsequent handling, processing, and environmental conditioning. The aim of this talk is to highlight the fundamental variables that drive this interfacial mechanical performance through the exploration of model material systems that cover a range of material performance including plastically deforming to brittle materials and hygroscopic to non-hygroscopic materials. In short, the interfacial strength of multi-layered tablets is derived from a combination of the interface architecture and material mismatch during compaction and unloading, the stresses invoked during ejection, and the mechanically and environmentally induced post compaction differential expansion. With this fundamental understanding in mind, experimental measurements are made that aim to predict multi-layer tablet performance at full-scale production through the use of hydraulic compaction simulators. Using this tool, the initial design space of a range of materials for multi-layer product is mapped out in terms of not only interfacial strength after compression but also under accelerated environmental conditions.

See more of this Session: Multi-Layer Units Manufactured by Compression
See more of this Group/Topical: Process Development Division

(346c) On the Structural Integrity of Bilayer Tablets Upon Ejection

Antonios Zavaliangos[1], Gerald Klinzing[1] and Adam Procopio[2], (1)Materials Science and Engineering, Drexel University, Philadelphia, PA, (2)Materials Characterization & Technology Assessment, Merck & Co., Inc., West Point, PA

Using a model system (microcrystalline cellulose, dicalcium phosphate) we explore the origins of processing difficulties of bilayer tablets. Differences in the mechanical behavior during compaction result in differential expansion in the two layers which in combination with the stress singularity of bimaterial interfaces and the geometrical characteristics of the interface may lead to cracking or splitting of bilayer tablets during ejection. The effect of tamping is reviewed and a process robustness map is presented. Preliminary results on the effect of differential moisture absorption in the long term structural integrity of bilayer tablets will also be presented.

See more of this Session: Multi-Layer Units Manufactured by Compression
See more of this Group/Topical: Process Development Division

(346d) Mechanistic Characterization of Bilayer Tablets .I: The Role of Pre-Compaction and Compaction Forces

Ilgaz Akseli[1], Admassu Abebe[2], Omar Sprockel[2] and Alberto Cuitino[1], (1)Mechanical and Aerospace Engineering, Rutgers University, Piscataway, NJ, (2)Biopharmaceutics, Bristol-Myers Squibb Company, New Brunswick, NJ

Solid oral dosage forms are the preferred route for many drugs and are still the most widely used formulations for new and existing complex-configuration tablets (e.g., bilayer tablets). Over the past decade or so, interest in bilayer tablets as an oral immediate-release/controlled-release system has increased substantially. However, during the production of such tablets, lack of sufficient bonding and adhesion at interfaces between the adjacent layers can compromise the mechanical integrity and performance of the final solid dosage form. A path forward to reduce the incidence of these issues in the production of bilayer tablets is to establish a rational strategy for the selection of compatible product formulations and manufacturing processes. This approach requires proper characterization of the interface of the bilayer tablets in representative pharmaceutical formulations.

In the reported study, bilayer tablets of the widely used excipient microcrystalline cellulose (MCC) in the form of Avicel 102 were formed with different pre-compaction (2kN, 4kN, 6kN, 8kN) and compaction (6kN, 10kN, 14kN, 18kN) forces to quantitatively characterize the strength (σ) of the interface and the adjacent compacted layers. In the reported experiments, using a developed tensile stress method, bilayer tablets of the same material were debonded to determine the fracture and axial tensile strength values. It was observed that when the first layer was compressed to a high density, the bonding with the second layer became difficult and it was not possible to produce intact bilayer tablets (Cases; 4kN vs. 2kN, 6kN vs. 2kN, and 8kN vs. 2kN) ($\sigma\_layer > \sigma\_interface$). It was demonstrated that the material response of the constrained MCC particles to an applied initial compression force within the initial compacted layer have a detrimental effect on the resistance to fracture of a bilayer tablet. The mechanism of failure at the interface or at the individual layers was also studied. It was shown that once the magnitude of the applied final layer compaction stress greatly exceeds the initial layer compaction stress (Cases; 2kN vs. 18kN and 4kN vs. 18kN), the tablet catastrophically fails at the first layer and not at the interface ($\sigma\_layer < \sigma\_interface$). The results of this study point out that the measurement of the interfacial strength provides insight into some of the major complications of bilayer tablets.

See more of this Session: Multi-Layer Units Manufactured by Compression
See more of this Group/Topical: Process Development Division

(346e) Mechanistic Characterization of Bilayer Tablets. II: The Effect of Layer Sequence and Fracture Surface Topography

Ilgaz Akseli[1], Admassu Abebe[2], Omar Sprockel[2] and Alberto Cuitino[1], (1)Mechanical and Aerospace Engineering, Rutgers University, Piscataway, NJ, (2)Biopharmaceutics, Bristol-Myers Squibb Company, New Brunswick, NJ

Bilayer tablets have some key advantages compared to conventional monolayer tablets. For instance, such tablets are commonly used to avoid chemical incompatibilities of formulation components by physical separation. In addition, bilayer tablets have enabled the development of controlled delivery of active pharmaceutical ingredients with pre-determined release profiles by combining layers with various release patterns, or by combining slow-release with immediate-release layers. However, these drug delivery devices are mechanically complicated to design/manufacture and harder to predict their long term mechanical properties due to the poor mechanical and compression characteristics of the constituent materials in the compacted adjacent layers, elastic mismatch of the layers, insufficient hardness, inaccurate individual mass control, cross contamination between the layers, reduced yield, and their tendency to delaminate at the interface between the adjacent compacted layers during and after the various stages of production downstream of the compaction process. Therefore, the major problem, that has to be overcome, is to understand in detail the sources of these problems in micro- and macro-scales and to develop remedies to solve them during solid dosage delivery design.

Bilayer tablets of the widely used excipients microcrystalline cellulose (MCC) and pregelatinised starch using unlubricated and lubricated punch and die sets were manufactured. The objectives of this presentation were three-fold: (i) to explore the effect of tableting conditions on the interfacial and individual layer strengths of bilayer tablets; (ii) to assess different fracture patterns related to the mechanical and fracture surface properties of compacted adjacent layers; and (iii) to determine the influence of layer sequence on the mechanical performance of bilayer tablets. In the reported experiments, using a developed tensile stress method, bilayer tablets were debonded to determine the fracture and axial tensile strength values. Fracture surfaces and topographic profiles of the debonded tablets were studied. Different fracture patterns, namely, clear layer break, half-half break, cap-shape break, and clear interface break were observed as a function of various initial and final compaction pressures. X-ray micro-computed tomography (μCT) also was utilized to examine the influence of localized density distribution on the delamination phenomena of bilayer tablets. The tensile strength values of individual layers after debonding were also determined and compared. It was observed that the bilayer tablets produced in a lubricated die fail at the interface while the tablets produced in an unlubricated die fail as half-half, cap-shape and clear layer breaks. Also it was shown why the layer sequence plays such a key role in the mechanical integrity of the bilayer tablets. It was found that, using the same compaction process, the tensile strength of the tablets with starch as the initial and MCC as the final layer were substantially higher than those of MCC as the initial and starch as the final layer. Fracture patterns and topographic profiles coupled with the X-ray μCT have provided a great insight into the effect of tableting conditions, individual layer properties, cross-contamination between adjacent layers, layer sequence, and stress distribution in bilayer tablets.

See more of this Session: Multi-Layer Units Manufactured by Compression
See more of this Group/Topical: Process Development Division

Innovative Tablet Technology for Pharmaceutical Drug Delivery

Rich Ferreira, Korsch America Inc., South Easton, MA

Extended Abstract

There is an increasing trend in the Pharmaceutical Industry for more innovative drug delivery technologies, and multi-layer tablet design is certainly at the forefront. New equipment technologies are being developed to support this trend, including bi-layer, tri-layer, four-layer, and five-layer tablet presses. In addition, other drug delivery platforms, including tablet in tablet - or compression coated tablets, and micro-tablets are also gaining in popularity.

This presentation examines the current state-of-the-art in tablet press technology for multi-layer, tablet-in-tablet, and micro-tablet applications, and reviews developments in the control system technology for these specialized machines.

Presentation Outline:

1. Introduction: Advantages of multi-layer tablet technology
2. Review of single layer process and principles governing weight control
3. Multi-layer principles for tablet compression machines
    a. Layer delamination
    b. Cross contamination of layers
    c. Production yield losses
    d. First layer filling principles
    e. Subsequent layer filling principles
    f. Tamping force measurement and instrumentation
    g. In process layer sampling systems
    h. In process layer sampling with hardening
    i. Tablet press control system features for precision weight control
4. State-of-the-art tablet compression equipment for multi-layer tableting.
    a. Development presses
    b. Production presses
    c. Compression core coating: Tablet-in-a-Tablet
    d. Specialty machines for unique multi-layer applications
5. Conclusion: Review of key machine principles for multi layer tableting

(355a) Real-Time Monitoring of Nucleation and Growth Rates in An Industrial Semi-Batch Sugar Crystallizer

Terry P. Redman, METTLER TOLEDO AutoChem, Columbia, MD, Greg Golovanoff, Amalgamated Sugar Company, Boise, ID and Wayne Genck, Genck International, Park Forest, IL

The process of boiling sugar cane syrup to produce refined sugar crystals was first performed in India around 300 AD. Although sugar manufacturing has evolved dramatically since that time, the basic methods of sugar crystallization in vacuum pans have been unchanged for almost two hundred years. Still, even in one of the oldest industrial processes, there is room for improving process efficiency and maximizing production capacity. As with many batch and semi-batch crystallization processes, the formation of large quantities of fine crystals – through nucleation events – can significantly reduce downstream separation efficiency and overall throughput.

A METTLER TOLEDO FBRM® crystallization monitor (model D600) was installed directly in a White Pan semi-batch crystallizer at The Amalgamated Sugar Company (Twin Falls, ID). Data were collected from 35 consecutive batches to monitor normal batch operation, to characterize batch-to-batch variability and to provide direct real-time analysis of how modifications to the batch recipe would impact the final crystal product.

In-line monitoring of the crystal population with FBRM permitted the calculation of relative nucleation and growth rates that were used to assess batch to batch variability and were able to pinpoint periods of intense secondary nucleation. The identification of factors and disturbances that promoted nucleation enabled adjustments to the operation of the sugar crystallizer that minimized nucleation events in subsequent batches - as confirmed with the inline particle size measurement.

See more of this Session: PAT for Crystallization Development and Manufacturing
See more of this Group/Topical: Separations Division

(355c) Apply PAT to the Definition of Design Space to Enable the Control Strategy for Crystallization Processes

George Zhou[1], Aaron Moment[1], Angela Spartalis[1], Sean Harrington[1], Shane Grosser[1], Marguerite Mohan[1], Naijun Wu[1], Ganeshwar Prasad[2], Tseng-En Hu[1], Paul Fernandez[1] and John Higgins[3], (1) Global Pharmaceutical Commercialization, Merck & Co., Inc., Rahway, NJ, (2)Global Technology Operations, Merck & Co., Inc., Barceloneta, PR, (3)Global Pharmaceutical Commercialization, Merck & Co., Inc., WEST POINT, PA

Process Analytical Technologies (PAT) have been widely used in the development and optimization of active pharmaceutical ingredient (API) processes, especially crystallization processes. In this talk, applications of a PAT-enabled crystallization platform to an API process will be described. Online IR has been applied to monitor the concentration and online FBRM to monitor the particle size distributions of APIs for the development and optimization of a crystallization process. As an important extension to QbD DOE work, to be able to dial in a reliable seeding point, a detailed solubility map of an API in a three-solvent system will also be illustrated. Furthermore, as a critical part of the QbD control strategy, an online NIR method has been established for controlling the composition of the process stream at the seeding point to ensure an efficient and robust crystallization process. The corresponding results of implementation at factory scale will also be highlighted. This approach has significantly enhanced our capability to explore the design space of crystallization processes and improved our ability to maintain the crystallization performance under unexpected disturbances and scale-up variations.

See more of this Session: PAT for Crystallization Development and Manufacturing
See more of this Group/Topical: Separations Division

(355d) Automated Crystallization Platform: Integrating Hardware, Software, and PAT to Expedite the Process of Crystallization Development

Amanda Rogers[1], Jacob Albrecht[1], James Vernille[1], Jose Tabora[1], Francesco Ricci[2], Mitsuko Fujiwara[3] and Richard D. Braatz[4], (1)Process Research and Development, Bristol-Myers Squibb Co, New Brunswick, NJ, (2)Princeton University, Princeton, NJ, (3)University of Illinois at Urbana-Champaign, Urbana, IL, (4)Department of Chemical Engineering, Massachusetts Institute of Technology, Cambridge, MA

In the absence of in situ monitoring techniques, crystallization development can prove labor intensive, requiring frequent sampling to monitor process variables such as solution concentration, crystal size distribution, etc. Frequent sampling for off-line analysis is inconvenient and may disturb the system, affecting the accuracy and limiting the completeness of the data collected. To this end, platforms have been developed that integrate process analytical technology (PAT) with hardware and software to enhance the efficiency and accuracy of data collection during crystallization studies [e.g., see Refs. 1 and 2]. This technology has been implemented at BMS for the study of both antisolvent and cooling crystallizations. Real-time readings from Attenuated Total Reflection-Fourier Transform Infrared (ATR-FTIR) spectroscopy and Focused Beam Reflectance Measurement (FBRM) are used to control solvent composition and temperature profiles for the crystallization process. In addition, automated experiments have been conducted to determine the solubility and metastable zone width, and monitor supersaturation throughout a recipe-driven antisolvent addition or cooling profile. This presentation will discuss results from the study of several pharmaceutical crystallizations, including those for late-stage active pharmaceutical ingredients (APIs). The productivity enhancements realized through the use of PAT, computer-driven hardware, and automated data analysis and visualization will also be described.

1. G. X. Zhou, M. Fujiwara, X. Y. Woo, E. Rusli, H.-H. Tung, C. Starbuck, O. Davidson, Z. Ge, and R. D. Braatz. Direct design of pharmaceutical antisolvent crystallization through concentration control. Crystal Growth & Design, 6, 892-898, 2006.

2. A. Cote, G. Zhou, and M. A. Stanik, Novel crystallization methodology to ensure isolation of the most stable crystal form, Organic Process Research & Development, 13, 1277-1283, 2009.

See more of this Session: PAT for Crystallization Development and Manufacturing
See more of this Group/Topical: Separations Division

(355e) Characterization of Flash Crystallization Process Using PAT

Barbara Wood, School of Chemical and Bioprocess Engineering, University College Dublin, Dublin, Ireland and Brian Glennon, Chemical and Bioprocess Engineering, University College Dublin, Dublin, Ireland

The objective of this project is to investigate flash crystallization. Flash crystallization is not well defined within available literature so the combination of in situ PAT such as FBRM, ATR-FTIR and PVM to characterize the process are key. Flash crystallization is a technique which utilizes a change in reactor pressure to drive crystallization. The overall aim is to produce a small narrow particle size distribution which is comparable to a typical micronization step. The adiabatic flash of solution causes a fast temperature change - a strong driving force for nucleation. The solution is also atomized by a nozzle upon entering the reactor which further promotes the formation of small particles by impeding growth. Adipic acid was chosen as the model system for this study. Flash crystallization was compared to various cooling crystallization strategies as well as different solvent systems. The use of FBRM ATR-FTIR and PVM were essential in the development of such comparisons. A flash crystallization process was developed incorporating a 2L jacketed vessel and vacuum pump with pressure control system. The pump maintains the system pressure as well as controlling the flow of solution to the reactor resulting in a dynamic system. The effect of temperature, pressure and flow rate as well as nozzle configuration and solvent selection on the final CLD are investigated. Solubility design space information was obtained which allowed for the development of cooling crystallizations. The use of extreme cooling profiles for the production of small particles with a narrow size distribution were examined, as well as more traditional cooling strategies. Various cooling rates and profiles were assessed for their impact on the systems' metastable zone width and particle dimension. Inline ATR-FTIR was used for assessing the dissolved concentration/supersaturation during the crystallization process. FBRM, PVM and SEM analysis show that the particles obtained via flash crystallization are much smaller in size, with a decrease in the FBRM mean square weight from 185 micron to 40 micron evident, when compared with the developed cooling crystallization. Additionally, the obtained CLDs are narrower in shape when compared with the developed cooling crystallization. Ethanol and water solvent systems were examined using traditional cooling strategies with the ethanol system resulting in a smaller CLD.

See more of this Session: PAT for Crystallization Development and Manufacturing
See more of this Group/Topical: Separations Division

A Guide to Scale-up of Batch Crystallization from Lab to Plant

Faster Optimization and Troubleshooting with Process Analytical Technology

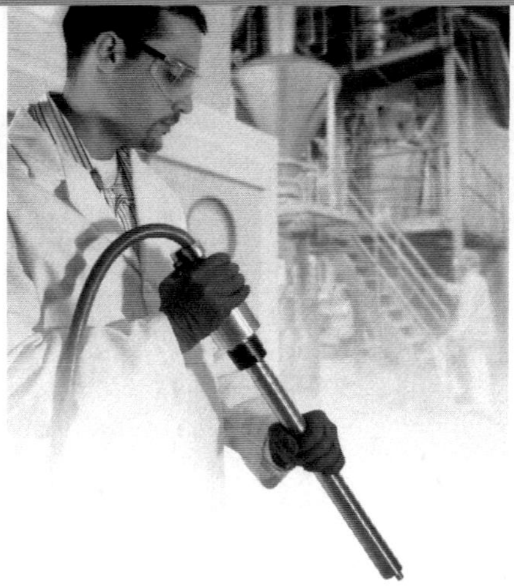

Real-Time Measurement of the Crystal Population

Terry P. Redman, Benjamin Smith, Mettler-Toledo AutoChem, Inc.
Mark Barrett, PhD, Solid State Pharmaceutical Cluster (SSPC) Ireland

Crystallization is a critical processing step in the manufacture of many fine chemical and pharmaceutical products. It can serve simultaneously as a purification step and as a separations unit operation. The design of a robust crystallization process can assure the isolation of a product in the desired size, shape and form required.

At the same time, crystallization is a notoriously complicated process where the final crystal product is a function of thermodynamics as well as kinetic and physical phenomenon. There are two traditional ways of dealing with this complexity of crystallization – either crash the solids out of solution and deal with the results downstream, or tune down the crystallizer to avoid common problems. Neither of these situations is optimized for assuring product quality or maximizing production yield and throughput. This has fueled a push towards real-time monitoring of crystallization with Process Analytical Technology (PAT) that can directly measure critical parameters such as the crystal size and shape distribution, the crystal form, and even the level of supersaturation.

Real-time monitoring of crystallization is shown to provide many benefits leading to improved methods for process development, optimization and scale-up. This paper reviews case studies examining many of these benefits, including the use of Process Analytical Technology (PAT) for:

- Providing detailed process knowledge that enables real improvements in yield, throughput and profitability of batch crystallization
- Eliminating downstream bottlenecks by improving filtration/dryer performance
- Speeding up identification and characterization of critical operating parameters for increased R&D productivity
- Enabling the design of more robust processes to assure batch-to-batch repeatability and consistent meeting of crystal specifications
- Identifying disturbances and undesirable events in real time to help ensure product quality

METTLER TOLEDO

Introduction

Crystallization is a critical process for the purification and isolation of chemical compounds in the manufacture of many fine chemical and pharmaceutical products. The results of the crystallization step have far reaching impacts on overall process efficiency and final product quality. It is also a very difficult process to effectively optimize and control. Crystallization is inherently complicated simply by being a process involving the creation and formation of solid particles. Timothy A. Bell (of DuPont Engineering Research and Technology), wrote a review of the challenges of scaling-up particulate processes where he stated:

> "Studies by the Rand Corporation in the 1980s identified substantial differences in the scale-up and start-up performance of plants processing particles versus those processing liquids or gases. These differences were inevitably unfavorable. Particulate process plants take longer to start up and are less likely to achieve desired production rates... These problems generally relate to an inadequate understanding of the behavior of particle systems. Many of these behaviors are sensitive to process scale or process history in ways that would not be expected by engineers familiar only with liquid or gas systems."[1]

Tim went on to identify crystallization as one of the most complicated particulate processes to scale-up (or scale-down) due to the significant impact of typical scale-up parameters – such as agitation – on the size distribution of the crystal product.

The implications of a well-designed crystallization process are considerable:

- Critical product parameters – such as product dissolution rates – and critical process parameters– such as powder bulk density, compressibility, and flowability – can be directly modified and controlled through effective crystallization process design.
- Achieving a target crystal size specification can significantly reduce cycle time – sometimes by an order of magnitude – by eliminating bottlenecks in filtration and drying to improve downstream process efficiency.
- Improving the repeatability of the crystallization process – by reducing batch to batch variability – improves the ability to design and fully utilize downstream processing equipment by avoiding the need to overdesign capacity for filtration, drying and milling equipment.
- Producing crystals that can be easily isolated from the mother liquor and used directly in the final product, or compressed directly into a granulated or tableted product, can avoid additional unit operations such as milling or wet granulation that can significantly reduce overall yield.

For all these reasons, crystallization continues to be an area in which incremental improvement in efficiency and yield can often have a significant impact on the overall production efficiency and product profitability. The use of Process Analytical Technology for optimization, scale-up, and ultimately control can help to quickly develop and capture the benefits of a more efficient crystallization process.

Authors

Terry Redman, MSc, MBA
Terry_Redman@mt.com

Benjamin Smith, BSC
Ben_Smith@mt.com

Mark Barrett BE, PhD

What Makes Crystallization Such a Complex Process?

The efficiency and profitability of a crystallization process is directly tied to the crystal size distribution produced in the crystallizer vessel. Problems with solids bulk density, flowability, crystal size and shape can often be directly related to the operations of the crystallization step. Missing the crystal size distribution specifications may result in costly rework. The production of excessive fines in the crystallizer vessel can dramatically cut production yields and can reduce throughput due to serious bottlenecks in downstream processes such as filtration and drying. Excessive milling of crystal product can result in further yield losses and result in potential dust hazards. Therefore, an engineered crystal size distribution that meets particle size specifications repeatedly, and avoiding excessive downstream modification through milling and sieving, can dramatically improve overall production efficiency and profitability.

> **"Crystallization is notoriously difficult to scale-up..."**
>
> Timothy A. Bell
> DuPont Engineering R&D[1]

Crystallization, however, is extremely difficult to directly transfer from the laboratory to pilot and production scale. Scale-up difficulties are compounded by the importance of both thermodynamic and kinetic properties in determining the final crystal size distribution.

Supersaturation, the thermodynamic driving force of crystallization, is a critical parameter in determining the final crystal population. In a laboratory vessel that is relatively well-mixed, supersaturation may be effectively constant throughout the vessel. At a larger scale, there are undoubtedly gradients of supersaturation throughout the crystallizer – due to the manner in which the supersaturation is created (most often by cooling or anti-solvent addition), and due to the mixing configuration (including parameters such as the vessel dimensions, baffles, impeller type, and agitation speed) which determines how effectively the supersaturation is dispersed throughout the vessel. The introduction of the supersaturation gradient plays a very significant role in the difference between laboratory-scale and full-scale crystallization.

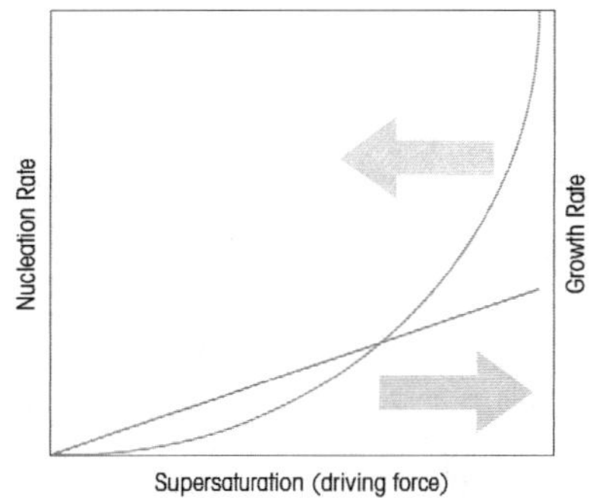

Figure 1. Nucleation and growth rates are both functions of supersaturation. The nucleation rate tends to exhibit a higher order exponential relationship, so as the supersaturation increases the nucleation tends to dominate. Promoting crystal growth requires operation within a limited range of supersaturation.

An undersaturated solution (negative supersaturation) dissolves the crystals present – resulting in crystal shrinkage and extinction. Due to their high ratio of surface area to volume, fine crystals can sometimes be completely redissolved with relatively little change to the largest crystals. Fines destruction (a controlled preferential dissolution of fines) is sometimes used as a means to control the final crystal size distribution or improve filterability of the final product.

Crystallization is further complicated by the fact that it is a multiphase system. The solid crystal product is very often a different density than the liquid phase (mother liquor). Crystals that are denser than the mother liquor have a desire to settle to the bottom of the crystallizer, and the mixing required to keep a liquid based system well-mixed is no longer sufficient to keep the solids suspended. Although a first instinct response might be to increase the agitation speed or to add baffles, one must also be aware of the possibility of crystal breakage and attrition due to the increased energy input. Attrition can be a significant cause of fine crystals or a source of secondary nucleation which will dramatically alter the final crystal size distribution and causes a number of scale-up headaches.

The kinetics of crystallization adds additional complexity to scale-up. Crystallization kinetics is commonly simplified to two parameters: nucleation (birth of new crystals) and growth (rate of increase of crystal dimensions). (More complex crystallization models may also include rates of dissolution, breakage, attrition, and agglomeration to more fully predict the population balance in the crystallizer – but these are usually situations we will work to avoid in a practical crystallization process.)

The nucleation and growth rates are primarily functions of supersaturation. At relatively low levels of supersaturation, growth tends to dominate. Nucleation rates, however, have a higher order relationship with supersaturation – so that if supersaturation reaches a high enough level, nucleation will dominate the crystallization process.

In addition, the presence of crystals – and the related crystal size distribution – is another critical factor in determining how the supersaturation is consumed and therefore the combination of the the final crystal product. This is because the quantity of crystals present – and specifically the viable crystal surface area available for growth – determines the rate at which supersaturation can be consumed by the existing crystal population.

If avoiding nucleation is desirable, the crystallizer can only generate supersaturation at a rate that the existing crystal surface area can handle at the corresponding growth rate. If the amount of surface area is insufficient to handle the generated supersaturation, then the overall supersaturation level will rise, eventually to the point where nucleation becomes a significant factor. If the amount of surface area is more than the current growth rate will sustain, the supersaturation will drop (implying that the crystallizer is running at less than full capacity.)

If that is not complicated enough, the fact that you have potential gradients of supersaturation and gradients of crystal size distribution throughout the full-scale vessel, also means that you have potential gradients of nucleation and growth rates making the final crystal product extremely difficult to predict from simple kinetic models based on laboratory data.

There have been two traditional ways of dealing with this complexity of crystallization – either crash the solids out of solution and deal with the headaches of the downstream processing bottlenecks, or simply tune down the crystallizer so far that it runs at a low enough level of supersaturation that problems of nucleation are generally avoided. Clearly, neither of these situations is optimized for maximum production yield and throughput, and this in part has fueled the recent push towards real-time monitoring of crystallization using Process Analytical Technology that can directly measure critical parameters such as the crystal size and shape distribution, the crystal form, and even the level of supersaturation.

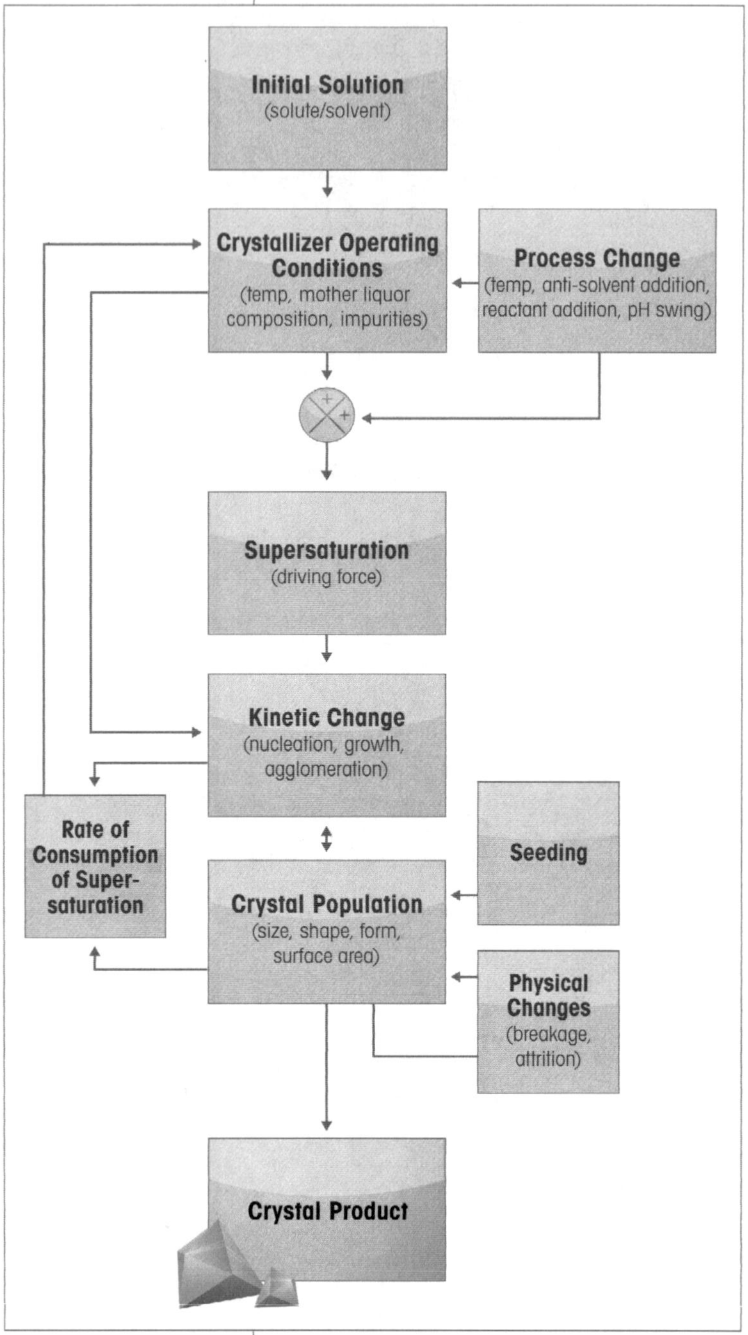

Figure 2. A moment in time in a batch crystallizer - visualizing a complex process

Supersaturation or the Crystal Size Distribution: Which Measurement is More Important?

In an ideal world, you may want to directly measure the crystal population within the crystallizer (a critical product quality attribute) and measure the supersaturation which is driving the process (a critical process parameter). It is good to know that today's advanced Process Analytical Technology allows you to measure both of these critical parameters in real time. But where should you begin – especially if budget constraints limit you to implementing only one advanced measurement.

The crystal population, as discussed previously, is often the product itself. An online measurement of the crystals gives you the possibility of control and the possibility of assurance that the crystals meet final product specifications before actually being discharged from the crystallizer.

Supersaturation monitoring and control can provide an optimal path to your final product. However, without actual crystal population information, this only makes sense if you have tight control of the crystallizer vessel, a precise supersaturation measurement, and a very reliable model of the system (i.e. where you can predict nucleation and growth as a function of supersaturation with reasonable accuracy throughout the operating range).

In laboratory-scale R&D this is certainly achievable. However, in a larger scale crystallizer there are limitations and complications that make the control of the crystallizer based solely on supersaturation very difficult. As discussed previously, gradients in temperature (and therefore supersaturation) and solids concentration throughout the vessel can have a dramatic impact on the crystal population. Therefore, to compensate for these potential gradients – without measuring the crystals themselves – the controlled level of supersaturation has to be tuned down to limit the maximum level of supersaturation that might occur.

And that's not really control, just avoidance.

If you manage to control the crystal product using only supersaturation, chances are you are over-tuned to stay far away from conditions that might promote nucleation. This likely means the production rate is not optimized.

From the viewpoint of process control, you can think of the two measurements as examples of feedforward and feedback control. Supersaturation gives you the ability to predict what is going to happen (feedforward), but you need a near-perfect model and a high level of measurement precision for this to work. In some batch crystallization applications, this has been shown to be successful using mid-IR (such as ReactIR™ ATR-FTIR) as the measurement of supersaturation.

Measuring the crystals in process with FBRM® real-time measurement (allowing feedback control) is much more reliable for dealing with disturbances. However, as with any feedback control, you actually depend on a slight disturbance before you can respond and correct the system.

An analogy of supersaturation control is simply trying to get from one location to another using a map (your model of the process) and a compass (your supersaturation measurement). To get to the desired endpoint, you need to know your starting point (clear liquor concentration) and you need to know what effect each step will have along your path. If your map is correct and your measurement has sufficient accuracy, you should reach your destination. But if there are changes to the terrain (such as the presence of impurities), or if your measurement is less than perfect, you can drift off course.

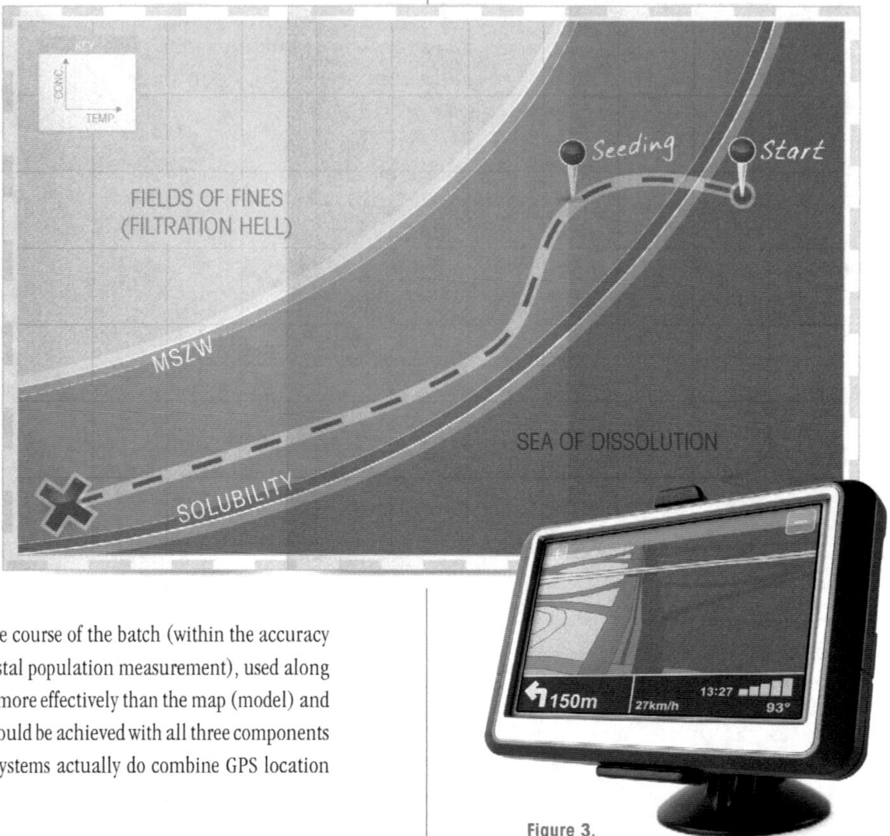

Figure 3.

Measurement of the crystal population, following the same analogy, is the addition of a global positioning system (GPS). It tells you exactly where you are throughout the course of the batch (within the accuracy of the measurement of course.) The GPS (crystal population measurement), used along with the map (process model), will guide you more effectively than the map (model) and compass (supersaturation). The best results would be achieved with all three components (and one can note that all GPS navigation systems actually do combine GPS location with a map and a compass).

And in a similar way, many of the best examples of batch crystallizer control actually use both FBRM® crystal population measurement and supersaturation measurement in their models and control algorithms (see academic research from the research groups of Prof. Richard Braatz (UIUC, USA)[2,3,4], Prof. Sohrab Rohani (Western Ontario, Canada)[5], Prof. Marco Mazzotti (ETH, Switzerland)[6], and Prof. Brian Glennon (UCD, Ireland)[7]). These advanced model-based controllers using neural networks and fuzzy logic rely on measurements of both supersaturation and the crystal population for optimum results.

If you have to choose one method of advanced Process Analytical Technology, measuring the crystals themselves is the best option for control AND assurance that the product will meet specifications. Supersaturation is very valuable in understanding and modeling the system, but it provides limited monitoring and control capability without real-time confirmation of the crystal size distribution.

This initial guide (Part 1) focuses on the *in situ* measurement of the crystal population, and the role these measurements play in the scale-up and optimization of pilot and manufacturing scale batch crystallizers. A separate guide (Part 2) will follow to provide an overview of supersaturation measurement and its applications in the scale-up and optimization of batch crystallization.

Selecting the Right Instruments using the 3 R's of Process Analytical Measurement: Robustness, Repeatability, and Reproducibility

A starting point for any effective attempt to optimize and control a process is the identification and reliable measurement of critical process parameters that directly influence product quality and process efficiency. Whether the measurements are online or offline, direct or inferred – there must be a connection between what is measured and what is desired in the final product. With processes being pushed for further efficiency and productivity, reliable real-time measurements become even more critical for enabling real-time decision making and automated control.

Selecting an online measurement to be used as the basis of scale-up requires consideration of a number of issues related to measurement and instrument precision and sensitivity. These can be described in terms of robustness, repeatability and reproducibility.

Robustness

Robustness (generally meaning "strong, reliable, healthy...") has a number of definitions that apply in this context:

Measurement robustness: In terms of instrumentation, a robust measurement is one which is both sensitive (able to detect significant changes in the underlying measured parameter) and precise (repeatable with a high signal to noise ratio)

Instrument robustness: The instrumentation itself is considered robust if it is constructed to operate reliably under expected process conditions that can include dust, solvent exposure, vibration, etc. once the process is scaled-up to the pilot or manufacturing plant. An important element in instrument robustness may include the designed ability to operate safely in hazardous environments that can include the presence of explosive dust or flammable vapors. Designs to meet hazardous area classifications such as Class 1, Div 1 or ATEX are commonly used to confirm compatibility with these hazardous environments.

Process robustness: In terms of process design and process operation, a robust process means that the system is designed to produce product that meets required specifications in spite of normal process variability[8]

Repeatability

Repeatability simply means that given the same inputs, the same result will consistently occur. This can also refer to both the measurement and the process.

Measurement repeatability:

Instrument repeatability is one definition of measurement precision. Given the same underlying measured system, it is desirable that the instrument provides the same measurement. This is often proven with calibration verification and operational qualification. All instrumentation has an element of random error (noise) that contributes some random variability to the measurement. A high level of precision corresponds to a low level of noise under normal operating conditions. To be able to track real-time changes in a process, the level of precision must be high relative to the actual changes in the underlying process.

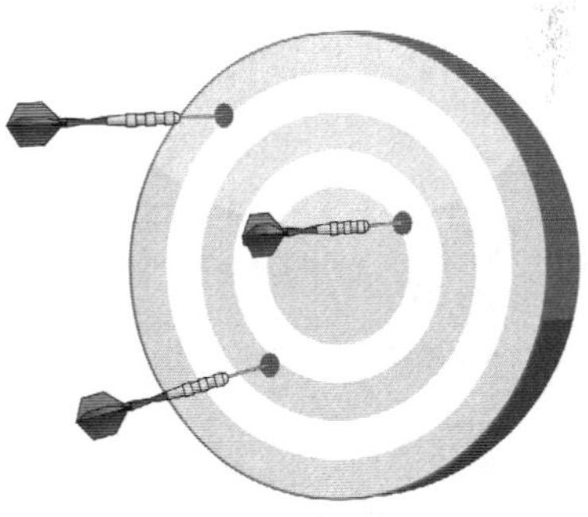

Poor precision (showing random error)

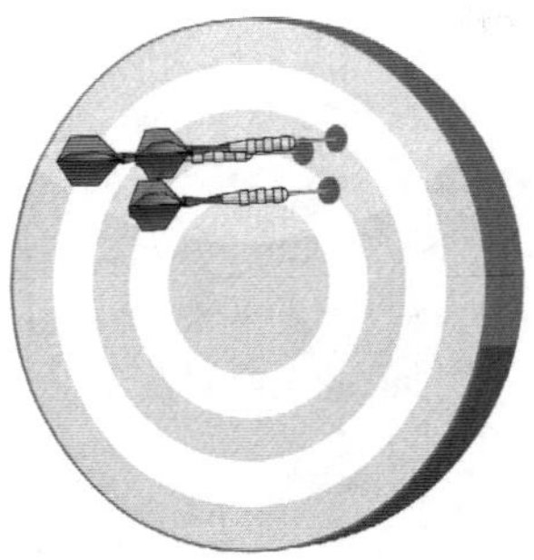

Good precision/poor accuracy (showing bias)

Figure 4. With Process Analytical Technology – precision is generally much more important than accuracy.

A lack of accuracy (bias) can be easily corrected for. A lack of precision (random error) will have a greater impact on the ability to track the process with adequate sensitivity.

Note that accuracy is less critical for inline measurement, provided there is adequate precision and sensitivity in the measurement. Measurement bias (the systematic difference between the measured variable and the variable of interest) can be corrected for by proper calibration and modeling.

Process (batch-to-batch) repeatability: Batch-to-batch repeatability is an important element of effective process design, indicating that the process is being operated in a robust region of the design space and is capable of handling random variability. Process Analytical Technology and real-time monitoring and control can greatly enhance batch-to-batch repeatability by providing the tools and handles to identify and compensate for some of this normal process variability.

Reproducibility

Reproducibility is a measurement of consistency from one system to another. Again, this is important to consider from both the measurement and the process viewpoint.

Instrument to instrument reproducibility: As processes are scaled-up from lab to plant, or transferred from one site to another site, the ability to reliably provide the same measurement at different scales and in different locations is a critical aspect of process optimization and control. In most cases this requires multiple instruments to achieve. A measurement technology that provides reproducible measurements of critical process parameters permits a more direct scale-up or transfer of the process from one location to another.

Process reproducibility: Ultimately, it is the reproducibility of the process that is of primary concern. In that sense, measurement reproducibility and process reproducibility go hand in hand. Real-time monitoring of the process can confirm that the process is progressing as designed, or can identify the points in time where the process is varying from the desired process - allowing faster identification of the scale-up factors that are affecting the process and providing the ability to rapidly re-optimize the process to take into account the effects of scale or equipment variations.

A Review of Selected Case Studies from Industry and Academia: Crystallization Scale-up Using Focused Beam Reflectance Measurement (FBRM®) and Particle Vision Microscope (PVM®)

METTLER TOLEDO FBRM® and PVM® has proven to be an extremely versatile tool for monitoring the crystal size distribution during development, scale-up and for monitoring at production scale. Probes designed for crystallizer vessels from 50mL to full production scale all employ the same robust method of measurement (see Appendix A for a review of the measurement technique). This consistency of measurement enables process chemists and engineers to obtain precise and sensitive real-time measurements of the crystals as they actually exist in the process at any scale. Real-time, inline monitoring of the crystal population has been shown to provide many tangible benefits that are discussed in this review.

We will look at the use of FBRM® for:

- Providing detailed process knowledge that enables real improvements in yield, throughput and profitability
- Eliminating downstream bottlenecks by improving filtration and dryer performance
- Speeding up identification and characterization of critical operating parameters for increased R&D productivity
- Enabling the design of more robust processes to assure batch-to-batch repeatability and consistent meeting of crystal specifications
- Identifying disturbances and undesirable events (such as oiling out) in real-time to enable process corrections to ensure product quality

Improving Throughput and Yield in Production

The crystal size distribution plays a critical role in process efficiency and product quality. For that reason, crystallization is almost always designed to produce a product of a defined crystal size distribution. This may include specifications for average crystal dimension (mean, median, mode), width of distribution (standard deviation, COV, d10, d90), or simply meeting a cut size (% of crystal mass less than or greater than a specified dimension.)

The ability to monitor the crystals throughout the process gives a wealth of dynamic information that can be used to understand how the final crystal population has been formed (through primary nucleation and growth, secondary nucleation, agglomeration, breakage) and how the crystal product is impacted by changes in critical process variables (such as agitation, operating temperature, cooling rate, anti-solvent addition rate.) This higher level of process understanding enables a more robust process design that reliably meets targets for crystal size, product yield, and purity.

Process Optimization Using FBRM® Provided These Direct Benefits to Excella

- Stable process with narrow particle size distribution (see figure below)
- Total cycle time reduction of **16hrs**
- Yield improvement by **10%**
- Cost reduction by **20%**

Andre Ridder, Excella[9]

Final Particle Size Distribution for Consecutive Batches

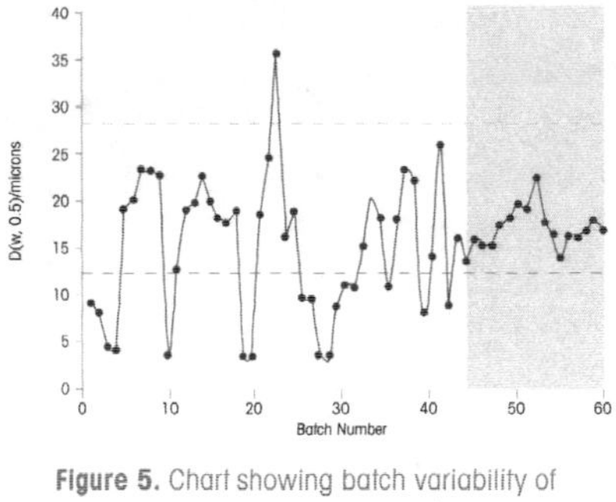

Figure 5. Chart showing batch variability of final particle size distribution. Optimization of the crystallization step dramatically reduced batch variability and eliminated batch failures due to off-spec crystal size distribution.

Andre Ridder, Presented at 15th International Process Development Forum, Annapolis, USA (2008)[9]

Better crystallization design – providing a more reliable and more repeatable process – allows greater control and allows optimized operation for maximizing crystal production rates while minimizing downstream process problems that are common when off-spec and non-optimized crystal product create bottlenecks in filters and dryers.

Excella Pharma Source, a European contract manufacturer, has reported on significant benefits achieved through process optimization at the plant scale with online FBRM® measurements.[9] Due to variability with incoming starting materials, they were experiencing significant process variability – leading to a number of batch failures and requiring a secondary crystal washing step. The specific benefits achieved (see sidebar) include a 20% reduction in operating cost and a 10% increase in crystal yield through crystallization optimization and elimination of the washing step. In addition, the repeatability of the process was drastically improved - essentially eliminating batch failures caused by off-spec crystal size distribution.

Eliminate Downstream Bottlenecks by Improving Filterability

One of the most common problems with poorly optimized batch crystallization is the impact it has on filtration and drying rates. Poorly filtering crystals can cause bottlenecks in the manufacturing process and can result in significant mother liquor holdup – resulting in high levels of impurities or the need for additional crystal washes which reduce yield.

There have been many examples of the relationship that the crystal size distribution can have on filtration rates[10, 11] and on the real benefits of using in-process crystallization monitoring with FBRM® to troubleshoot and optimize crystallization processes specifically to address filtration issues.[12, 13, 14]

Accelerate Development and Scale-up Timelines

The application of any Process Analytical Technology is based on the concept that real-time, in-process measurements will provide better information for process development, optimization and control. The benefits of Process Analytical Technology – as opposed to the alternative methods of running blind or taking process samples for offline analysis – are that the real-time and often data-rich information provides a dynamic understanding of the process. Understanding the dynamics of the process and measuring the real-time response to changes in operating conditions enables much faster process characterization and optimization. Direct measurement of the process during scale-up permits immediate confirmation that the process is performing as designed – or provides the tools for fine tuning the process if the scaled-up process is sensitive to the differences in mixing and heat transfer at the increased scale.

The use of FBRM® for monitoring batch crystallization has been shown to routinely deliver these benefits[15, 16, 17] – allowing FBRM® users to streamline development and scale-up to achieve an optimized process with significantly reduced time and materials.

Benefits Achieved Through Optimization of Crystallization to Minimize Filtration Issues:

- Particle attrition in crystallizer and filter dryer reduced, eliminated slurry handling problems
- Mean particle size doubled
- Eliminated filter blinding and need to interrupt campaign for periodic washing of filter plates
- Reduced process cycle time by three hours (10% reduction)
- Reduced operator labor by 120 hours per month
- Increased monthly throughput by 20% (increased from 20 to 24 batches)

Kaz Wood-Kaczmar, GlaxoSmithKline[12]

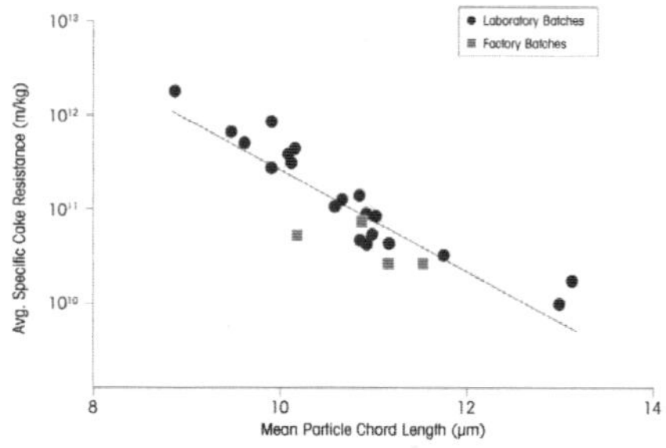

Figure 6. Correlation of FBRM® Chord Length Data to Filtration Rate. A direct correlation between crystal dimensions measured with FBRM® and downstream filterability is straightforward.[10]

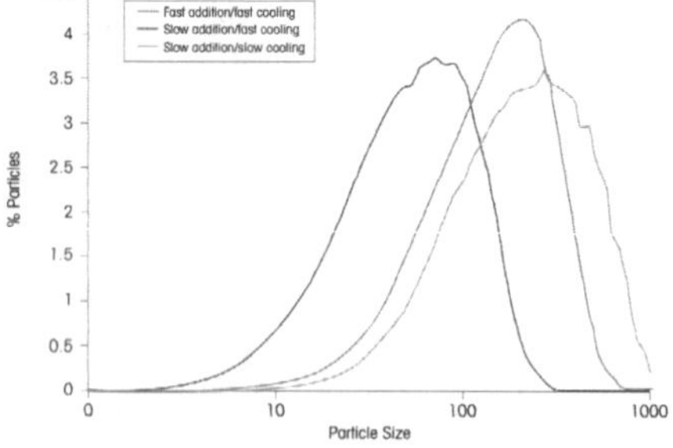

Figure 7. The above figure, from a Pfizer Case Study[17], demonstrates that real-time, in-process measurements make it simple to characterize the direct effect of critical operating variables on the crystal population. Design of Experiments (DoE) or optimization routines can be used to quickly design a robust process for scale-up.

Design More Robust Processes to Assure Batch-to-Batch Repeatability

As indicated, when processes are sensitive to scale-up variations, Process Analytical Technologies providing a sensitive measurement of critical process parameters can be used to diagnose and eliminate the source of these problems.

A case study, published by researchers at Sepracor[18], illustrates how FBRM® information was valuable in eliminating a problem that was leading to batches that were failing the crystal size specification at their Contract Manufacturer (CMO). After a number of batch failures at the plant scale, scale-down experiments were performed in the laboratory to understand the reasons for the batch failures. With the use of FBRM® a problem with secondary nucleation was identified after seed crystals were added. Scale-up of seeding protocol tends to be complicated due to the difficulty of ensuring consistency in seed source, seed size distribution, and dispersion of seeds throughout the vessel. Measuring the secondary nucleation event in real time (Fig. 8b), it was determined that the temperature (and level of supersaturation) at which secondary nucleation occurred was a direct function of the shear rate at the impeller (Fig. 8c). Optimization of the agitation and seeding protocol allowed them to eliminate the secondary nucleation and avoid any further failed batches. In fact, they eliminated batch failures in subsequent production runs at the CMO site, and they dramatically improved overall batch efficiency (the average centrifugation time was reduced from 7.5 hours to 2.2 hours per batch and they eliminated the need for manual discharge of the centrifuge contents.)

This example shows how Process Analytical Technology, such as METTLER TOLEDO FBRM®, contributes to the mapping of the operational design space and in determining and setting the critical operating variables to ensure a robust process that can deliver the necessary quality, yield and batch repeatability.

Optimization of Seeding Protocol Provided the Following Benefits to Sepracor

- The centrifugation and drying operations were drastically improved, resulting in considerably lower cycle times
- The robustness of the process also appears to be improved: in 2006, 39 batches under these conditions were operated with zero failures, while 18 batches were operated using the original conditions with three failures

Patrick Mousaw et al
Sepracor[18]

Parameter	Original	10x seeding
Secondary nucleation induction time	average: 37°C range: 26-43°C	average: 44°C range: 42-45°C
Centrifugation time	average: 7.5h range 3-12h	average: 2.2h range: 1-4h
Manual discharge of centrifuge required?	yes	no
No. of failed batches ÷ no of batches in 2006 campaign	3/18	0/39

Figure 8a.

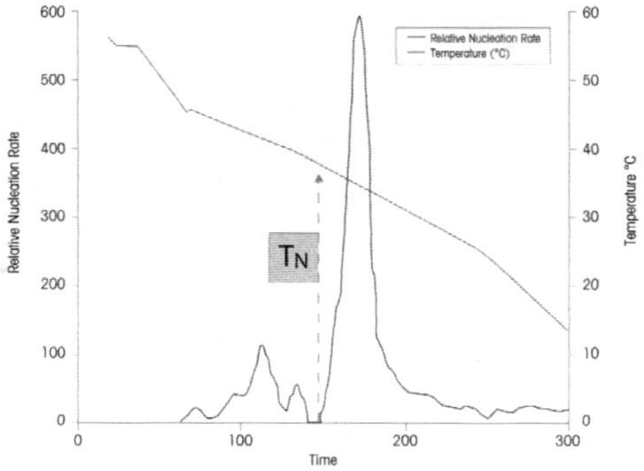

Figure 8b. Detection of Secondary Nucleation

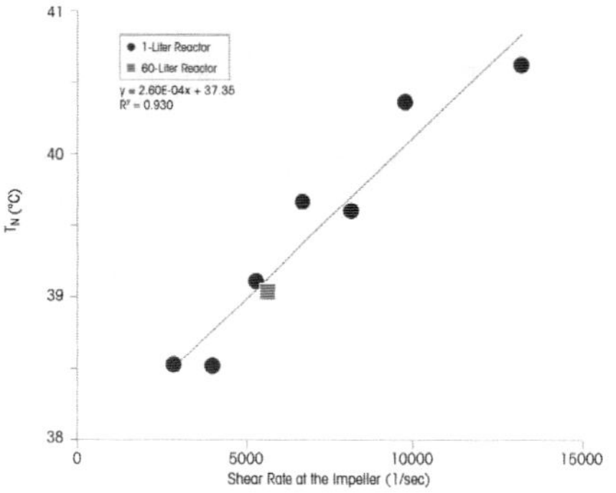

Figure 8c. Secondary Nucleation as a Function of Shear Rate

Reference for Figures 8a, 8b, 8c: Mousaw et al, 2008[18] (c) 2008, American Chemical Society (used by permission)

Immediately Identify Disturbances with Real-time Monitoring

Disturbances and unexpected events will happen, but they are unpredictable by definition. The design of a robust process will often take into account the expected variability in the process and will hopefully account for most of the risk of batch failure.

Even so, as a crystallization process is scaled-up or transferred to a different set of equipment, there are many possible sources of variability that can have disastrous effects. One of the side benefits of Process Analytical Technology is that it can often provide a window into the unexpected and can actually help eliminate the unexplained.

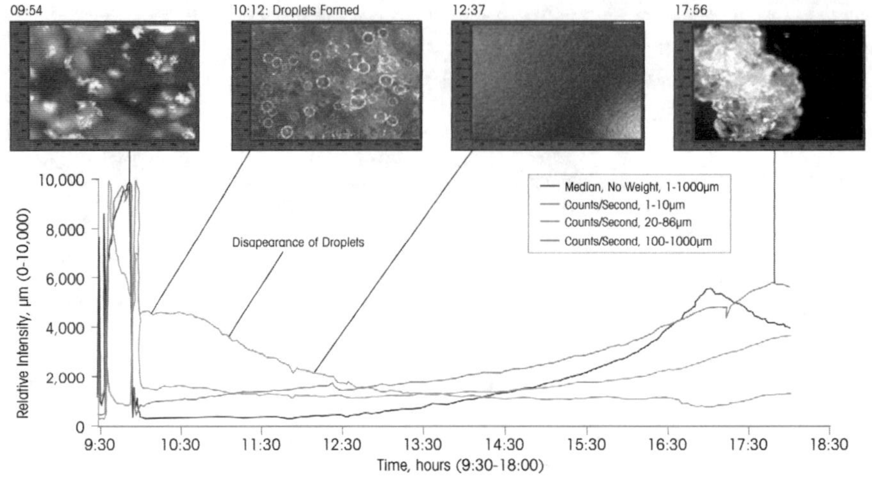

Figure 9a. A Case Study by Bristol Myers Squibb[19] highlights how PVM® real-time imaging and FBRM® were used together to investigate process deviations. Seed crystals were shown to dissolve and a phase separation occurred during normal operation. The process was subsequently modified to avoid phase separation and ensure a repeatable crystal product.

In many cases, a batch failure requires an analysis of the root cause of the failure. This analysis can be time consuming and costly. With the appropriate measurement of critical process parameters in place, early detection of process deviations can dramatically speed the discovery and alleviation of disturbances that can cause potential failures.

Many case studies have highlighted the ability to use FBRM® and PVM® technologies to identify non-ideal batch behavior and to quickly investigate the root cause of unexpected batch failures.[19, 20] The application of Process Analytical Technology also provides a faster path to further optimization of the process to avoid further problems.

Figure 9b. *A picture is worth a thousand measurements.* In some instances[20], PVM® in-process Imaging has been shown to detect undesirable polymorph crystals at concentrations below the detection limits of more traditional measurements such as Raman or FBRM®.

Appendix A:
Focused Beam Reflectance Measurement (FBRM®)

Measurement for optimization in real time – FBRM® is a highly precise and sensitive technology which tracks changes to particle dimension, particle shape, and particle count. Over a wide detection range from .5 to 3000µm, measurements are acquired in real time while particles are forming and can still be modified enabling process optimization and control. No sampling or sample preparation is required – even in highly concentrated (70% and higher) and opaque suspensions.

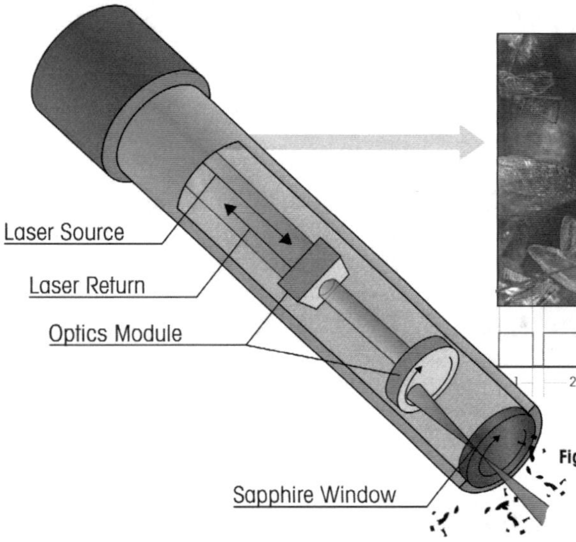

Figure 10.

Figure 11.

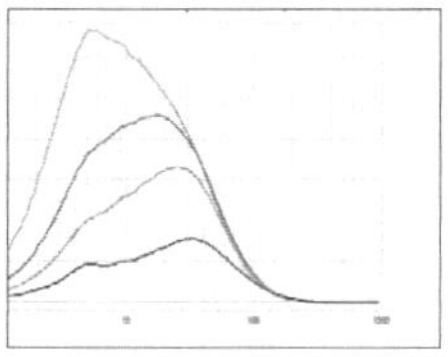

Figure 12. Chord Length Distributions

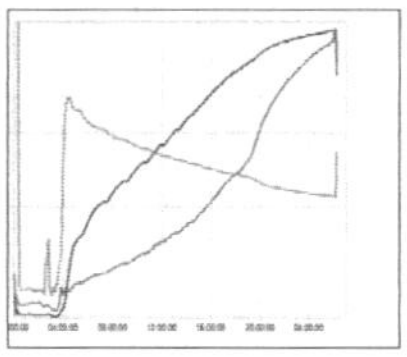

Figure 13. Trended Statistics

How does FBRM® work?

The FBRM® probe is immersed into a dilute or concentrated flowing slurry, droplet emulsion, or fluidized particle system. A laser is focused to a fine spot at the sapphire window interface (Figure 10). A magnified view shows individual particle structures will backscatter the laser light back to the probe (Figure 11). These pulses of backscattered light are detected by the probe and translated into Chord Lengths based on the simple calculation of the scan speed (velocity) multiplied by the pulse width (time). A chord length (a fundamental measurement of particle dimension) is simply defined as the straight line distance from one edge of a particle or particle structure to another edge. Thousands of individual chord lengths are typically measured each second to produce the Chord Length Distribution (CLD) (Figure 12). The CLD is a "fingerprint" of the particle system, and provides statistics to detect and monitor changes in particle dimension and particle count in real time (Figure 13).

Unlike other particle analysis techniques, with FBRM® measurement there is no assumption of particle shape. This allows the fundamental measurement to be used to directly track changes in the particle dimension, shape, and count.

Importance of a Fixed Focal Spot

In Focused Beam Reflectance Measurement of concentrated suspensions, the user should be aware that having a fixed focal spot at the probe window has been shown to provide optimum results.[21]

Technologies that use a varying focal depth will have inconsistent and often unpredictable results due to the inability of light to penetrate even small distances (< 1mm) at relatively low concentrations (1000ppm). The result of an oscillating focal position is a measurement zone that shows extreme variation with concentration (Figure 14). Particles that are close to the probe window effectively interfere with portions of the measurement where the focal position is located away from the window. In extreme cases, this can actually cause a reduction in particle count with increasing concentration. At higher concentrations, the loss of resolution will be even further magnified.

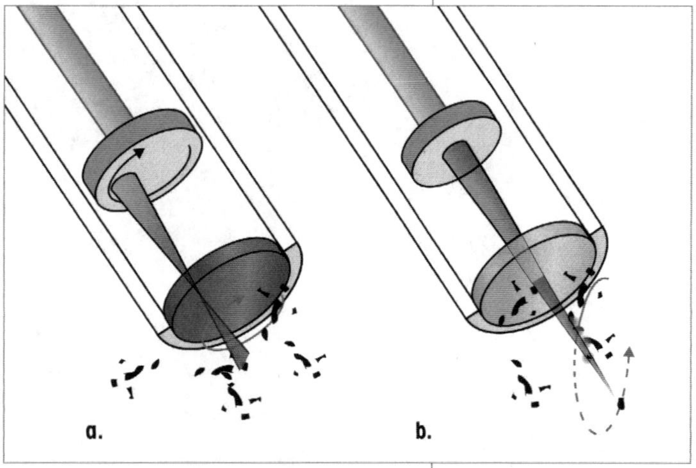

Figure 14. a. Ideal FBRM® measurements are made with a focal position that scans at the window surface to minimize the affect of concentration on the measurement

b. A variable or oscillating focal position loses measurement sensitivity as concentration increases, the same way as transmittance turbidity saturates at relatively low concentration

Appendix B:
Particle Vision Microscope (PVM®)

Vision for understanding and optimization – PVM® is a real-time probe based vision tool which provides instant critical insight into crystal, particle, and droplet systems. PVM® enables chemists and engineers to detect and understand process changes that could take months to discover with traditional offline microscopy techniques.

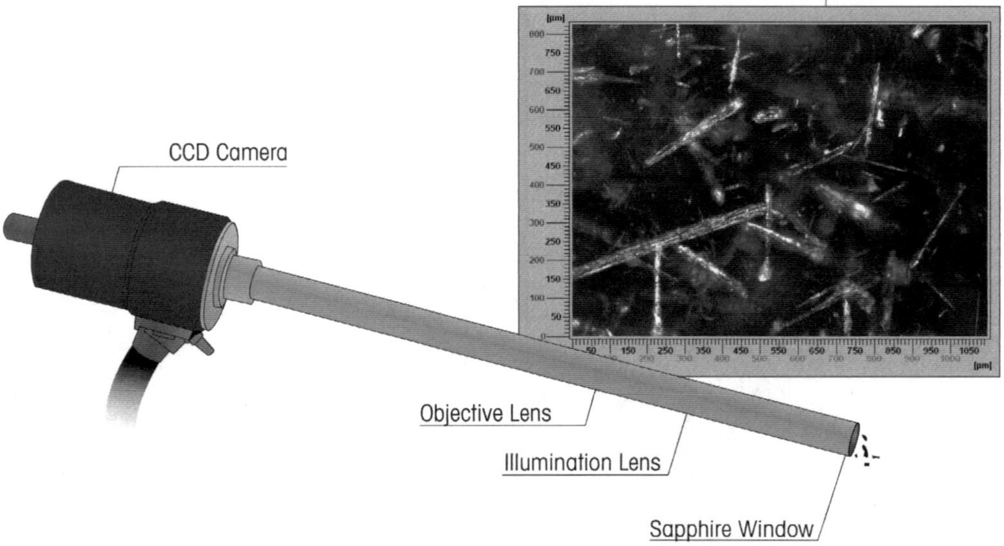

How does PVM® work?

PVM® uses a high resolution CCD camera and internal illumination source to obtain high quality images even in dark and concentrated suspensions or emulsions. With no calibration needed and easy data interpretation, PVM® quickly provides critical knowledge of crystal, particle, and droplet behavior.

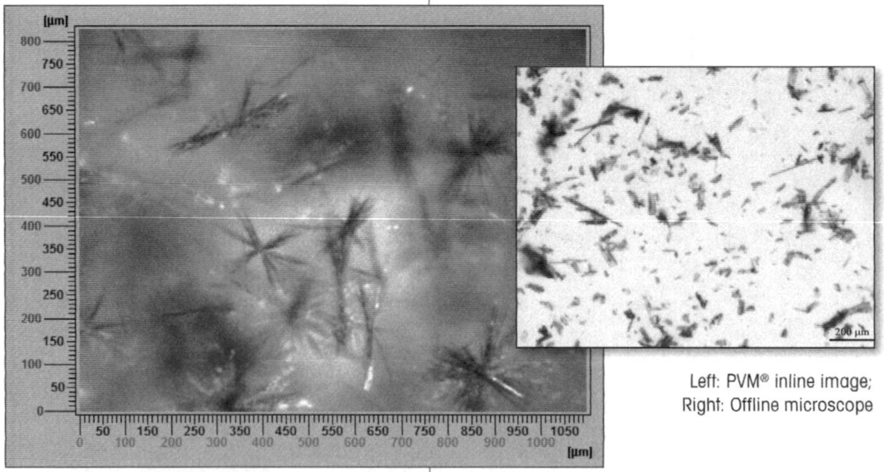

Left: PVM® inline image;
Right: Offline microscope

References

1. T. A. Bell, **Challenges in the scale-up of particulate processes—an industrial perspective**, Powder Technology, 150 (2): 60-71 (2004).
2. X. Y. Woo, Z. K. Nagy, R. B. H. Tan, R. D. Braatz, **Adaptive Concentration Control of Cooling and Antisolvent Crystallization with Laser Backscattering Measurement**, Crystal Growth & Design, 9 (1): 182-191 (2009).
3. Z. K. Nagy, M. Fujiwara, J. W. Chew, and R. D. Braatz. **Comparative performance of concentration and temperature controlled batch crystallizations**. Special Issue in Honor of Dale Seborg, J. of Process Control, 18 (3-4): 399-407 (2008).
4. T. Togkalidou, H.-H. Tung, Y. Sun, A. Andrews, and R. D. Braatz. **Parameter estimation and optimization of a loosely-bound aggregating pharmaceutical crystallization using in-situ infrared and laser backscattering measurements**. Ind. Eng. Chem. Res., 43 (19):6168-6181, (2004).
5. M. Trifkovic, M. Sheikhzadeh, and S. Rohani, **Multivariable real-time optimal control of a cooling and antisolvent semibatch crystallization process**, AIChE J., 55 (10): 2591-2602 (2009).
6. C. Lindenberg, M. Krättli, J. Cornel, J. Brozio, and M. Mazzotti, **Design and Optimization of a Combined Cooling/antisolvent Crystallization Process**, Crystal Growth & Design, 9(2): 1124-1136 (2009).
7. D. O'Grady, M. Barrett, E. Casey, B. Glennon, **The effect of mixing on the metastable zone width and nucleation kinetics in the anti-solvent crystallization of benzoic acid**. Chemical Engineering Research and Design, Transactions, 85 (A7):945-952 (2007).
8. M. Glodek, et al., **Process Robustness – A PQRI White Paper**. Pharmaceutical Engineering, November/December 2006 (on-line exclusive) (2006).
9. A. Ridder, **How to Control Particle Size on Plant Scale: High Shear Wet Milling or Traditional Optimization**, Proceedings of the 15th International Process Development Forum, Annapolis (2008).
10. B. Johnson, C. Szeto, O. Davidson, and A. Andrews, **Optimization of Pharmaceutical Batch Crystallization for Filtration and Scale-up**, AIChE Annual Meeting, Los Angeles (1997).
11. T. Redman, J. Ward, **The Effect of Particle Size on Downstream Processing**, AIChE Annual Meeting, Indianapolis (2002).
12. K. Wood-Kaczmar, **The Applications of FBRM® to Resolve Batch Processing Problems in Drug Manufacture, A Case History**, Proceedings of 2001 Lasentec User Conference, Barcelona (2001).
13. R. Beck, D. Malthe-SZrenssen and J.-P. Andreassen, **Polycrystalline growth in precipitation of an aromatic amine derivative and L-glutamic acid**, Journal of Crystal Growth, 311 (2): 320-326 (2009).
14. S. Kim, B. Lotz, M. Lindrud, K. Girard, T. Moore, K. Nagarajan, M. Alvarez, T. Lee, F. Nikfar, M. Davidovich, S. Srivastava, and S. Kiang, **Control of the Particle Properties of a Drug Substance by Crystallization Engineering and the Effect on Drug Product Formulation**, Org. Process Res. Dev., 9 (6): 894-901 (2005).
15. M.D. Argentine, T.M. Braden, J. Czarnik, E.W. Conder, S.E. Dunlap, J.W. Fennell, M.A. LaPack, R.R. Rothhaar, R.B. Scherer, C.R. Schmid, J.T. Vicenzi, J.G. Wei, J.A. Werner*, and R.T. Roginski, **The Role of New Technologies in Defining a Manufacturing Process for PPAR** Agonist LY518674, Org. Process Res. Dev., 13 (2): 131–143 (2009).
16. F. Sistare, L. St. Pierre-Berry, and C.A. Mojica, **Process Analytical Technology: An Investment in Process Knowledge**, Org. Process Res. Dev., 9 (3): 332-336 (2005).
17. J. Åhman, M. Birch, S.J. Haycock-Lewandowski, J. Long, and A. Wilder, **Process Research and Scale-up of a Commercialisable Route to Maraviroc (UK-427,857), a CCR-5 Receptor Antagonist**, Org. Process Res. Dev., 12 (6): 1104–1113 (2008).
18. P. Mousaw, K. Saranteas, and R. Prytko, **Crystallization Improvements of a Diastereomeric Kinetic Resolution through Understanding of Secondary Nucleation**, Org. Process Res. Dev., 12 (2): 243-248 (2008).
19. S. Desikan, W.P. Davis, J.E.Ward, R.L. Parsons Sr., and P.A. Toma, **Process Development Challenges to Accommodate A Late-Appearing Stable Polymorph: A Case Study on the Polymorphism and Crystallization of a Fast-Track Drug Development Compound**, Org. Process Res. Dev., 9 (6): 933–942 (2005).
20. A.T. Andrews, P. Fernandez, M. Midler, R. Osifchin, S. Patel, Y.-K. Sun, H.H. Tung, J. Wang, and G.X. Zhou, **On-line Monitoring of Crystallization of a Polymorphic Compound Using FBRM®-PVM®-Raman™**, 2002 Lasentec User Forum, Charleston (2002).
21. A. Ruf, J. Worlitschek and M. Mazzoti, **Modeling and Experimental Analysis of PSD Measurements through FBRM®**, Part & Part. Syst. Char., 17 (4): 167-179 (2001).

www.mt.com/particle

Terry Redman, MSc, MBA
Terry_Redman@mt.com

Benjamin Smith
Ben_Smith@mt.com

(355g) Direct Measurement of Length and Orientation of Microfibers in Solution

Taslima Khanam[1], Nurur Rahman Mohammad[1], Arvind Rajendran[1], Vinay Kariwala[2] and Anand K. Asundi[3], (1)School of Chemical and Biomedical Engineering, Nanyang Technological University, Singapore, Singapore, (2)Division of Chemical and Biomolecular Engineering, Nanyang Technological University, Singapore, Singapore, (3)School of Mechanical and Aerospace Engineering, Nanyang Technological University, Singapore, Singapore

Size measurement of non-spherical objects represents a fundamental scientific challenge. Most particle sizers report volume averaged distributions which are converted to size distributions assuming that the particles are spherical. For non-spherical particles, this assumption is questionable and hence the conversions result in erroneous results. In the recent times, imaging based techniques are being deployed to measure the lengths of non-spherical particles, e.g. micro fibers. However, fibers pose an important challenge also for imaging based systems. Using conventional imaging tools, it is possible to only measure the projected lengths of fibers and not their real lengths, as the angle of tilt is seldom measurable.

In this work, we present the use of a novel measurement technique using digital holography. In digital holography, a hologram of the object is stored digitally. Subsequent numerical processing yields reconstructions focused at particular depths of the scene. Thus, 3D information can be obtained using a single hologram by reconstructing it at different depths. The use of digital holography has been demonstrated for the measurement of particle size and shape measurements [1, 2]. The proposed method is first applied to experimentally recorded holograms of single fibers of known length and orientation placed on a mount, which can be fixed at a pre-determined off-axis tilt. The hologram of the fiber is captured and processed by a customized image process algorithm which provides the location, orientation and the length of the fiber. The measured length and tilt of the microfiber are found to be in excellent agreement with their true values. It is worth noting that the image analysis algorithm is automated, which provides the same accuracy for microfibers with different lengths (150 and 1320 microns) and different tilts along the optical axis (0-60°). The proposed method is applied to the fiber suspension, where the length, position and orientation of the multiple fibers are determined without any a-priori information. Finally the results from holography are compared with those obtained by microscopy which show an excellent match confirming the potential of digital holography for measurement of micro fibers. The potential of this technique for use as an online monitoring tool for crystallization will be discussed.

1.E. Darakis, T. Khanam, A. Rajendran, V. Kariwala, T. Naughton, A.K. Asundi, Microparticle characterization using digital holography, Chem. Engg. Sci. 65:1037-1044, 2010.

2.M. Kempkes, E. Darakis, T. Khanam, A. Rajendran, V. Kariwala, M. Mazzotti, T. Naughton, A.

Asundi, Three dimensional digital holographic profiling of micro-fibers, Opt. Express, 17:2938-2943,2009.

See more of this Session: PAT for Crystallization Development and Manufacturing
See more of this Group/Topical: Separations Division

(381a) Application of a Synthesis and Design Methodology Incorporating Process Intensification

Philip Lutze, Department of Chemical & Biochemical Engineering, Denmark Technical University, Lyngby, Denmark, Rafiqul Gani, CAPEC, Department of Chemical and Biochemical Engineering, Denmark Technical University, Lyngby, Denmark and John M. Woodley, Department of Chemical & Biochemical Engineering, Technical University of Denmark, Lyngby, Denmark

In recent years, Process Intensification (PI) has attracted considerable academic interest as a potential means of process improvement, to meet the increasing demands for sustainable production. PI aims to benefit processes without sacrificing product quality by increasing efficiency, reducing energy consumption, costs, volume, and waste as well as improving safety. A variety of intensified operations and equipment has been developed in academia and industry. Potentially, this creates a large number of options to improve the process but to date only a limited number have achieved implementation in industry, such as reactive distillation, dividing wall columns and reverse flow reactors [1].

Some of the reasons for this are that the currently available methods for design and synthesis for process intensification to find the set of feasible solutions within which the optimal can be found are neither simple nor systematic and take considerable resources. Hence, a process synthesis tool to achieve PI would potentially assist in the generation and evaluation of PI options. This leads to the development of a general PI synthesis and design methodology in which redundant intensified options are systematically removed by checking against predefined constraints through a decomposition approach of the optimization problem where the lower level steps employ simple and easy calculations, while the higher level steps employ more rigorous and detailed calculations. Such a methodology is reliant on a set of tools as well as structured knowledge, which is provided in this work through retrieval from a computer-aided knowledge base where relevant information obtained through a literature survey is classified in terms of known intensified processes and the principles on which the intensification are based, different methods for achieving PI and the tools employed to achieve them.

In this contribution the developed systematic methodology together with supporting methods and tools is applied to improve a production plant of hydrogen peroxide. Different steps of this process have been under investigation for improvement using PI equipment previously [2,3] but a systematic generation and comparison of different PI options to improve the design has not been tackled to date and this is therefore seen as a good example. Starting point is a base-case design of a cyclic anthraquinone process [4,5]. It comprises the catalytic hydrogenation of an anthraquinone solution [6] in a fixed-bed reactor [7], followed by oxidization of the hydrogenated solution with air in a bubble-column reactor [8] to produce hydrogen peroxide. Hydrogen peroxide is separated via extraction with water and further purified through distillation to achieve a concentration of 70 wt.% while the working solution is recycled back to the hydrogenation reactor [4,5].

[1] Harmsen, J. Process Intensification in the petrochemicals industry: Drivers and hurdles for commercial implementation. Chem Eng Process 2010; 49: 70-73. [2] Turunen, I. Process for producing hydrogen peroxide. US5637286 (1997). [3] Meili, A. Distillation plant for producing hydrogen peroxide. US5171407 (1992). [4] Eul, W., Moeller, A., Steiner, A. Hydrogen Peroxide. in Kirk-Othmer Encyclopedia of Chemical Technology, 5th Edition, 2007. [5] Goor, G., Glenneberg, J., Jacobiin, S. Hydrogen Peroxide. in Ullmann's Encyclopedia of industrial Chemistry, 7th Edition, 2010. [6] Ranbom, W. Hydrogen peroxide process. US4394369 (1982). [7] Porter, D.H. Production of hydrogen peroxide by anthraquinone process in the presence of a fixed bed catalyst. US3009782 (1961). [8] Liebert, M., Delle, H., Kabisch, G., Apparatus for the production of hydrogen peroxide. US3880596 (1975).

See more of this Session: Advances in Process Intensification I
See more of this Group/Topical: Process Development Division

NEW OPTIMIZATION-BASED APPROACH TO PROCESS SYNTHESIS – TOWARDS THE FULL INTEGRATION OF PROCESS DESIGN, OPERATION AND CONTROL

Nikola Nikacevic [1], **Adrie Huesman**[1], **Paul Van den Hof**[1], **Andrzej Stankiewicz**[2]

[1] *Delft Center for Systems and Control, Delft University of Technology, Delft, The Netherlands, n.nikacevic@tudelft.nl*
[2] *Process & Energy Department, Delft University of Technology, Delft, The Netherlands*

Conceptual process design still follows a traditional procedure which consists of three sequential stages, as depicted in Figure 1. In the first stage, the operation mode is selected according to the particular process and manufacturing demands. The decision is strictly based on heuristic rules, for instance batch operation is preferable for small capacities and multi-products while continuous operation is recommended for bulk production of a single product. In the second stage, the process is designed according to the process' physical and chemical characteristics, desirable production rate and quality and related economical issues. Although in this step equipment dimensions are commonly calculated using mathematical models, this phase also heavily relies on heuristics. In most cases predefined process units for a specific application are used and a selection of units is completed according to this practical method. In the third phase, a control system is designed for the purpose of maintaining the specified product quality and quantity, as well as process safety. This procedure is well established in practice; it offers ready-made mathematical models and technical solutions. However, it does not allow the interaction of design, operation and control and therefore it is generally suboptimal from both economical and environmental perspectives.

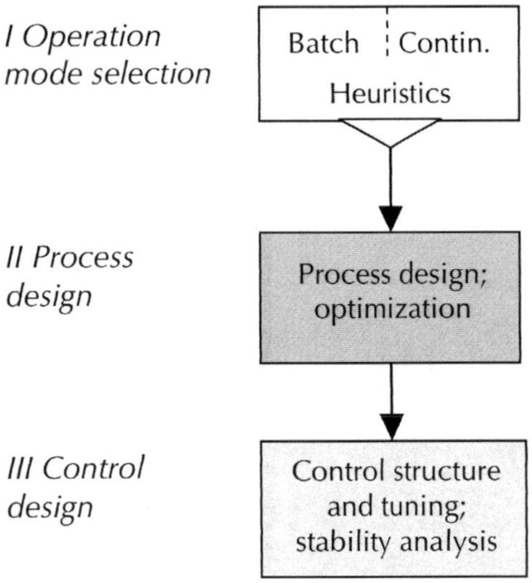

Figure 1. Traditional process synthesis

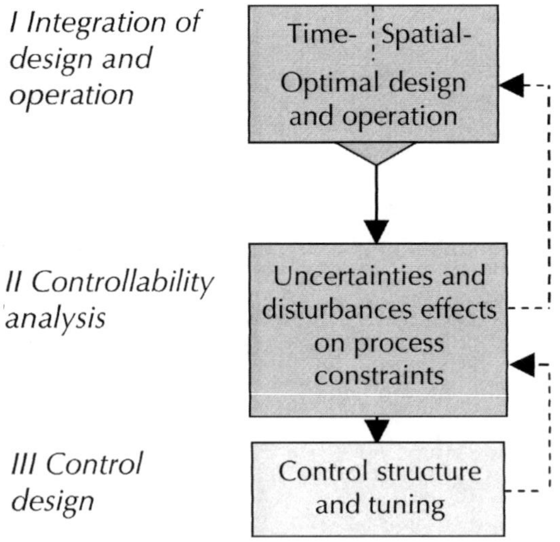

Figure 2. New optimization-based process synthesis

A new approach to chemical process synthesis is going to be presented and illustrated on an example. The concept is based on dynamic optimization and it also contains three stages as presented in Figure 2. In contrast to traditional approach this concept provides strong interactions between process operation, design and control. In the first stage, the complete integration of process design and operation should be attained by means of dynamic optimization, based on simple first principles mathematical models. Optimization objectives are economical and they may include environmental issues. Optimization results should provide optimal process design and optimal operation. Proposed process synthesis concept is it is out of the limits of traditional unit operation approach and it is determined by exploration of phenomena and manipulation of fluxes and driving forces. Therefore, this concept promotes the exploitation of process intensification principles and methods for process design. It also explores the possibilities for actuation improvement for optimal operation and control.

In the second stage (see Figure 2) controllability study for the optimal process solution (achieved in the first phase) is to be performed, by means of robust optimization techniques. In this analysis, the effects of model uncertainties, specifically physical parameters uncertainties, and process disturbances on the process constraints are to be examined. This analysis may lead to a broadening of operational values for actuation variables, in order to avoid violation of the constraints. In the third phase, the control system will be conceptually designed, similarly to the traditional approach. It is reasonable to assume that existing advanced control systems, like model predictive control, will function adequately for the proposed optimal designs and realization of the optimal operation. In this presentation, the first and most important stage of new process synthesis concept will be addressed, while the second stage will be tackled in the following paper.

The methodology for attaining the integration of process operation and design (stage I in Figure 2), could be generalized as follows. First, after a chemical problem is given (with known reaction mechanism and properties) an economic objective function is modeled, which includes the revenues for a production of desired product and costs for: a) separation, b) investments, c) energy, d) environmental issues. The problem is then systematically analyzed to consider as much as (physically) possible structural and operational enhancements, in order to maximize fluxes and driving forces within the process. Formulation of a single unique mathematical model which incorporates different process structure options and ways of operation is not attainable at the present. Therefore, the proposed method implies the decomposition of a generic problem into two or more sub-problems to be subjected to an optimization (Figure 2). One type of general model will generate more spatially-oriented solutions, operated often in continuous regime, both steady and unsteady state. The second form of model will result in more timely-oriented solutions, operated more frequently in discontinues or semi-continuous mode. After the concurring problems are selected and models are set, all degrees of freedom (control variables) need to be listed and classified between time-variable or time-invariant. Apparent physical constrains are to be incorporated in the optimization set-ups, as usual. After dynamic optimizations (MIDO) are performed, the target function values will be compared for a selection of the more beneficial solution, which would further be treated in stages two and three.

Proposed process synthesis methodology, precisely its first phase, is illustrated on an industrially relevant example – methane steam reforming (MSR) accompanied with water gas shift reaction (WGS). The goals are to have: a) high conversion of methane; b) efficient separation of hydrogen; c) high driving forces for reaction and separation; d) low energy costs; e) low capital costs and f) compact design.

To overcome the equilibrium limitations and achieve high conversion of methane, a functional integration of catalytic reactor with H_2 selective membrane is a very good process intensification option. A membrane will also provide immediate separation of a product from the undesired product and allow the use of lower temperatures. The alternative could be an integration of catalytic reaction with *in situ* adsorption; however this option was not consider at present, in order to have manageable and illustrative example. To attain high driving forces for the membrane separation, as well as for the reactions, consecutive order of process functions in one unit (reaction-separation-reaction-...) is considered, aside to fully integrated system. Without going into technical details, the spatially-distributed continuous systems could be realized as a tubular (or plate) catalyst bed reactor with membrane-wall annular (or sandwich) compartment for the H_2 permeate, either fully integrated or in consecutive order. The other, time-distributed class of systems could be designed as a rotating reactor with several cylindrical layers of catalyst bed and membrane-wall permeate compartments in between.

Alongside mentioned structural intensification, the actuation improvements should be examined systematically and rigorously within the optimization. The first reaction (MSR) is highly endothermic and requires high temperatures and the second reaction (WGS) is slightly exothermic and requires moderate temperatures. In order to attain optimal temperature and concentration profiles and their accurate control, different actuation possibilities are explored – optimal heat supply and optimal steam feeding, distributed in space and/or in time for all of the systems analyzed. The optimal (time and/or space) heat supply, with high energy efficiency could possibly be realized by using electromagnetic fields as heat sources, like microwaves or radio waves. A potential for selective heating (difference in material's dielectric properties) of catalyst pellets and membrane walls allows very good energy utilization in comparison to classical heating systems. Further, optimal control in space and time could be achieved effectively due to electromagnetic fast dynamic characteristics. For the continuous system, unsteady state (periodic) operation is also examined, as another possibility for operational improvement.

Economical target function comprises a maximal production of hydrogen with a minimal costs for catalytic and membrane surfaces and energy consumption. The dynamic models are simplified using common assumptions in order to attain convergence during the optimizations with a number of constraints and control variables. For all analyzed systems the dynamic optimization results provide optimal design parameters and optimal operation procedures, satisfying process constraints. The comparison of objective functions between two analyzed systems suggests the optimal process solution. The results of the analyzed systems are compared to the results of a conventional process performed in continuous catalytic packed-bed reactors with additional membrane separation unit. The results demonstrate significant improvements in economic performance, which is motivating for further development of the presented concept.

(381d) Coupled Isothermal Continuous Preferential Crystallization (CIC-PC) – A Novel Promising Technology for Gaining Pure Enantiomers with High Productivity

Andreas Seidel-Morgenstern, Chemical Engineering, Otto von Guericke University, Magdeburg, Germany and **Martin Peter Elsner**, Physical and Chemical Foundations of Process Engineering, Max Planck Institute for Dynamics of Complex Technical Systems, Magdeburg, Germany

The recognition of differences in the pharmacological activity of enantiomeric molecules has induced the necessity to administer them as isolated enantiomers in order to ensure the desired optimum therapeutic effect. Driven by the becoming more and more restrictive policy of regulatory authorities such as US Food and Drug Administration and the EU Committee for Proprietary Medical Products the production of pure enantiomers is considerably increasing in several industrial branches, e. g. pharmaceutical, agrochemical, food industries as well as in cosmetic and fragrance industries [1].

An attractive process for gaining pure enantiomers from racemic mixtures is the so-called preferential crystallization (PC) which has been up to now usually investigated in a discontinuous operation mode [2-4]. According to the literature [5] there is still a deficiency and therefore some mandatory need for action in a systematic investigation of a continuous process in order to gain access to higher productivities being much more attractive for an industrial production. For elucidating the general principle of such a simple isothermal process (SIC-PC, ***S**imple **I**sothermal **C**ontinuous **P**referential **C**rystallization*) one might consider a suspension crystallizer revealing MSMPR characteristics, i.e. a perfectly mixed tank (concerning both phases), which is continuously fed with mother liquor possessing a racemic composition. Of course, some suspension has to be continuously withdrawn from the crystallizer. By a continuous supply of homochiral seed crystals of the preferred target enantiomer E(p) the preferential crystallization of just this preferred enantiomer is initialized (i.e. crystal growth of the seed crystals and possibly secondary nucleation of those crystals), if the crystallization takes place within the metastable zone where spontaneous, uncontrolled primary nucleation is kinetically inhibited. During a starting-up phase which strongly depends on the properties of the system as well as on the process parameters the concentration of the goal enantiomer E(p) is decreasing until a steady state is reached where the composition is determined by the residence time τ. Due to the different kinetic mechanisms and their inherent different time constants for the phase transformation (liquid -> solid) a different depletion of the supersaturation for each enantiomer can be realized by an appropriate choice of the process conditions. As long as a critical residence time τ_{crit}, where primary nucleation may appear, is not exceeded the concentration of the undesired counter enantiomer E(c) remains all the time constant. This reveals also the benefit of this continuous process in comparison to the batch one: a choice of the "right" process conditions allows a constant production of the goal enantiomer at a high purity level. Above all should be mentioned that relatively short residence times, assuring highly pure product, entail relatively low yields which can be improved and optimized. Therefore, an upgraded arrangement has been investigated theoretically which might even provide a better performance where two continuous suspension crystallizers are coupled via the liquid phase, i.e. the crystal free mother-liquor is exchanged between these two vessels (cf. Fig. 1). Because of this exchange, the liquid phase shows a higher overall concentration of the preferred enantiomer in that vessel in which the preferred enantiomer was seeded even if the chosen residence time exceeds the critical one. The supersaturation level which creates the crystallization driving force is higher during the whole process in comparison to the case without an exchange (DIC-PC, ***D****ecoupled* ***I****sothermal* ***C****ontinuous* ***P****referential* ***C****rystallization*). Additionally, the concentration of each counter enantiomer in the liquid phase decreases in both vessels. For the borderline case of infinite exchange flow rate, racemic composition is maintained in the fluid phase of both vessels. The described effect of decreasing the counter enantiomer concentration makes the probability for primary nucleation lower. This corresponds to higher product purity at higher residence times and enhances therefore the productivity. This more attractive and effective operation mode using two continuous crystallizers coupled via their liquid phases has been investigated theoretically considering as a case study the separation of the enantiomers of the amino acid threonine in water based on data and modified models published by our group before [6-8]. The influence of specific process parameters, as e. g. residence time, stirrer speed, crystallization temperature, subcooling, size distribution as well as mass of the seeds, has been analyzed and will be presented in this contribution. Some selected results in Fig. 2 depict the high potential of the coupled continuous mode in comparison to the decoupled one.

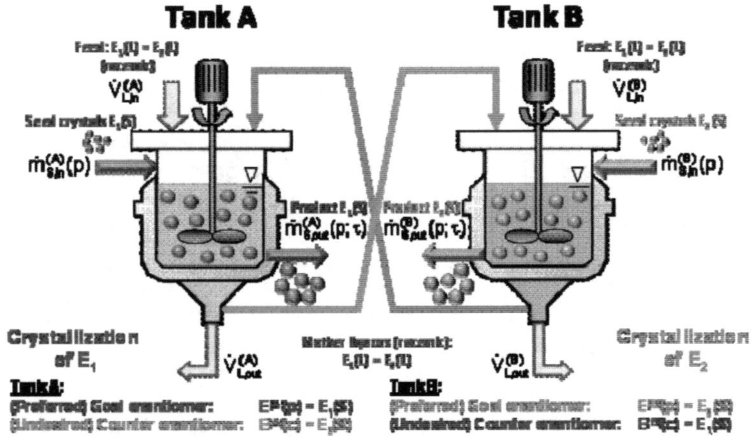

Fig. 1: *Arrangement of two continuous crystallizers coupled via the liquid phase (CIC-PC). Without an exchange of the (crystal-free) mother liquors the figure illustrates the DIC-PC alternative.*

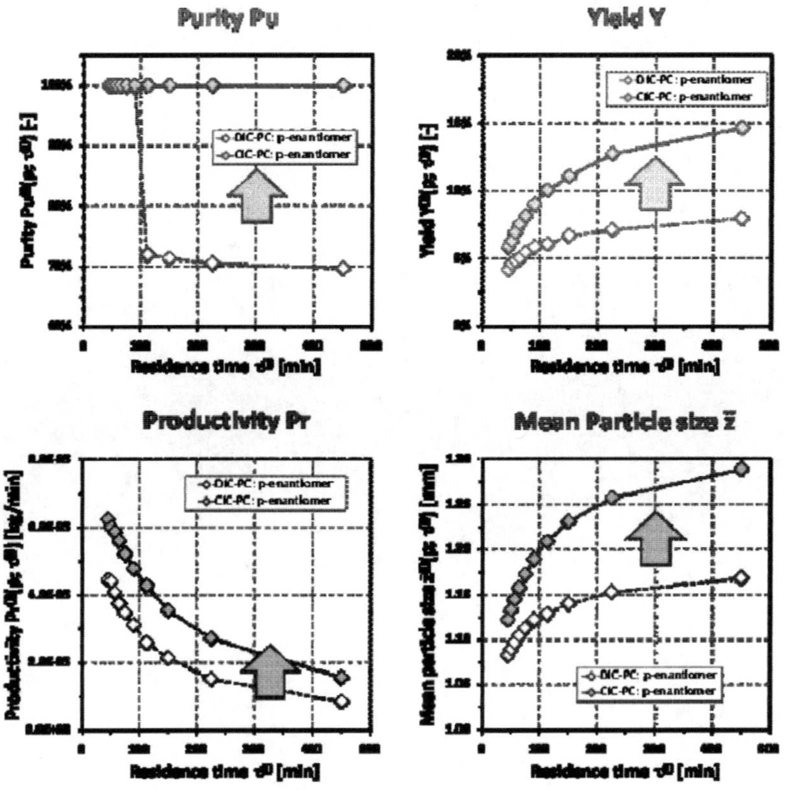

Fig. 2: *Selected parameter study for both, decoupled and coupled continuous PC of threonine in H_2O. In any case the simultaneous exchange of the mother liquors in the CIC-PC mode has a positive impact on all parameters being of interest for PC.*

[1] Caner, H.; Groner, E.; Levy, L.; Agranat, I. (2004): *Trends in the development of chiral drugs.* Drug Discovery Today. 9 (3), 105-110

[2] Jacques, J.; Collet, A.; Wilen, S.H. (1994): *Enantiomers, racemates and resolutions*, Krieger, Malabar

[3] Elsner, M.P., Fernández Menéndez, D., Alonso Muslera, E., Seidel-Morgenstern, A. (2005): *Experimental study and simplified mathematical description of preferential crystallization*, Chirality **17** (S1), S183-S195

[4] Lorenz, H., Perlberg, A., Sapoundjiev, D., Elsner, M.P., Seidel-Morgenstern, A., (2006) *Crystallization of enantiomers*, Chem. Eng. and Proc. 45(10), 863-873

[5] Coquerel, G. (2007): *Preferential crystallization.* In: Novel optical resolution technologies. K. Sakai, N. Hirayama, R. Tamura (Eds.). Topics in Current Chemistry **269**, 1-51

[6] Elsner, M.P.; Ziomek, G.; Seidel-Morgenstern, A. (2007): *Simultaneous preferential crystallization in a coupled, batch operation mode. Part I: Theoretical analysis and optimization.* Chem. Eng. Sci. **62** (17), 4760-4769

[7] Angelov, I.; Raisch, J.; Elsner, M.P.; Seidel-Morgenstern, A. (2008): *Optimal Operation of Enantioseparation by Batch-Wise Preferential Crystallization*. Chem. Eng. Sci. **63**(5), 1282-1292

[8] Elsner, M.P.; Ziomek, G.; Seidel-Morgenstern, A. (2009): *Efficient separation of enantiomers by preferential crystallization in two coupled vessels*. AIChE Journal **55**(3), 640-649

See more of this Session: Advances in Process Intensification I
See more of this Group/Topical: Process Development Division

(381e) Permeation, a Process Intensification Tool in Methyl Acetate Synthesis

Thomas Winkler, Susanne Lux and M. Siebenhofer, Department of Chemical Engineering and Environmental Technology, Graz University of Technology, Graz, Austria

The major goal of process intensification is to overcome limitations of product yield and selectivity of chemical reactions, mass transfer and heat transfer through new or different combination of technologies and/or process steps. Speeding up reaction kinetics of equilibrium reactions and elimination of azeotropes by adaptive configuration of reaction and separation steps has been basic strategy of this process intensification project. The synthesis of methyl acetate via esterification of acetic acid with methanol served as a welcome model process. The equilibrium reaction is heterogeneously catalyzed. Methyl acetate forms low boiling azeotropes with methanol and with water. And the bulk product methyl acetate is produced in large quantities based on the established Eastmann Kodak process. In this research project alternative process routes for synthesis and isolation of methyl acetate were investigated. Selective permeation may contribute to azeotrope breaking. When esterification is run under acid limiting conditions and methanol is removed by permeation a maximum yield of product without azeotrope formation is achieved. After investigating several membranes for separation selectivity and flux the process concept with acid limited reaction was designed and validated.

See more of this Session: Advances in Process Intensification I
See more of this Group/Topical: Process Development Division

(381f) Characterization of Microwave Assisted Melting of Metals

Ramanathan Srinivasan, Tanmay Basak and Chandrasekaran Sekaran, Chemical Engineering, Indian Institute of Technology Madras, Chennai, India

Although metals do not normally absorb microwave, it has been shown that certain forms of metals can be heated using microwave. In this work, the heating and melting of metals such as lead , aluminum and copper using microwave is characterized. Microwave absorbing materials were used to assist the heating process. In situ temperature measurements show that the heating rate is very high and is temperature dependent. The dependence of heating rate on the power level and the load was measured. Up to 750 C, the heating rate is a linear function of temperature. Beyond 750 C, the heating rate decreases with temperature. Based on the dependence of dielectric loss of different materials, a suitable combination of susceptor material is proposed so that a high heating rate can be obtained at all temperatures. The microwave heating is also four times more energy efficient compared to conventional heating and methods to improve the efficiency further are also discussed

See more of this Session: Advances in Process Intensification I
See more of this Group/Topical: Process Development Division

(420a) High Throughput Experimentation in Process Research Utilizing the Protasis "One Minute NMR" System

Carrie Brockway Wager, Chemical Research and Development, Pfizer, Inc., Groton, CT and Mark Zell, Analytical Research and Development, Pfizer, Inc., Groton, CT

In our hands, rapid reaction screening via high throughput NMR analysis has been indispensible. Several reasons include, the short cycle time for repeat analysis, the ability to get qualitative and quantitative information without reaction standards and no sample preparation or diluent needed. The presentation will focus on our work in implementing and fine-tuning the analytical system, as well as, the application to several projects in process research.

See more of this Session: PAT and High Throughput Technologies for Process Development and Control
See more of this Group/Topical: Process Development Division

(420b) Microfluidic Platforms for Pharmaceutical Screening

Michael R. Thorson[1], Sachit Goyal[2], Benjamin R. Schudel[2], Yuchuan Gong[3], Geoff G.Z. Zhang[3] and Paul J. A. Kenis[4], (1)Department of Chemical & Biomolecular Engineering, University of Illinois at Urbana Champaign, Urbana, IL, (2)Department of Chemical & Biomolecular Engineering, University of Illinois at Urbana-Champaign, Urbana, IL, (3)Solid State Sciences, Abbott Laboratories, North Chicago, IL, (4)Chemical and Biomolecular Engineering, University of Illinois at Urbana-Champaign, Urbana, IL

Efficient screens for crystal formation and polymorph identification are an expensive and critical step in active pharmaceutical ingredient (API) development. Microfluidic applications have the potential to improve high throughput screens for crystallization and polymorph identification by reducing reaction volumes, improving control in both mixing and evaporation and drastically increasing the number of conditions that can be screened [1].

Presently, macroscale approaches use on the order of 5 μl of CD per condition tested [2]. Here we present a microfluidic platform that screens for crystallization of candidate drugs (CD) while using as little as 50 nl of solution per well with as many as 96 wells on one platform. This platform employs three modes of operation to control supersaturation and thereby induce nucleation and growth of CDs: evaporation, temperature control and free interface diffusion. These three methods were chosen because of their propensity for inducing nucleation and their wide use on the macroscale [3]. By utilizing these techniques for control of supersaturation in microfluidic networks, we generate large combinatorial combinations of different crystallization conditions with respect to CD concentration and precipitant concentration and composition, enabling improved efficiency in crystal screening and polymorph identification. Additionally, the platform has been modified such that analysis of the crystals can be obtained with raman spectroscopy without removing crystals from the wells (e.g. on-chip).

This presentation will cover the design, fabrication, and application of the microfluidic platform. The utility will be demonstrated with model compounds such as Acetaminophen, Naproxen, and Sulfathiazole.

References

[1] C. Hansen, E. Skordalakes, J. Berger, S. Quake, Proc. Nat'l Acad. Sci., 99, 2002, 16531-6. [2] S. Talreja, D.Y. Kim, A.Y. Mirarefi, C.F. Zukoski, P.J.A. Kenis, J. Appl. Crystallography, 2005, 38, 988-995. [3] C. Gardner, O. Almarsson, H. Chen, S. Morissette, Computers and Chemical Engineering, 28, 2004, 943–953; E. Ware and D. Lu, Pharmaceutical Research, 21, 2004, 177-184. [4] B. Schudel, C. Choi, B. Cunningham, P. J. A. Kenis, Lab on a Chip, 2009 9(12): 1676-1680.

Methodological Aspects of On-Line Probes in Batch & Semi Batch Stirred Tanks

This presentation considers the impact of probe location and hydrodynamic conditions using Focused Beam Reflectance Measurements (FBRM) as an example probe, as it is one of the most common and reliable techniques of on-line particle sizing.

On-line monitoring of essential process parameters is an inherent element of PAT. Adequate data interpretation recorded by on-line probes requires understanding of the physical processes involved in measurements. The factors affecting operation of on-line probes might be divided into two groups: those related to the physics of the measurements that underlie the operation of the probe and those related to the conditions in which a probe operates.

The complex flow structure in stirred tanks results in a non-uniform distribution of dispersed phase, temperature and concentration gradients, etc., consequently, changing the location of the probe and intensity of stirring, results in different probe readings. Ignoring such phenomena might lead to misinterpretation of the data and invalidation of the entire PAT system.

A deep investigation of the operational features of a probe is often available in the literature and/or from the manufacturer. Various issues associated with optical and mathematical aspects of these measurements are presented in the literature. Factors affecting results of FBRM measurements include shape of particles, concentration effects leading to possible overlapping of particle projections, and aggregation of small particles.

In order to avoid such disturbance and focus on purely hydrodynamic effects, spherical polymer beads were chosen for this study.

Polyvinylchloride (PVC) beads were used as a model of solid particles with a narrow, almost monodispersed, particle size distribution (PSD) with a median of 100 microns. A mixture of two sizes of copolymer of styrene with di-vinyl benzene (DVB) beads (200 and 500 microns) was used as a model of solid particles with a wider PSD. No abrasion or aggregation of particles was detected in the study and water was used as a liquid phase throughout.

All experiments were conducted at room temperature in a 3 L cylindrical glass laboratory reactor equipped with a 4-blade pitch stirrer. The impeller stirring speed was varied from 50 to 400 rpm. Turbulent flow, i.e. a Reynolds number greater than 3000, was maintained throughout the study.

Measurements were conducted using an FBRM D600 L probe running at a laser beam rotation rate of 2 m/s, and the measurement time was kept at 30 seconds in all experiments. The radial position of the probe was fixed at 60 mm from the reactor center. Three vertical positions of the probe were used with a window located at 20, 80 and 120 mm from the liquid surface in the quiescent reactor i.e. before the start of stirring. Hydrodynamic simulation was carried out using VisiMix (Turbulent 2K8 package).

Experiments conducted with monodisperse PVC particles show that there is no appreciable difference in PSD through the entire range of stirring speeds and probe locations. This confirms the absence of significant particle aggregation through the entire range of stirring speeds. Polymer beads used in this study are heavier than water, therefore, if the stirring speed is decreased one would expect the concentration of particles in the lower part of the reactor to increase and consequently the concentration of particles in the higher part of the reactor to decrease. Conversely, by increasing the stirring speed, one would expect the total number of detected particles to increase in the higher part of the reactor and decrease in the lower part. The phenomenon is observed for different locations of the probe at a constant stirring speed. The total number of detected particles, however, increases as the stirring speed is increased at all probe locations.

Movement of solid particles in the stirred tank might be considered as the superposition of a vertical and tangential flow. The vertical flow is governed by sedimentation or floating of particles and the circulating flow of the liquid phase. The total number of particles is proportional to the flow of particles against the window i.e. to the tangential speed of the liquid.

Good agreement between FBRM trends and an estimation conducted with the data from Visimix was achieved assuming the FBRM has a constant depth of particle detection.

In contrast to monodisperse particles, analysis of the wide PSD measured in different locations and different stirring speeds illustrates a shift in the median of the measured distribution. The total number of counts increases with stirring speed.

The change of PSD with stirring speed might be explained by a consideration of a difference in sedimentation rate of particles of different sizes. A combination of large and small particles having significantly different rates of sedimentation leads to significant impact of stirring speed on the detected PSD.

Our experiments show that even for such a simple case as non-aggregated, spherical particles suspended in a Newtonian liquid in a completely turbulent stirred tank, quantitative interpretation of the results of FBRM should be supported by hydrodynamic modeling.

One of the conclusions, which might be drawn, is that a widely used approximation of laboratory reactors by a perfect mixing model is not acceptable for interpretation of FBRM data and should be replaced by a more detailed study of hydrodynamics if a quantitative interpretation of the data is intended. VisiMix software provides an easy to use and accurate tool for such tasks.

Assessment of stirring conditions at which the non-uniformity of solid particle distribution in the reactor might be neglected could be carried out with the help of hydrodynamic simulation. Such simulation requires knowledge of the maximum equivalent particle size, the densities of liquid and solid phases, the viscosity of reaction mixture, and the design of the reactor and impeller. Measurement of settling velocity for the largest particles in independent experiments is highly desirable.

If realization of uniform stirring conditions is not possible due to technical capabilities of the reactor and/or sensitivity of product to shear, then PSD measured by FBRM at a particular location and stirring speed should be supported by hydrodynamic modeling of the reactor.

(420d) Jacket Controlled RC1 Heat Transfer Scale Down

Edward M. Davis, Chemical Process Development, Eli Lilly, Indianapolis, IN

Mathematical models of dose controlled reactions based on RC1 heat flow (Qr) data are quite common and well developed however mathematical predictions for non-dose controlled reactions from Qr data are quite a different matter. Even dose controlled predictions can be tricky if one is attempting to optimize the large scale reaction in such a fashion that reaction temperature (Tr) increases. Reaction rate, viscosity, and mechanistic behavior are influenced by the batch temperature change but these are seldom accounted for in mathematical models. Presented here is a method for mimicking the heat transfer of a large vessel by manipulating Tj in the RC1 so as to directly observe and measure any changes in Tr and their resulting effect on the reaction. The approach described is based solely on widely accepted scale-up first principles and is derived mathematically as well as empirically tested.

See more of this Session: PAT and High Throughput Technologies for Process Development and Control
See more of this Group/Topical: Process Development Division

(420e) Simulation and Optimization for Product Loss Reduction in a Chemical Plant

Xiongtao Yang, Qiang Xu and Kuyen Li, Dan F. Smith Department of Chemical Engineering, Lamar University, Beaumont, TX

The hydrogen/methane separation section is the coldest section in an ethylene plant. This section includes mainly two facilities, chilling train and de-methanizer. Chilling train consists of countercurrent multi-stream heat exchangers (cold box) and separation drums. The front-end of this facility is used to liquefy C2+ components from the charge gas and feed them to the de-methanzier. The back-end of chilling train is to split the left charge gas of hydrogen and methane as well as provide cooling duty for the liquefaction in the front-end. De-methanizer is employed to separate methane from C2+ components as the top stream, which is directed to the fuel gas system. If ethylene effluents from the de-methanizer top, it will be accounted as the product loss. Obviously, the hydrogen/methane system is a highly integrated and energy-concentrated cryogenic separation system. Its operation strategy should be optimized to effectively balance the ethylene product loss and the energy consumption. This work employs rigorous simulation, optimization, and principle component analysis method to identify the best operating conditions of the hydrogen/methane system, so as to reduce the ethylene product loss with the consideration of energy consumption. A virtual study shows that the optimized operation based on the developed methods can save millions of dollars per year compared with the original plant operations.

See more of this Session: PAT and High Throughput Technologies for Process Development and Control
See more of this Group/Topical: Process Development Division

(420f) Ultrasound Induced Destruction of Liposomes Using Vaporizing Emulsion Droplets

James R. Lattin, Chemical Engineering Dept, Brigham Young University, Provo, UT, Taylor Atkinson, Neuroscience, Brigham Young University, Provo, UT, Ghaleb A. Husseini, Chemical Engineering Department, American University of Sharjah, Sharjah, United Arab Emirates and William Pitt, Chemical Engineering, Brigham Young University, Provo, UT

Several groups have reported the encapsulation and delivery of Doxorubicin and other drugs from liposomes. More recently, ultrasound has been utilized as a method to release the encapsulated drugs[1-3]. This method of targeted delivery could be more useful with liposomes that are designed to be more susceptible to ultrasound without compromising stability, allowing for a faster and more complete release of drugs when the ultrasound is applied. In this study perfluorocarbon emulsion droplets were encapsulated inside of liposomes to increase the rate and amount of liposome destruction with ultrasound. Upon insonation, the perfluorocarbon emulsion droplets vaporize and disrupt the liposomal membrane as they expand (Figure 1).

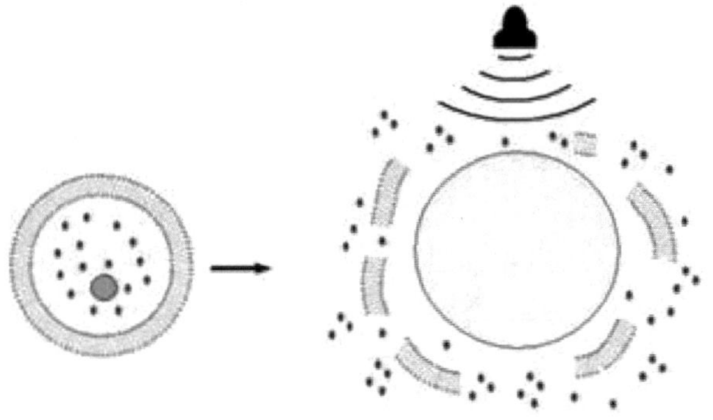

Figure 1: Emulsion droplets and calcein are encapsulated inside of DPPC liposomes. Upon insonation, the emulsion droplets vaporize, rupturing the liposomes and releasing their contents as they expand.

Emulsion droplets were formed by mixing perfluorohexane, DPPC, and water. A 20-kHz ultrasound probe was used to shear the solution, and the resulting emulsion was extruded through a 0.1 µm filter to reduce the size of the stabilized droplets to approximately 100 nm. Emulsion droplets were encapsulated inside of liposomes using a "vesosome" technique described by Kisak, et al[4]. Dipalmitoylphosphatidylcholine (DPPC) liposomes were formed by film hydration and extruded through a 0.1 µm filter. The 100 nm liposomes were stirred while ethanol was added dropwise to a final concentration of 3M, causing them to fuse into bilayer sheets. Ethanol was removed from the sheet solution by centrifugation and resuspension of the pellet in excess PBS. This process was repeated 3 times. Concentrated calcein was added to the sheet solution with or without emulsion droplets and samples were heated to 50ºC for at least 20 minutes to reform closed vesicles. The resulting vesicles were extruded through a 0.8 µm filter. The average size was determined to be 800 nm by dynamic light scattering. Unencapsulated calcein was removed from the outside of the vesicles by size exclusion chromatography using a Sephadex G25 column. A third sample was prepared by mixing an emulsion suspension and a liposome suspension, thus placing the emulsion droplets only on the outside of the liposomes.

The liposome solutions were diluted in a sample cuvette and the amount of calcein released from the liposomes was quantified using a fluorometer. A 20-kHz ultrasonic probe was immersed directly into the sample cuvette through an opening in the fluorometer. Excitation and emission wavelengths were set at 488 and 525 nm, respectively. At the conclusion of each experiment, Triton X-100 was added to the cuvette to destroy the liposomes, thus recording the fluorescence level of 100% release. Data was normalized by fractional calcein released. Sample data is shown in Figure 2.

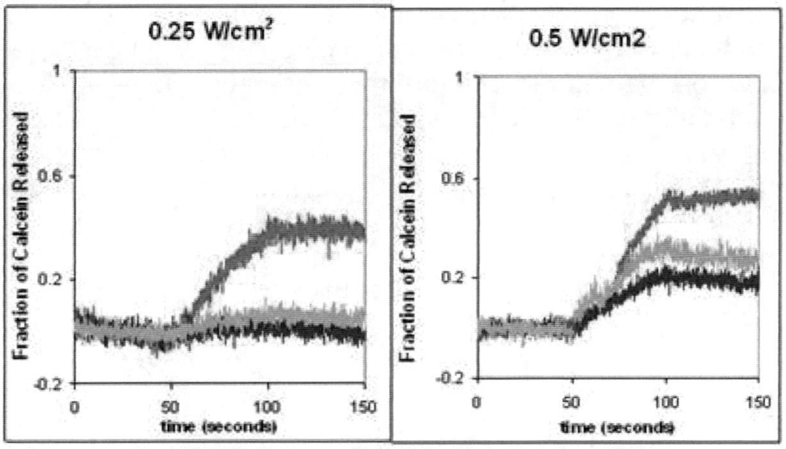

Figure 2: Fluorescence of calcein released from liposomes (----), liposomes with exterior emulsion droplets (----), and liposomes with encapsulated emulsion droplets (----). Ultrasound was applied from 50 to 100 seconds.

The presence of internal emulsion droplets increased the amount and rate of calcein release compared to conventional liposomes, particularly at low intensities. External emulsion droplets also increased the amount and rate of calcein release from liposomes, although not to the level of encapsulated emulsion droplets. Further work is required at varied ultrasonic parameters and with different liposomal formulations to optimize ultrasonically-mediated release from liposomes and to describe ultrasonic thresholds for vaporization of perfluorocarbon liquids.

References

(1) Ning, S.; Macleod, K.; Abra, R.; Huang, A. H.; Hahn, G. M. *Int. J. Radiation Oncology Phys.* **1994**, *29*, 827-834.

(2) Vyas, S. P.; Singh, R.; Asati, R. K. *J Microencapsul* **1995**, *12*, 149-154.

(3) Unger, E. C.; McCreery, T. P.; Sweitzer, R. H.; Caldwell, V. E.; Wu, Y. *Invest Radiol* **1998**, *33*, 886-892.

(4) Kisak, E.T., et al. *Current Medicinal Chemistry*, **2004**. 11(2): 199-219.

See more of this Session: PAT and High Throughput Technologies for Process Development and Control
See more of this Group/Topical: Process Development Division

Improvement in recuperation process of acid by-product using Six Sigma

Paulo C. Meneghetti and Theodore P. Kircher, Momentive Performance Materials, Willoughby, OH

Abstract

This paper describes the use of Six Sigma methodology to help improve recovery of a by-product in a manufacturing process; specifically, the recovery of by-product hydrofluorosilicic acid (HFS) from a silica purification process. The primary tools used were the process map and the cause and effect matrix, followed by statistical analysis. Using load cell readings of the reaction vessel, a gage R&R was performed, allowing process improvement to be quantifiable. Process baseline accounted for an average of 4,875 pounds of HFS recovered per batch, and change-implementation increased average recovery to 4,975 pounds/batch. The increase of at least 100 lbs/batch was attributed mainly to two factors (out of at least 9 factors): length of recovery time and gas displacement (introduction of gas during pump out). Quality of product (sand) was not affected since impurity levels of key elements, measured via inductively coupled plasma mass spectrometry (ICP-MS), were not statistically different. Engineered controls for pump time should reduce further variation in total weight recovered, stabilizing the process even more.

1. Introduction

Six Sigma is a project-driven management approach to improve an organization's products, services and processes, by continually reducing defects in the organization [1-7]. It can allow companies to dramatically improve their bottom line by designing and monitoring everyday business activities in ways that minimize waste and resources while increasing customer satisfaction [2]. Going back to the mid 1980s, applications of Six Sigma methods helped many organizations sustain their competitive advantage by integrating their knowledge of process with statistics, engineering and project management [1, 3]. Numerous books and articles have been published providing the main concepts and benefits of the Six Sigma methodology [2, 4, 5]. Since then, Six Sigma has been successfully applied in many manufacturing organizations such as General Electric, Boeing, DuPont, Toshiba, Seagate, Allied Signal, Motorola, Kodak, Honeywell, Texas Instrument, Sony, etc [1]. The results of a recent survey of process management tools indicates that Six Sigma yields the most improvement results, as shown on Table 1 [6, 7].

Table1. Rating of process improvement techniques [6, 7]

Process Improvement Tool	Impact (%)
Six Sigma	53.60
Process Mapping	35.60
Root Cause Analysis	33.50
Cause and Effect Analysis	31.30
ISO 9001	21.00
Statistical Process Control	20.10
Total Quality Management	10.30
Malcolm Baldridge criteria	9.80
Knowledge Management	5.80

The term Sigma Quality Level refers to an indicator of process goodness. Lower Sigma quality level means greater possibility of defective products while higher Sigma quality level means smaller possibility of defective products within process [8,9]. Improving a highly variable process to a minimum level of defect (6 sigma = 3.4 defects per million opportunities) requires leadership, infrastructure, and appropriate tools and methods.

As described in this paper, Six Sigma methodology for process improvement was used to improve the recovery of a by-product. The Six Sigma steps are described in detail along with the phases of the project. The goal of the project, maximized recovery of by-product weight, should translate into an increase of productivity for the company, contributing to its business success and profitability.

2. Six Sigma methodology

As discussed in the literature, Six Sigma is about improving profitability, although improved quality and efficiency are immediate by-products of Six Sigma [2]. The Six Sigma project discussed in this paper focused on the recovery of hydrofluorosilicic acid (HFS) from silica purification process. The Six Sigma approach follows the structure DMAIC methodology shown in Figure 1 (Define, Measure, Analyze, Improve, and Control phases) [2, 4, 8].

Phase	Step	Description
Define	A	Identify Project CTQs
	B	Develop Team Charter
	C	Define Process
Measure	1	Select CTQ characteristic
	2	Define Performance Standards
	3	Evaluate Measurement System on Y
Analyze	4	Establish Process Capability
	5	Define Performance Objectives
	6	Identify Variation Sources
Improve	7	Screen Potential Causes
	8	Discover Variable Relationships
	9	Establish Operating Tolerances
Control	10	Define and Validate Measurement Systems for X's
	11	Determine New Process Capability
	12	Implement Process Control

Figure 1: 12-step Six Sigma process [2, 4, 8].

2.1 Define phase

This phase defines the processes that contribute to the functional problems [2]. When evaluating the entire sand purification process, it was observed that the acid recovery process required attention. The goal for the project was to improve the recovery of HFS, which would translate to a more efficient process, less scrap, and increased productivity.

In Six Sigma approach, adequate metrics for evaluating project success should be established. These metrics encompass ratio volume/cost, labor cost, tool cost, downtime, rolled throughput yield and others. In this case, increased yield recovery, or additional pounds, obtained from the process translate into saleable product and increase in profitability [8].

In the beginning of the project, historical data were available, which allowed the identification of the critical-to-quality characteristics (CTQ) – i.e. production of HFS (lbs. recovered). In accordance to the DMAIC methodology, in this phase CTQs should be determined, project/process goals should be defined and cross-functional teams identified including individuals with appropriate knowledge. Additional tools are used as the project proceeds to the measurement phase.

2.2 Measurement phase

The measurement phase is concerned with selection of one or more product characteristics to be improved, mapping the respective process, making certain the measurement system is workable, making the necessary measurements and recording the results [2, 4]. One approach to connect the voice of the customer (VOC) with the actual process is to perform a high level CTQ flowdown correlating key customer output to key process output as shown in Figure 2. In this case, the big picture output (Big Y) is variable cost productivity (VCP). The business was looking for an increase in productivity, and as the process was examined through the HFS acid, the production was the process output (y) that is measurable and can lead to the increase in productivity. Tied to the production process is the pump out process, which will be examined in greater detail next, in order to capture its key inputs.

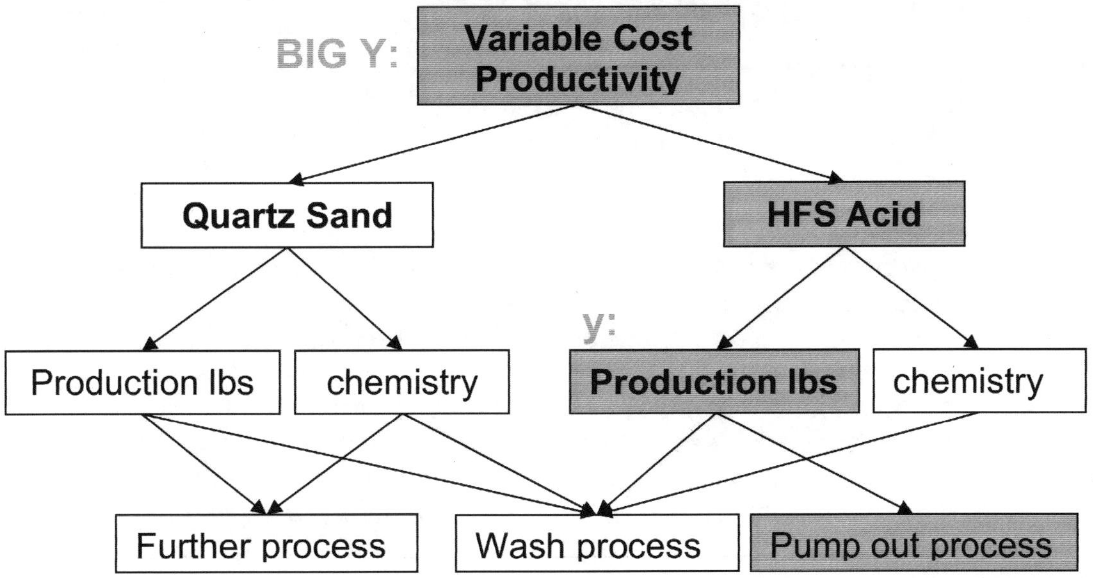

Figure 2. CTQ Flowdown

In general, at this stage, a process map is generated to help visualize and identify the key steps of the process, the input and output variable of a process and opportunity for improvements. Due to the proprietary nature of the project, the operations prior to the acid "pump out" process are not shown. Figure 3 shows the basic pump out process.

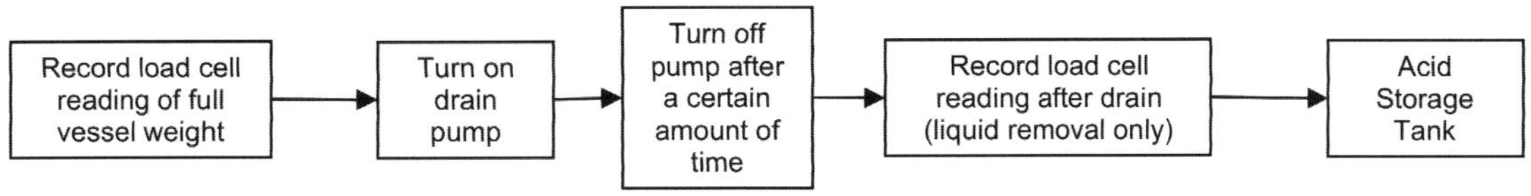

Figure 3. Process map of pump out process

The measurement system analysis (MSA) is the next step in this phase, which will assess how much variation is associated with the measurement system and compare it to the total process variation or tolerances [10]. In this project, the production weight of HFS was the key process output variable (KPOV) measured using a load cell mounted underneath the reaction vessel. The load cell reads weight from 100 lbs to 20,000 lbs. Using a calibrated scale, 2 operators and 10 samples of roughly 1,770 lbs, 10 measurements were taken by each operator at the calibrated scale and at the load cell with each sample. Table 2 shows the average of each of the 10 samples measured with the calibrated scale, and then via the load cell. The variation from the load cell and the calibrated scale was very small, and the total standard deviation divided by the average was almost the same (1.15%) for calibrated scale and load cell.

Table 2. Calibrated scale versus load cell data for 10 samples.

	avg lbs	Stdev	stdev / avg
Calibrated Scale	1770.5	20.3	1.15%
Load cell	1767.3	20.2	1.14%

To check the adequacy of the measurement system, a GR&R was performed. GR&R simultaneously quantifies the amount of error in the measurement system as well as the amount of variation from one operator to another, estimated as a % of Tolerance. Reproducibility refers to amount of variation from operator to operator. Repeatability refers to the amount of variation attributable to the equipment or gage. Due to the nature of the process each sample used was different for each operator. Once the sample was measured at the load cell, it was consumed and not available for other operator to do a measurement. Thus, a short GR&R was performed and resulted in an 18% GR&R. This is a good gage since typical rule of thumb is that the %GR&R should be less than 30%.

$$GR\&R\% = \frac{100 * 5.15 * \text{Standard Deviation of Measurement}}{\text{Upper Specification} - \text{Lower Specification}}$$

2.3 Analyze Phase

The first part of this phase is to determine the viability of the current process (benchmarking), determine how well it is performing, and identify the performance objective [2]. In collecting the baseline data for this process substantial variation was observed, which pointed to potential room for improvement. Statistical analysis software Minitab was used to perform a capability study on the historical data, which is shown in Figure 4. The data ranged from 3200 lbs to 5200 lbs recovered per production batch. The lower specification limit was set to 4800 lbs, and the average run produced 4875 lbs. The defect per million opportunities (DPMO) was 138,461.

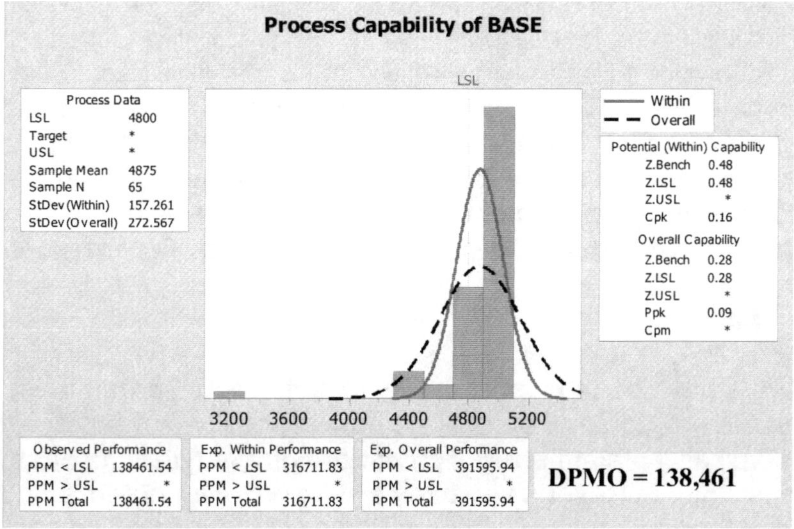

Figure 4. Process Capability Analysis for baseline (historical project data)

Based on the results from the historical data, the performance objectives were created to obtain a realistic and much improved process. The target weight to be recovered was 100 lbs more than current average and was set to 4,975 lbs. In addition, the variability of the process would be reduced (raising the sigma level by at least 2 sigma).

The team used a cause and effect diagram to identify the key process input variables (KPIVs) for further study. Since one main output was desired (production weight), besides identifying the input variables, the team ranked each variable according to ease of impact in the process (1 = highest impact) and also implementation (1 = easiest to implement/fix). Figure 5 below shows the C&E diagram (similar to fishbone).

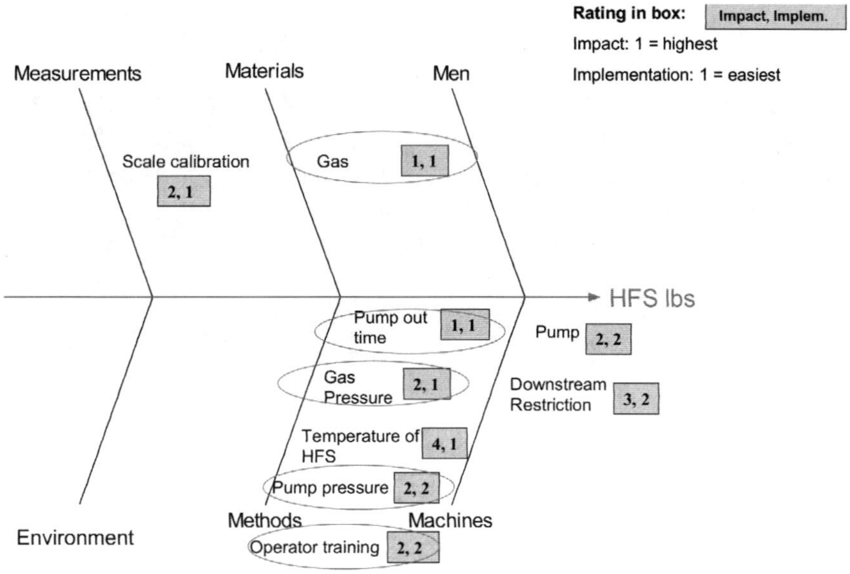

Figure 5. Cause and Effect diagram for recovery of HFS weight (HFS lbs).

From the cause and effect diagram, at least 9 key inputs were identified and scored in terms of impact to the key output variable (HFS acid production lbs) and ease of implementation. A couple of factors were ruled out immediately: scale calibration and pump. With an acceptable GR&R, the scale calibration was removed. Keeping the pump constant and new also eliminated this variable. The resulting 7 factors were examined individually in the next phase to determine their significance in achieving the project goal. The factors were: gas, gas pressure, pump out time, temperature of acid, pump pressure, operator training and downstream obstruction.

2.4 Improvement Phase

Once the KPIVs have been identified, the next step is to screen these variables, select the best possible solution and predict the new capability of the process and/or value stream [10]. Typically, a planned Design of Experiment (DOE) is one of the more common approaches when numerous variables exist along with potential interactions. In this project, prior to elaborating a DOE, the 7 variables identified earlier were tested one by one.

The hypothesis testing covers a large range of data types (continuous or discrete) for both KPIVs (Xs) and KPOV (Ys). The most common type of testing is Analysis of Variance (ANOVA) applicable to discrete Xs, continuous Ys, normal distribution, and two or more factors [4]. Some of the data in this project were discrete Xs and continuous Ys, but were not normally distributed. Therefore, the median was analyzed instead of the mean (Mood's Median Test) [4].

Variables Gas and Gas Pressure were evaluated together. First, a simple on and off experiment was used to determine the effect of the gas pressure, and then a variation in the pressure (high and low) was tested to evaluate effect in the production of HFS. As the acid was being drained from the closed reaction vessel, the Gas variable was simply a possibility of adding an element to the vessel to balance its negative pressure. Figure 6 shows 3 plots: normality plot (non-normal data), test of equal variances (non-equal variance), and box plots for each setting with Gas or no Gas. The addition of Gas produced a mean shift increase in the KPOV while dramatically reducing the variance. The box plots show that there was a large difference in using Gas versus not using Gas. Analyzing the median via Mood's median test reveals that the factor was significant; P-value is less than 0.05 (Figure 7). The second part of the Gas test (high or low pressure) was less significant to the output; P-value was 0.2 (not shown).

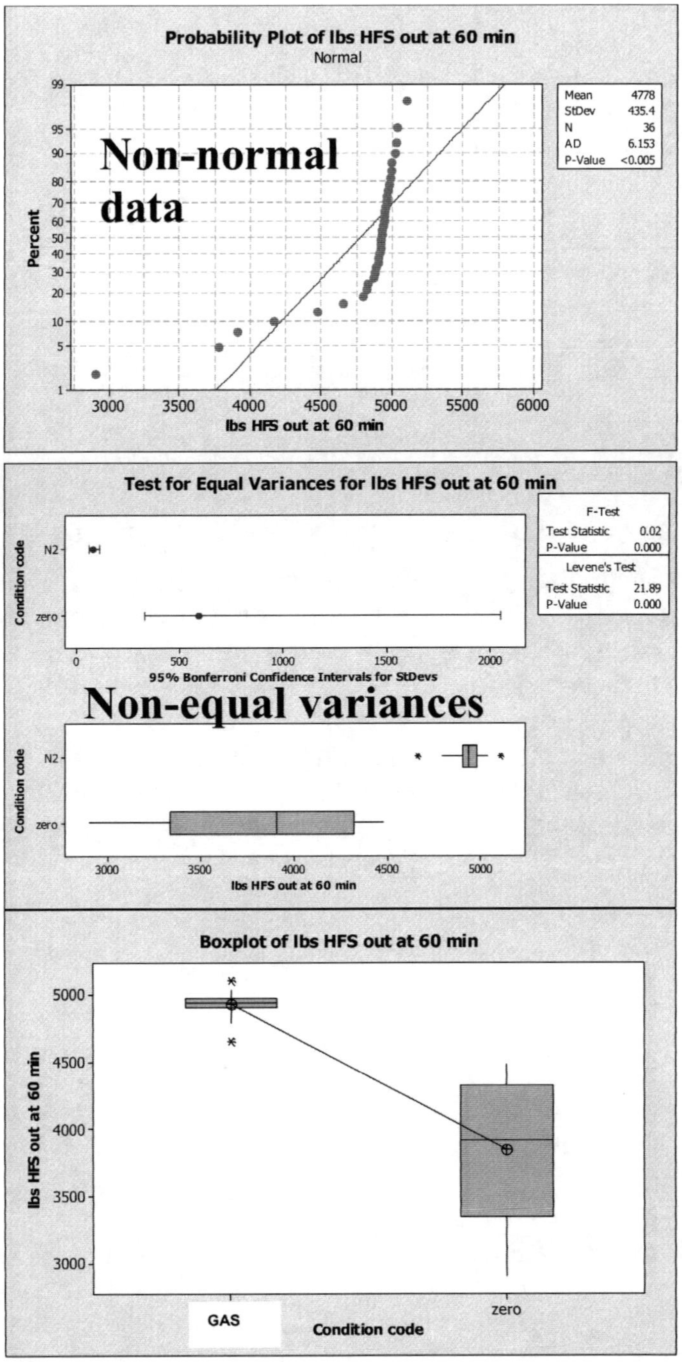

Figure 6. Normality plot (non-normal data), test of equal variances (non-equal variance), and box plots for each setting with Gas or no Gas.

```
Mood Median Test

Mood median test for lbs HFS

Chi-Square = 5.81   DF = 1   P = 0.016

                                Individual 95.0% CIs
Conditio  N<=   N>  Median   Q3-Q1  -+---------+---------+------
  GAS     13    18   4935     75                              +)
zero       5     0   3910    982  (--------------+---------)
                               --+---------+---------+---------+---
                               3000      3600      4200      4800
Overall median = 4925
```

Figure 7. Results of Mood's Median Test

Pump out time was the other KPIV significant to KPOV acid production. First, four different pump out times were observed: 45, 60, 75 and 90 minutes. However, the shortest time, 45 minutes, was not sufficient to achieve even baseline data (4,875 lbs), and 90 minutes was too long for the production process. Thus, the times tested were for 60 vs. 75 minutes. The data were normal, with equal variances, and ANOVA showed a P-value of 0.026, indicating statistically different means when using 75 minutes vs. 60 minutes. At 75 minutes the mean was 4,972 lbs. and at 60 minutes the mean was 4928 lbs. Figures 8 and 9 depict these results. The other factors (temperature of HFS, pump pressure, operator training) were not significant to the KPOV and results are not shown.

```
One-way Analysis of Variance

Analysis of Variance
Source    DF       SS       MS       F       P
Factor     1    29178    29178    5.19   0.026
Error     60   337397     5623
Total     61   366575
                          Individual 95% CIs For Mean
                          Based on Pooled StDev
Level      N     Mean    StDev  ------+---------+---------+-------
60min     31   4928.2     82.5  (-------*-------)
75min     31   4971.6     66.6            (-------*-------)
                                ------+---------+--------+--------
Pooled StDev =    75.0             4920      4950      4980
```

Figure 8. ANOVA results for pump out time

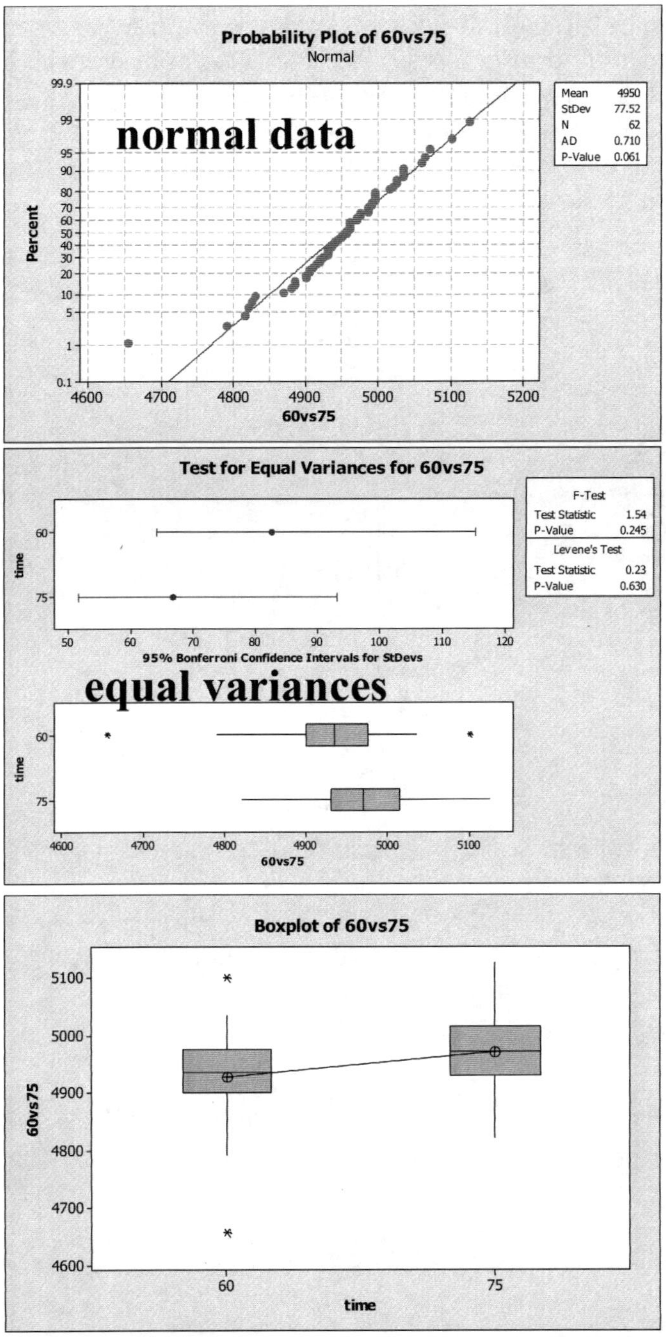

Figure 9. Normality plot (normal data), test of equal variances (equal variance), and box plots for each setting with pump out time (60 vs. 75 minutes).

2.5 Control phase

With the key input variables identified (Gas and pump out time), the objective of the final phase of the project was to implement the accepted solution (Gas & 75 minute pump out time) and ensure that the gains obtained from these changes could be controlled [10]. The baseline data were compared with the data containing the new variables setting, and a statistically significant difference was observed. The data were not normal, requiring a Mood's Median test. Figure 10, with the results, showed a P-value of 0.004 (below 0.05), indicating a significant difference in the medians. In terms of the means, the average production weight with the new variables was 4973 lbs versus a historical 4,875 lbs (an increase of 98 lbs).

```
Mood Median Test: HFS lbs versus Type

Mood median test for HFS lbs
Chi-Square = 8.16   DF = 1   P = 0.004

                     Individual 95.0% CIs
Type  N<=  N>  Median  Q3-Q1   +---------+---------+---------+-----
BASE   44  21  4930.0  100.5   (----------*----------)
FINAL  45  55  4975.0   73.8                    (--*---)
                       +---------+---------+---------+-----
                     4900      4925      4950      4975

Overall median = 4965.0


A 95.0% CI for median(BASE) - median(FINAL): (-75.0,-10.0)
```

Figure 10. Mood's median test for comparing Final settings to baseline settings

A new process capability was plotted and observed under the new conditions. The variation was reduced, and an increase of 2 sigma (Z) was observed in the new process. The DPMO level was reduced from 138,461 to 9,523 (Figure 11).

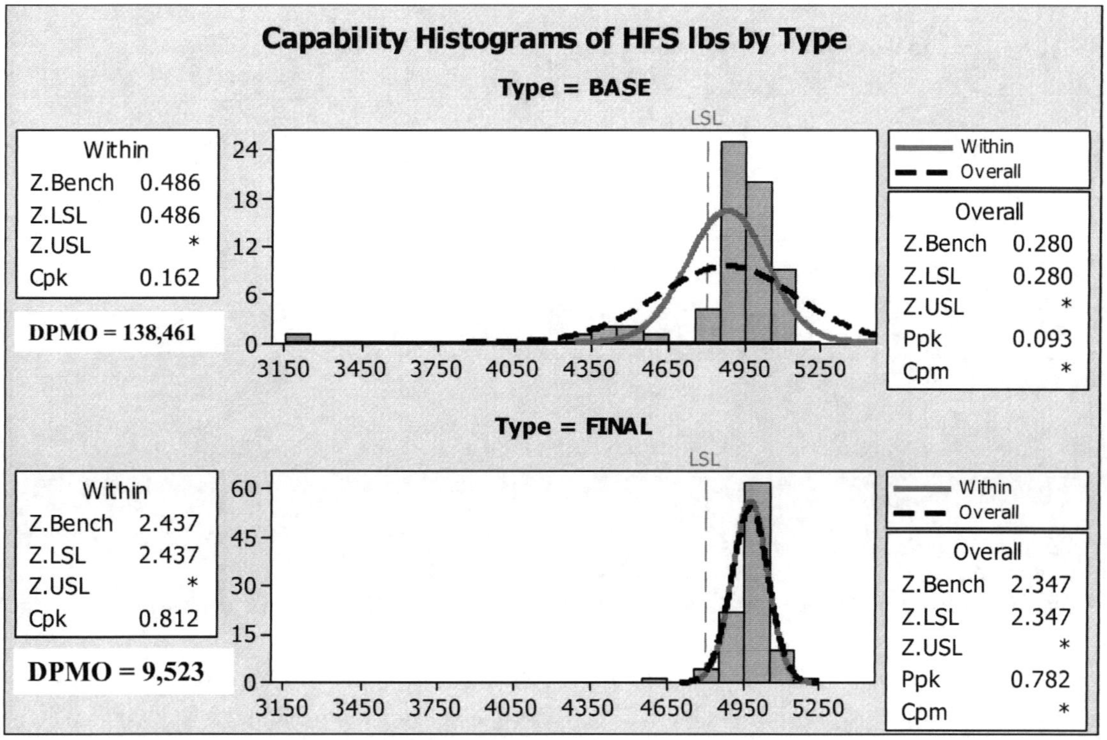

Figure 11. New process capability after changes, compared to baseline capability.

The key deliverable in the control phase was to ensure that a control plan was created that ensured the process and products will consistently meet customer requirements [8]. In this example, the final process was a function of the gas being used during the pump out process and pump out time (75 min) since these settings yielded the highest acid production (HFS lbs). In order to keep the process in control, manufacturing process instructions were revised to account for these changes, and the operators were trained. A long-term goal for this project is to adopt an engineering control that will set the pump out time with automatic on/off controls to further reduce variations in the process. Figure 12 is a control chart with the data before changes (Base) versus after the changes in the project (January 2010), clearly showing the reduction in variability of the process and the shift in mean to a higher value with the new process.

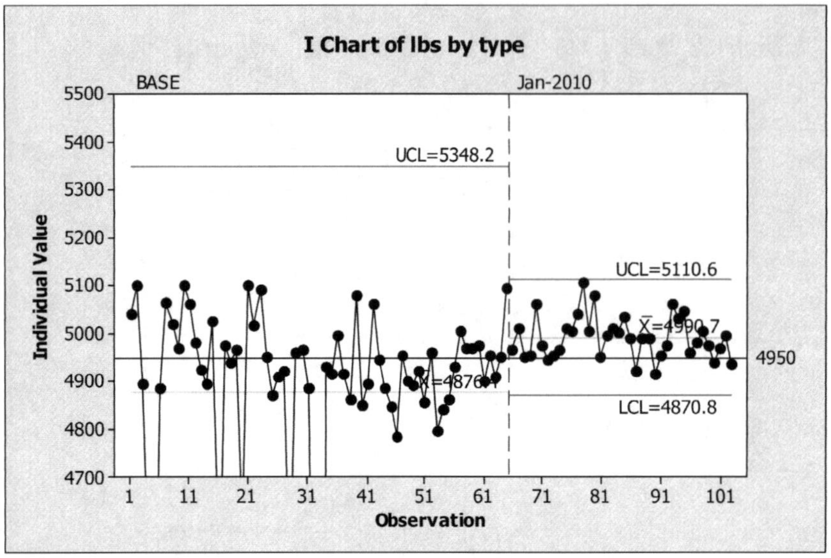

Figure 12. Control chart showing data before and after changes were implemented

Another part of the project not described in detail (due to proprietary reasons): is the quality of the actual sand, which is the main product of the reaction vessel. Any changes to the HFS process regarding process improvement to maximize production of acid must be evaluated against changes in the sand chemistry. When changing a process, it may not be obvious to the operator that the actual product is changing with the KPIVs change. It is easy to assume that increase in productivity will not affect the quality of the final product. To verify this assumption, the sand impurity elements were tested via ICP-MS. Results showed no difference in elements before and after the change (within 1 sigma variation).

3. Discussion and Conclusion

Six Sigma seeks to improve profitability with simultaneous quality improvement and efficiency [2]. It is an effective method to determine the greatest process needs and the softest points of the process. It also provides measurable indicators and adequate data for analytical analysis [8].

In this example the use of six sigma DMAIC methodology resulted in increased HFS acid production (recovery process) by an average of 100 lbs per batch. It also reduced the variability of the process. A well-defined project and team was set up from the start. The measurement system showed that the GR&R was reliable and the load cell was accurate to record the necessary differences that would be observed. Taking advantage of an expert team and tools such as cause & effect diagram, 9 KPIVs were identified, but only 2 proved to be statistically significant to the KPOV (acid production). With the new settings for the project implemented, the resulting benefits translated to significant improvement in the variable cost productivity. Furthermore, the quality of the main product (sand) was not affected by the increase in productivity related to the by-product (HFS acid).

Implementing six sigma methodology allows a company to change/improve their business approach. Simply changing the mentality on how to solve a problem can carry through the company so that a better solution is obtained. Whether it is with the development of new products or solving current problems, the methodology provides close acquaintance with all phases of the process while Six Sigma tools enable proper decisions and the most significant improvements [8].

References

1. Kwak, Y. and Anbari, F. (2006), "Benefits, obstacles, and future of six sigma approach," Technovation, 26, pp. 708-715.
2. Harry, M. and Schroeder, R. (2000), Six Sigma: The Breakthrough Management Strategy Revolutionizing the World's Top Corporations, Doubleday, New York, NY.
3. Anbari, F.T. (2002), "Six Sigma Method and Its Applications in Project Management," Proceedings of the Project Management Institute Annual Seminars and Symposium [CD], San Antonio, Texas. Oct 3-10, Project Management Institute, Newton Square, PA.
4. Pande, P., Neuman, R., and Cavanagh, R. (2000), The Six Sigma Way Team Field Book, McGraw Hill, New York, NY.
5. Chowdhury, S. (2001), The Power of Six Sigma, Dearborn Trade Publishing Co., Chicago, Illinois.
6. Kumar, U., Nowicki, D., Ramirez-Marquez, J., and Verma, D. (2008), "On the optimal selection of process alternatives in a Six Sigma implementation," Int. J. Production Economics, 111, pp. 456-467.
7. Dusharme, D. (2010), "Six Sigma Survey: Big Success . . . What About Other 98 Percent?" Quality Digest, accessed on 14th September 2010 at http://www.qualitydigest.com/feb03/articles/01_article.shtml
8. Sokovic, M., Pavletic, D., and Krulcic, E. (2006) "Six Sigma process improvements in automotive parts production" Journal of Achievements in Materials and Manufacturing Engineering, 19, pp. 96-102.
9. Breyfogle III, F., et al. (2001), Managing Six Sigma, John Wilet & Sons, Inc., New York, NY.
10. Hsieh, C-T., Lin, B., and Manduca, B. (2007), "Information Technology and Six Sigma Implementation," Journal of Computer Information Systems, pp. 1-10.

(437a) New Developments in Heat Transfer Scale Down and in-Situ FT-IR for Safe Scale up From the Lab to the Plant

Leen Schellekens, Mettler-Toledo AutoChem, Inc., Columbia, MD

"In today's world, the chemical and pharmaceutical industry faces major challenges, for instance: globalization, environmental and safety regulation, and shortening product life cycle and increase time pressure for commercial releases. Breakthroughs in process operations and modeling are necessary to achieve energy and material efficiency gains supporting shorter development and scale-up times and improving the environmental, safety and economical profile of chemical processes . Those breakthroughs are made possible thanks in part to the use of process analytical technologies and reactor systems optimally designed for real time in situ process monitoring and accurate control of processes, both in lab and plant scale. These breakthrough innovations are designed to safely develop a chemical synthesis route in less time, ensuring a robust and economic process that can be efficiently and safely scaled to manufacturing. This paper will present case studies exploring the use of heat transfer scale down in the RC1e calorimeter to simulate a plant reactor's heat removal and will demonstrate the use of in situ FT-IR as real time chemistry monitoring technique from lab to plant.

See more of this Session: Advanced On-Line Analytical and Optimization Tools in Pilot Plants
See more of this Group/Topical: Particle Technology Forum

(437b) Optimization and Scale-up of 4,4'-Diamino-3,3'-Azoxyfurazan (DAAF)through the Use of Real-Time Analytical Tools

Kimberly Hanson[1], Shilpa Amato[1], Jerry S. Salan[2] and Matt Jorgensen[2], (1)Chemical Development Branch, Indian Head Division, Naval Surface Warfare Center, Indian Head, MD, (2)Nalas Engineering Services Inc., Salem, CT

ABSTRACT

Introduction: The Joint Insensitive Munitions Technology Program (JIMTP) has invested in the development of explosive formulations using 4,4'-diamino-3,3'-azoxyfurazan (DAAF). Los Alamos National Laboratory (LANL) and Indian Head Division, Naval Surface Warfare Center (IHD NSWC) collaborated in the synthesis and scale-up of DAAF as well as a formulation effort that has evaluated DAAF as an IM explosive. The synthesis to DAAF was modified1 and results in pure product of sufficient particle size for LANL's formulation efforts. The synthesis utilizes commercially available reagents including water, Oxone® and aqueous buffer. The process was evaluated in detail with relevant factors being identified including pH of the aqueous media during the reaction and temperature. The process was also developed with an eye towards continuous manufacture based on the engineering challenges including pH controlled feedback loop and low concentration of product produced per reactor volume. The investigators will share the details of the engineering assessment, laboratory experimentation to obtain process understanding, process and equipment designs and product quality. An overview of costs associated with future manufacture of DAAF will also be presented.

1 US Patent 6358339 - Use of 3,3'-diamino-4,4'-azoxyfurazan and 3,3'-diamino-4,4'-azofurazan as insensitive high explosive materials

See more of this Session: Advanced On-Line Analytical and Optimization Tools in Pilot Plants
See more of this Group/Topical: Particle Technology Forum

(437c) Energetic Material Process Development Using Scale-Down Experiments

Jerry S. Salan, Nalas Engineering Services Inc., Salem, CT

Chemical process development is a challenging and exciting task for the chemical engineering profession. Utilization of in-situ tools combined with offline analytics afford the proper degree of process understanding to safely scaleup energetic materials from gram scale to multi-kilogram scale. Energetic materials pose inherently different challenges when compared to production of fine chemicals or pharmaceuticals. Energetic materials are not forgiving and require the ultimate respect by the chemical engineer as the processes are transitioned from the laboratory to pilot scale. Case studies on scaling up energetic materials utilizing various chemical engineering principles will be presented.

See more of this Session: Advanced On-Line Analytical and Optimization Tools in Pilot Plants
See more of this Group/Topical: Particle Technology Forum

(437d) Application of in-Situ Technology in 2,6-Diamino-3,5-Dinitropyrazine-1-Oxide (LLM-105) Recrystallizations

Andrew G. Pearsall[1], Bradley Sleadd[1], Tod F. Ricks[1] and Shilpa Amato[2], (1)Chemical Development, Naval Surface Warfare Center, Indian Head, MD, (2)Chemical Development Branch, Indian Head Division, Naval Surface Warfare Center, Indian Head, MD

The increased knowledge gained through the use of in-situ technologies is invaluable in determining the best method of obtaining the desired particle size and morphology of 2,6-diamino-3,5-dinitropyrazine-1-oxide (LLM-105) in recrystallization processes. Extensive experimentation performed at the Indian Head Division, Naval Surface Warfare Center (IHD NSWC) has shown the advantages and the usefulness of having this technology in understanding the recrystallization of LLM-105 from several different solvent systems. Differing methods of inducing nucleation have also been studied. A focused beam reflectance and measurement (FBRM) probe measures the particle size counts and distributions of LLM-105 crystals, and offers valuable insight. In addition, a particle vision and microscopy (PVM) probe allows real time images to be collected of the solid material in the slurry. Several case studies will be presented that highlight the knowledge and value gained from the use of these probes.

See more of this Session: Advanced On-Line Analytical and Optimization Tools in Pilot Plants
See more of this Group/Topical: Particle Technology Forum

(437e) Taking the Min to the Max: A Case Study in Small Scale Process Development Using On-Line Reaction Calorimetry and in-Situ Particle Characterization

Cheryl Teich, Core R&D/ Process Chemistry & Development, The Dow Chemical Company, Spring House, PA, Aaron Sarafinas, Core R&D / Engineering Sciences, The Dow Chemical Company, Spring House, PA and Paul Morton, Core R&D / Analytical Sciences, The Dow Chemical Company, Spring House, PA

Many specialty and fine chemicals are produced in batch processes and require fast assessment and development of manufacturing plans to speed time to market and profit. It is unlikely that any one specialty product could justify construction of a dedicated pilot plant, and its construction timeline would further decrease that likelihood. The subject case study demonstrates the breadth, depth, and quality of development and design data that can be obtained for fast, multi-phase chemistry in a small scale reaction calorimeter operated over a range of defined mixing conditions and equipped with in-situ particle characterization.

When a small-volume specialty was transferred to a different plant, the new reactor produced smaller size particles, impacting its downstream processing as well as the Plant's primary operating area. Initial discussions with the Plant centered on "fixing" the downstream equipment, but the focus moved quickly to the reactor and consideration of the most economical design for the low-volume product. A directed experimental program, incorporating macro, meso, and micro mixing considerations and study of the solid-liquid equilibrium was conducted in a METTLER TOLEDO RC1e™ reaction calorimetry equipped with a METTLER TOLEDO FBRM® in-situ particle characterization system. The program results led to an understanding of what "broke" the process and to a low cost design option, and demonstrate how the lab-scale equipment can be used as an effective pilot plant.

See more of this Session: Advanced On-Line Analytical and Optimization Tools in Pilot Plants
See more of this Group/Topical: Particle Technology Forum

(437f) Real-Time Analysis of Process Conditions During a Super-Saturation Controlled Recrystallization

Tod F. Ricks, Chemical Development, Naval Surface Warfare Center, Indian Head, MD

As part of a parametric study in particle size control, 3-Nitro-1,2,4-triazol-5-one is being recrystallization under a condition of constant super-saturation level. A super-saturated solution is one in which the concentration of a solute dissolved in a solvent is higher than the saturation concentration of that solute in the same solvent at a given temperature. Super-saturation level is the difference in concentration between the supersaturated concentration and the equilibrium concentration (saturation concentration) at a given temperature. In order to maintain a constant super-saturation level, real-time analysis of process conditions within the reactor such as temperature, particle size, and dissolved energetic concentration are constantly measured and relayed back to the control system to automatically adjust reactor temperature along the desired cooling curve. This paper will discuss the results of how real-time analytical tools were used to control process parameters and effect particle size and distribution control.

See more of this Session: Advanced On-Line Analytical and Optimization Tools in Pilot Plants
See more of this Group/Topical: Particle Technology Forum

(442a) Intensification Options for Different Hierarchical Process Levels Illustrated in the Conceptual Design of a Novel Cyclohexanol Production Process

Hannsjörg Freund, Rakesh Kumar, Amit Katariya and Kai Sundmacher, Physical and Chemical Process Engineering, Max Planck Institute for Dynamics of Complex Technical Systems, Magdeburg, Germany

A chemical production process can be decomposed into a multi-scale structure of four hierarchical levels [1,2]. The most detailed level is the molecular level, at which phenomena on the scale of individual molecules are investigated. At the next level, molecule populations that build up a thermodynamical phase (phase level) are considered. In the process, the thermodynamical phase(s) are embedded into apparatuses, or – more abstract – into individual process spaces. This is the process unit level. Usually, the process consists of several such process units. The interconnection between the individual process units and thus the overall process flowsheet can finally be analyzed at the superordinated plant level. Options for process intensification arise from all of these levels.

In this contribution, the conceptual design of a novel reactive distillation process for the production of cyclohexanol from cyclohexene via indirect hydration is presented and compared to state-of-the-art processes. It is exemplarily illustrated that a certain measure for process intensification – here the use of a reactive entrainer as reaction partner – has an impact on all levels of the process: on the molecular level (new reaction route, new catalyst), on the phase level (eliminate the problem of limited solubilities), on the process unit level (use of a different reactor type with structured internals) and on the plant level (integration of reaction and separation leads to a reduced number of process units and thus to a simplified process flow scheme).

The process development performed includes the experimental identification of thermodynamic and kinetic data, feasibility studies by means of reactive residue curve map analysis [3], rigorous process simulations [4] and experimental validation via mini plant experiments [5]. It is demonstrated that the new process scheme features a very promising performance, i.e. high conversion in combination with high selectivity (nearly 100% conversion of cyclohexene and more than 99 mole-% purity of cyclohexanol based on rigorous process simulations).

[1] H. Freund, K. Sundmacher, Chem. Eng. Process. 47 (2008) 2051. [2] H. Freund, K. Sundmacher, Process Intensification, In: Ullmann's Encyclopedia of Industrial Chemistry, Wiley-VCH: Weinheim, (2010) accepted. [3] F. Steyer, H. Freund, K. Sundmacher, Ind. Eng. Chem. Res. 47 (2008) 9581. [4] A. Katariya, H. Freund, K. Sundmacher, Ind. Eng. Chem. Res. 48 (2009) 9534. [5] R. Kumar, A. Katariya, H. Freund, K. Sundmacher, Appl. Catal. A-Gen. (2010) submitted.

See more of this Session: Advances in Process Intensification II
See more of this Group/Topical: Process Development Division

(442b) Process Intensification in Reactive Distillation: Design of a Diabatic Microwave Heated Column for the Synthesis of n-Propyl Propionate

E. Altman[1], G. D. Stefanidis[1], T. Van Gerven[2] and A. Stankiewicz[1], (1)Process&Energy, TU Delft, Delft, Netherlands, (2)Department of Chemical Eng., Katholieke Universiteit Leuven, Leuven, Belgium

Process intensification (PI) is a good example of how innovation is driving development in chemical engineering. Completely new approaches are being reported every year to perform conventional chemical processes with improved efficiency. Although the application of electromagnetic radiation as an alternative energy form in chemical syntheses has been studied extensively and the literature regarding intensification effects is remarkably rich, no successful commercial implementations in chemical plants have been reported. In addition to reactions, distillation processes benefit from microwave radiation (MW) with improved molecular separations. We have reported that when the vapor-liquid interface is exposed to microwaves, the system can be disturbed from conventional equilibrium conditions and a positive deviation in phase composition can be achieved, as the vapor phase becomes richer in the component with the lower boiling temperature; the extent of separation improvement depends on the boiling points and the dielectric properties of the mixture components. Since the two processes (reaction and distillation) separately can be enhanced by microwave radiation, the synergy gained by the integration of both operations in terms of chemical equilibrium shift and heat exchange can be additionally improved. In this investigation, we take a step further with the design and construction of a laboratory-scale diabatic RD unit using MW energy (f = 2.45 GHz). As a test system the heterogeneously catalyzed reaction of n-propyl propionate (ProPro) from 1-propanol (ProOH) and propionic acid (ProAc) is used. The MW heated column comprises a modified version of an Armfield UOP3 distillation column (DN-50) coupled with a specially designed mono-mode MW cavity from Sairem. The research is divided in two parts. First, experimental results of three out of the six constituent binary systems (ProOH/ProPro, ProPro/ProAc, ProOH/ProAc) are compared to the conventional heated process in the diabatic column using glass sieve plates and structured packing transparent to MW. In the second part of the research screening for a suitable catalyst including a BEA zelite, an acidic MOF and metal oxides is performed. The selected catalyst is coated upon the structured packing and tried in reactive distillation experiments.

See more of this Session: Advances in Process Intensification II
See more of this Group/Topical: Process Development Division

(442c) Development of a Highly Integrated Reaction-Distillation Process for the Production of Methylal

Jan-Oliver Drunsel[1], Hans Hasse[1], Ulrich Schmidt[2], Mario Renner[2] and Sergej A. Blagov[2], (1) Laboratory of Engineering Thermodynamics, University of Kaiserslautern, Kaiserslautern, Germany, (2) INEOS Paraform GmbH & Co. KG, Mainz, Germany

Methylal is an important industrial chemical. It is used as a solvent or as an intermediate for other processes like the production of poly(oxymethylene)dimethylethers. In this work, a new process for the production of methylal from an aqueous, methanolic formaldehyde solution was developed. Several process alternatives were evaluated, the most promising of which is a highly integrated reaction-distillation process in which methylal is produced by a heterogeneously catalysed reaction using an acidic ion exchange resin. In order to investigate the kinetics of this reaction, experiments in a tubular fixed-bed reactor were carried out. The parameters of an activity-based pseudo-homogeneous model were fitted to the results of these experiments. As formaldehyde solutions are complex reacting mixtures, their true speciation had to be explicitly taken into account in the modelling. The new process yields highly pure methylal – in a concentration exceeding that of the methylal-methanol azeotrope. Almost complete conversion of formaldehyde is achieved. Also methanol and water are obtained in reasonably high purity. The new process will be presented and discussed in detail at the conference. It is shown, that the desired specifications can be achieved with reasonable effort. The process development and optimization is based on process simulations. These are extremely demanding for the present system as besides the heterogeneously catalysed formation of methylal the complex reactions in the formaldehyde solutions have to be taken into account - not only in the reactor model but also in the distillation model. Based on previous extensive work of our group both on thermodynamic properties of the complex system studied here and on the simulation of processes with formaldehyde containing reactive mixtures, a reliable model for the new process was developed. It was successfully validated by a comparison to distillation experiments in a laboratory glass column (diameter 50 mm, packing height 4 m) that were carried out in the present project.

See more of this Session: Advances in Process Intensification II
See more of this Group/Topical: Process Development Division

(442d) Chemical Synthesis with Extractive Product Isolation as An Approach for Process Intensification

Susanne Lux, Thomas Winkler and Matthäus Siebenhofer, Department of Chemical Engineering and Environmental Technology, Graz University of Technology, Graz, Austria

Chemical synthesis reactions such as esterification reactions may result in extended complexity due to formation of azeotropic mixtures. Stringent efforts for subsequent product isolation are required. Production of methyl acetate is a representative example for formation of highly complex mixtures during reaction. Methyl acetate forms low boiling azeotropic mixtures with the reactant methanol as well as the by-product water. The state-of-the-art process for production of methyl acetate is based on a reactive distillation process, the so called Eastman Kodak process. In principle the Eastman Kodak process is a dual step esterification process with the basic intention of avoiding azeotrope formation by supplying different excess reactant in both reactor sections. In this project technical approaches for production and isolation of methyl acetate below boiling point operation are investigated. The reaction is carried out in a packed bed plug flow reactor. Strongly acidic cation exchange resins are used as heterogeneous catalysts. Formation of the methyl acetate/methanol azeotropic mixture is avoided by metering acetic acid in excess. The azeotropic mixture methyl acetate/water can be overcome through selective extraction of methyl acetate from the multicomponent reaction mixture with consecutive distillative isolation, finally leaving a mixture of acetic acid and water for distillative separation. For technical as well as economic process comparison adsorptive separation of water and methanol from the reaction mixture was investigated too. During esterification the catalyst undergoes swelling in the multicomponent reaction mixture. For modeling purposes the composition and propagation of swelling of the resins was investigated by determining the adsorption characteristics of reactants and products. The project provides basic strategy for process design of esterification processes with the aim of developing new on-site-processes for stock quantities.

See more of this Session: Advances in Process Intensification II
See more of this Group/Topical: Process Development Division

(442e) Biodiesel Production Using An Intensified Spinning Disc Reactor

Zheyan Qiu, Chemical & Petroleum Engineering, The University of Kansas, Lawrence, KS and Laurence R. Weatherley, Chemical & Petroleum Engineering, University of Kansas, Lawrence, KS

Biodiesel, as an alternative fuel, has attracted more attention and been more widely accepted because it is made from renewable feedstocks. Usually biodiesel is synthesized from transesterification of vegetable oils or animal grease with alcohols in a stirred tank reactor (STR) in the presence of acid or base. Transesterification reaction is reversible and its reaction rate is limited by mass transfer between immiscible two phases. Hence, biodiesel production requires long residence time, high molar ratio of alcohol to oils and catalyst concentration, leading to high operating cost and energy consumption, and low production efficiency. In this paper, a novel intensified spinning disc reactor (SDR) with a stator, which has been developed to improve the mixing and mass treansfer between methanol and oil, thus increasing the product efficiency of biodiesel production via alkali-catalyzed transesterification is described. The modified SDR includes a stationary disc, which is coaxially spaced adjacent to a rotating parallel disc separated by only a fraction of a millimeter. The configuration may lead to intense, forced molecular inter-diffusion of liquid-liquid two phases caused by high shear rate. The performance of the new SDR has been investigated by examining the effects of gap sizes between the discs, rotational speed, flowrate of oil phase and designs of the rotating disc. Additionally, reaction conditions have been optimized to determine the values of molar ratio of methanol to oil, reaction temperature and catalyst concentration which result in the highest biodiesel conversion. Finally, the performance of the intensified SDR has been compared with that of a STR.

See more of this Session: Advances in Process Intensification II
See more of this Group/Topical: Process Development Division

(442f) Study On a Cryogenic Separation System Under Abnormal Situations Via Dynamic Simulation

Xiongtao Yang, Qiang Xu and Kuyen Li, Dan F. Smith Department of Chemical Engineering, Lamar University, Beaumont, TX

A cryogenic separation system, especially the chilling train and de-methanizer facilities, plays an important role in olefin plants. This system can chill and refrigerate the cracked gas from cracking furnaces to -160 °C, so as to separate the light components such as hydrogen, methane, and C2+ at different temperatures. As it closely relates to heat exchange at low temperatures, previous studies mainly focused on the energy use efficiency of this system. Because the cryogenic separation system operates at low temperature and high pressure, it features inherently un-safety, this paper will address its design and operational safety issues at dynamic situations, i.e. identify the safe operation capability of the cryogenic separation system under abnormal situations. In addition to the safety concerns, this paper will also study air emission problems during these abnormal situations. A large-scale dynamic simulation system integrating rigorous models of chilling train and de-methanizer facilities is developed for these study purposes. The abnormal situations such as system start-up, upsets, and shutdown will be simulated. They may cause off-spec product generations or system over-pressure events, which eventually lead to the flaring emissions. Dynamic simulated emission source profiles will be provided when these induced emission events arise. Based on the safety and environmental impact analysis for the current cryogenic separation system, design and operational improvements will be proposed and tested in the virtual dynamic environment. A case study is employed to demonstrate the efficacy of this study.

See more of this Session: Advances in Process Intensification II
See more of this Group/Topical: Process Development Division

(467a) Ethanol Production by Bioconversion of Softwood Prehydrolysates Supplemented with Pulp Mill Sludges

Li Kang[1], Sunghoon Yoon[1], Gopal A. Krishnagopalan[2] and Y. Y. Lee[3], (1)Chemical Engineering, Auburn University, Auburn, AL, (2)Chemical Engineering, Auburn University, Auburn University, AL, (3)Department of Chemical Engineering, Auburn University, Auburn University, AL

Eighty percent of the hemicellulose fraction of pulp mill feedstock is released into black liquor during kraft pulping process. In pulping process, the black liquor is burned to recover chemicals and generate steam and electricity. It is technically feasible to remove part of the hemicellulose before pulping keeping the quality of pulp unaffected. The hemicellulose portion of the pulp mill feedstock can be selectively converted to soluble sugars (termed as prehydrolysate) by water or mild acid hydrolysis. The prehydrolysate thus prepared contain pentose, hexose, and their oligomers. In this study, bioconversion of the prehydrolysate to ethanol was investigated using Saccharomyces cerevisiae. The prehydrolysate produced from wood also contain toxins, primarily organic acid, lignin and sugar degradation products. These toxins strongly inhibit microbial action. The prehydrolysate was therefore detoxified by over-liming and by addition of peroxide. The detoxified prehydroysate was further treated with pectinase to hydrolyze the mannose oligomers to monomer. After these treatments, the hysrolyzate was efficiently converted to ethanol. Use of mixture of prehydrolysate and pulp mill sludge as the fermentation feed has given significant increase of the ethanol concentration in the bioreactor over that of prehydrolysate fermentation.

See more of this Session: Innovations in Bio-Refinery Processes
See more of this Group/Topical: Process Development Division

(467b) Molecular Design of Solvents for 2nd Generation Biofuels Separation Processes

Ana Silveira[1], José Scilipoti[1], Martin Cismondi[2] and **Esteban A. Brignole**[3], (1)Chemical Engineering, PLAPIQUI - Universidad Nacional del Sur - CONICET, Bahía Blanca, Argentina, (2)Chemistry, Facultad de Ciencias Exactas y Naturales, Universidad Nacional de Córdoba, Córdoba, Argentina, (3) Plapiqui, PLAPIQUI - Universidad Nacional del Sur - CONICET, Bahía Blanca, Argentina

The second generation biofuels are based on the use of biomass , mainly from lignocellulosic sources, as raw materials for the production of biofuels and for the chemical and paper industry. Many challenging separations problems are faced for the conversion of cellulosic materials in fuels and chemicals. Even though there are different wood processing, in the present work we focus mainly on the process of detoxification of hydroxilates; that will be used for the production of ethanol and other chemicals by fermentation processes.

In this work, the ECOFAC molecular design of solvents program (Cismondi and Brignole [1]) is upgraded to deal with the separation of organic chemicals found in wood hydroxilates, that are toxic for the fermentation process. For this purpose, the last UNIFAC parameter data bank revision are updated with new parameters reported in the literature for the components of interest. Examples of selection of solvents for different toxic components are presented and a strategy, for solvent selection with multiple objetive optimization goals, is presented. The ECOFAC program is also updated for the prediction of environmental properties, among others: octanol – water partition coefficients and water solubilities of organic compounds.

The program also incorporates a more general molecular design procedure to generate families of lignocellulosic derivatives that are toxic for the fermentation process. In this way potential solvents can be screened in their capacity to separate these products. As a difference with a typical extraction process the raffinate is in this separation problem the product to be recovered and for this reason the primary solvent properties are more restrictive than in conventional extraction processes.

Another option under development for this particular type of application is the use of A-UNIFAC, an group contribution approach taking into account association.

References [1] M.Cismondi and E.A.Brignole, Molecular Design of Solvents: An efficient search algorithm for branched molecules. Ind. Eng. Chem. Res. 43(2004) 784-790

See more of this Session: Innovations in Bio-Refinery Processes
See more of this Group/Topical: Process Development Division

REACTIVE EXTRACTION AS A METHOD TO PRODUCE BIODIESEL FROM *JATROPHA CURCAS* SEEDS

Farizul H. Kasim, Adam P. Harvey, Newcastle University, Newcastle, UK

Abstract

Jatropha curcas (jatropha) produces oil that can be converted to biodiesel by transesterification. However, because of its high free fatty acid value (>2 %), acid catalysts are usually used in the process, rather than alkalis. Because of this, longer reaction times are required. Furthermore, jatropha oil contains phorbol ester, a carcinogen that can be exposed when the seeds are extracted for the oil. The application of reactive extraction as a method to produce biodiesel from jatropha, enables alkali based catalyst to be utilized, and shortens the reaction time without compromising on conversion and quality of biodiesel. Since seeds are used as raw material instead of oil, the phorbol ester exposure is greatly reduced. This study evaluates the parameters involved in reactive extraction of jatropha seeds. The molar ratio of alcohol to oil was found to be the most important parameter in this reaction. The yield in 1 hour was low at 100 molar ratio (2%) but then increased sharply when the ratio was increased to 200 (55%), 300 (75%) and 400 (82%). Only small increments were observed when the ratio was increased further to 500 (85%) and 600 (90%). Methyl ester yield was a strong function of seed particle size. Of 5 different sizes of particle size, the smallest (0.5 mm), produced the highest yield (86%) and the biggest (2.0-4.0 mm) produced the lowest (35%). On agitation speed parameter, low speed agitation (100 rpm) was unable to produce a high yield compared to high speed agitation (300, 400 rpm). The study on temperature meanwhile revealed that there was little change in yield when the reaction was performed at 30°C (90%), 40°C (89%), 50°C (91) and 60°C (98). Interestingly, it was shown that the process was complete relatively quickly, typically after 30 minutes. The methyl ester content of the samples demonstrate that high purity of methyl ester (<90%) can be obtained from this method. An implication of this is that reactive extraction could be a method of producing biodiesel from seeds with high free fatty acid contents.

Introduction

Current biodiesel production technology converts refined oils to biodiesel. The oils undergo a number of processes: mechanical crushing and solvent extraction (using hexane) to produce the raw oil, followed by degumming and esterification steps to clean up the oil prior to transesterification to biodiesel. If, instead of oil, oilseeds themselves can be converted directly to biodiesel, all of these steps can be missed out. This should greatly simplify the process, which should make it suitable for small-scale distributed production, as the capital cost and service requirements will be significantly reduced. This should allow oilseed farmers to realise more of the full value of their crop, and simultaneously reduce local biodiesel prices. The process is "greener" as hexane is no longer required.

The aim of this work is to develop a process for directly converting jatropha seeds to biodiesel, via "reactive extraction". The specific objectives are: to demonstrate that this can be done and to map out the parameter space affecting the process. Jatropha is of particular interest as it can grow on marginal or degraded land, so does not compete with food crops for land. Due to the lower land quality requirement, it does not require land clearance, as is so often the case with palm and soya. Another benefit in terms of the food vs. fuel debate is that it is inedible, so is not in the same market as the edible oils.

Materials and methods

Preparation of J. curcas Seeds

Jatropha seeds were provided by Indian Institute of Petroleum (IIP), Dehradun, India. The seeds were stored in dark air-tight containers. The oil content and free fatty acids (FFA) value of the seeds were determined prior to experiments and whole jatropha seeds (shell and kernel) were utilised. The seeds were ground using a blender (Kenwood BL440, UK) to reduce the particle size. A vibrator sieve-shaker (Retsch, Germany) was then employed to separate the ground seeds into different sizes. These seeds then were placed in a drying pan and were dried in a drying oven (Memmert, Germany), at 105°C for an hour.

Chemicals and Reagents

Methanol (99.8% purity) was purchased from Fisher Scientific, UK. Sodium hydroxide (NaOH), in pellet form was procured from Merck, Germany. Hexane and acetic acid (glacial) which were used in post-reaction steps were acquired from Fisher Scientific, UK as well. Heptane, which was used as a solvent in gas chromatography sample preparation and the internal standard, methyl heptadecanoate, were purchased from Sigma Aldrich, UK.

Reactive Extraction

Reactive extraction experiments were conducted in sealed, chemically resistant Schott bottles (250 ml). The bottles were filled with methanol to achieve the required pre-determined molar ratio and relevant amount of catalyst (NaOH) then was dissolved in the methanol. The mixture was then pre-heated to the reaction temperature in a programmable incubator shaker (IKA, Germany). Jatropha seeds (10 g) were put into the bottles after the mixture reached previously set temperature. The reaction was conducted in an incubator shaker with controlled temperature, agitation speed and time.

The bottle was taken out from the incubator and vacuum-filtered after the reaction. This will separate the meal and the liquid mixture. To ensure that the transesterification reaction stopped, the catalyst was neutralised using a small amount of glacial acetic acid. 2 mL of the sample was then analysed by gas chromatography (GC) to determine the yield. The remaining mixture was put in an evaporator flask, to evaporate the methanol. The rotary evaporator (Buchi, Switzerland), which was set at 55°C - 60°C evaporated the methanol and collected it in another connecting flask.10 mL of hexane then added to the remaining mixture. The solution then was transferred to separating funnel and was allowed to separate gravitationally. After 1 hour, two distinct layers, the biodiesel-rich upper layer, and the glycerol-rich lower layer appeared. The upper layer was yellowish in colour the lower layer, dark brown. The mixture was then washed with warm water to remove impurities from the upper layer. After the glycerol layer was drained, the hexane in the upper layer was evaporated using a hot plate in fume hood. The mass of the purified biodiesel was then recorded, and analysed by GC to determine percentage methyl ester. The solid part of the reaction (the meal) was heated in the oven at 100°C to remove the methanol. It was then weighed and recorded to determine the percentage of weight loss.

Table 1 listed the setting of all parameters studied in the experiments. The coloured boxes indicate the parameter which was varied for the particular experiment set.

Table 1: List of experiments conducted with respective parameters setting

Experiment	Parameters					
	Seed size (mm)	Mixing speed (rpm)	Reaction temperature (°C)	Reaction time (min)	Catalyst concentration (N)	Alcohol:oil molar ratio
Effect of particle size	<0.5-4	400	60	60	0.1	400
Effect of mixing speed	<0.71	100-400	60	60	0.1	400
Effect of reaction temperature	<0.71	400	30-60	60	0.1	400
Effect of reaction time	<0.71	400	60	10-60	0.1	400
Effect of catalyst concentration	<0.71	400	60	60	0.1-0.2	400
Effect of methanol to oil molar ratio	<0.71	400	60	60	0.1	200:1-600:1

Analysis

A flame ionisation detector (FID)-equipped HP5890 Series II (Hewlett Packard, USA) gas chromatograph fitted with BPX70 column, 30 m long x 0.32 mm ID x 0.25 μm film thickness (SGE, Australia) was used to analyse the samples. Helium was used as carrier gas at a pressure of 7 psi. The oven was maintained at 230°C for 30 minutes.

The data was acquired and processed using Clarity Chromatography Station for Windows (DataApex, Czech Republic). The software allowed integration of peaks to be performed on the chromatogram. For the calculation of FAME conversion as well as methyl ester content, the solvent peak was excluded from the chromatogram, and this was done by integrating the peaks eluded from 10 – 30 minutes.

The internal standard, methyl heptadecanoate solution was prepared prior to the sample preparation. 250 mg of sample was then weighed and placed in the vial. After that, 5 ml of methyl heptadecanoate solution was added to the sample. The mixture was mixed thoroughly using the MS1 Minishaker (IKA, Germany).1 μL of sample was then injected to the GC using a 5 μL microsyringe (SGE, Australia).

Yield and FAME Percentage Calculation

The yield was calculated by modifying the standard for calculating FAME (BS EN 14103:2003). Firstly, the mass of the liquid mixture after the separation of solid (meal) and liquid (mixture of methyl ester, methanol, sodium hydroxide, glycerol and unreacted triglycerides) was taken and recorded (w). Methanol was used to prepare the internal standard. After adding internal standard to the sample, the sample was then injected to the GC.

The ester content in the sample, C, expressed as a mass fraction in percent, can be calculated using Equation 1 below.

$$C = \frac{(\Sigma A) - A_{Ei}}{A_{EI}} \times \frac{C_{EI} \times V_{EI}}{m} \times 100\% \qquad \textbf{Equation 1}$$

where

$\sum A$ = total peak area from the methyl ester $C_{14} - C_{24:1}$
A_{EI} = peak area corresponding to methyl heptadecanoate
C_{EI} = the concentration, in mg/ml of the methyl heptadecanoate solution
V_{EI} = the volume, in ml of the methyl heptadecanoate solution being used
M = mass of the sample (mg)

The yield then was calculated by multiplying C with mass of the sample, w, as in Equation 2.

$$Yield(\%) = C(\%) \times w(g) \qquad \textbf{Equation 2}$$

FAME content is the percentage of methyl ester in the biodiesel rich phase of the sample. As in the yield calculation, the method and calculation was based on the BS EN 14103:2003 standard. In the sample preparation, heptane was used as a solvent to dissolved methyl heptadecanoate, the internal standard.

Results and Discussion

The results of all the experiments are as in Figure 1. In this study, the molar ratio of methanol to oil was varied from 100 to 600 (Figure 1a). No methyl ester was produced at the lowest molar ratio (100), even though 18.1% of mass was extracted from initial 10g of seeds. The yield of methyl ester was significantly higher at a ratio of 200, although there had been little increase in % mass loss (21.9%). The yield then increases steadily as the ratio increases. The yield at a molar ratio of 300 is 74.7% and the yield for 400, 500 and 600 are 81.9%, 85.7% and 86.9%, respectively. The mass loss levels off when the ratio reaches 400.

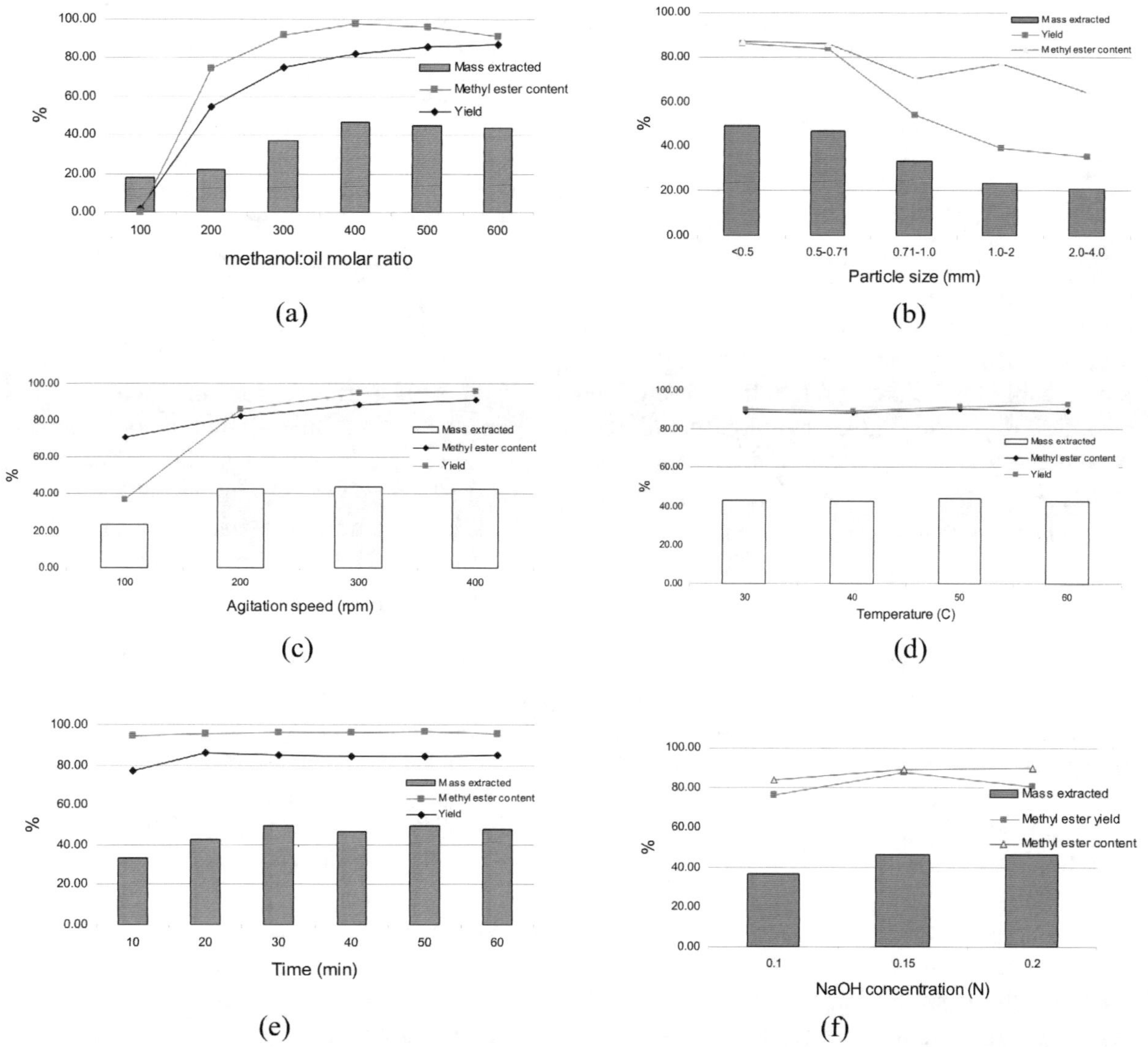

Figure 1. Methyl ester yield, methyl ester content and mass extracted on various experiments parameters (a) 100 to 600 molar ratio of methanol to oil.; (b) <0.5 mm to 2-4 mm particle size; (c) 100 to 400 rpm agitation speed; (d) 30°C to 60°C reaction temperature; (e) 10 minute to 60 minute reaction time and (f) 0.1N to 0.2N NaOH concentration. The other parameters set as in Table 1.

The results suggested that the amount of methanol must be very high to achieve an appreciable rate. This was presumably to drive the penetration of alkaline methanol into the seed, as observed in Ren *et al* [1]. Further increasing the excess of methanol (e.g.600) does not greatly increase the yield, but is undesired, as it will increase the load on downstream separation processes. This plateau has also been observed by Qian *et al* [2]. They also found, however, that the amount of methanol required was a function of the catalyst concentration. Haas *et al* [3] also found that the greater amount of alcohol used in the process, the lower the catalyst concentration needed to give a good yield. Above a molar ratio of 300, the methyl ester content is above 90%, but it exhibits a maximum around 400, indicative of significant amounts of triglyceride had been converted into methyl ester.

In Figure 1(b), the yield of methyl ester decreases with increasing particle size, when the particle size is larger than some threshold value of around 0.71 mm. No significant difference was determined between the yield for the smallest two particle size ranges, <0.5 mm and 0.5 – 0.71 mm, with the former yielding 86.1% conversion and the latter 83.7%. The largest particle size range (2–4 mm) produced the lowest yield, at 35.5%. This finding is consistent with those reported by Qian *et al* [2]. In terms of methyl ester content, only the two smallest particle sizes produced more than 80% of methyl ester in the sample. The other particle sizes produced lower methyl ester contents; with size 0.71-1 mm has 70.5% methyl ester, size 1-2 mm; 77.0% and 2-4 mm, 64.0%. No further analysis was done to determine the other compounds in the sample, but the most likely suggestion would be the presence of unreacted glyceride in the form of mono-, di- and triglycerides [4.5]. The chances of other compounds such as glycerol and polar lipids being present in the sample are small, as they will have been removed by settling and hexane washing.

It is noticeable for Figure 1(c) that the mixing intensity is not a mass transfer limiting factor once it reaches 300 rpm. The finding suggests that there is no point increasing the agitation beyond this level, as it will not lead to a significant improvement in yield. However, decreasing the speed to 200 rpm, and further to 100 rpm, decreased the yield (from 94.8%) to 85.7% to 37.2%. A similar pattern was also observed by Meher and co-workers in their work on alkali-catalyzed transesterification of *Pongamia pinnata* oil [6]. They attributed the low yield (< 85%) achieved with 180 rpm mixing intensity to incomplete reaction as opposed to 97% yield when 360 rpm and 600 rpm mixing intensity were applied.

There was no significant difference between the masses extracted at the 4 different temperatures (Figure 1d). There was also minimal change in the yield and methyl ester content. The results obtained, are in agreement with observations by Haas and co-researchers [3]. Both observed that triglyceride can be converted to FAME at both low and high temperature (23°C and 60°C in the case of Haas' group). However, they also found that at the lower temperature 2.4 times more methanol was required to achieve the same conversion as at 60°C. This phenomenon also has been reported elsewhere [2]. In contrast, temperature influences both the reaction rate and yield in conventional transesterification [7].

The results obtained from the reaction time experiments are presented in Figure 1(e). The mass extracted from the seeds during reactive extraction were similar (around 40%) starting from 30 minute onwards. This result mirrors the methyl ester yield, which also reached a plateau after 30 minutes. It is therefore very likely that the reactive extraction itself completed between 20 and 30 minutes. It also can be observed that reactions less than 20 minutes did not achieve high yields. This is in agreement with other findings, which report rapid increases in yield within the first 30 minutes of reaction [2,3].

The addition of NaOH, although in small quantity (0.1 N) give a significant effect in term of converting oil to methyl ester. Increasing the concentration to 0.15 causing the yield to increase from 76.2% to 87.8%. However further increment to the concentration (0.2 N) decrease the yield to 80.8%. It is also interesting to note that, with a further increase in NaOH concentration, an emulsion started to form, consequently reducing the yield. The most likely cause of the emulsion is the formation of soap which is a competing reaction in alkali-catalysed transesterification process. Interestingly, it is apparent from Figure 1(F) that through reactive extraction, the high free fatty acids content of jatropha oil had no negative effect to the yield.

Conclusion

This study was designed to determine the effect of various parameters on the reactive extraction of jatropha. Of the 6 parameters investigated (particle size, mixing speed, reaction temperature, reaction time, NaOH concentration and molar ratio of methanol to oil), 5 parameters have some effect on the yield of FAME. The only parameter which did not exhibit any significant change in 60 minute is the reaction temperature. Particle size and molar ratio of methanol to oil strongly affect the yield of methyl ester: it increases with increasing molar ratio to an optimum, and increases with decreasing particle size to a plateau. Investigation of the effect of mixing speed indicated that a certain minimum speed must be provided to minimise the immiscibility between oil and alkaline methanol, otherwise the FAME yield decreases. The reaction was shown to be complete after 20-30 minutes. This study demonstrated that for jatropha, as for other oilseeds, some form of alkaline catalyst, in this case NaOH, is required to achieve a meaningful yield when performing extraction with. Finally, there is an important limitation that must be considered. It is impossible to rule out the interaction between parameters as a consequence of the involvement of 6 different parameters in this reactive extraction reaction. Therefore, this one-at-a-time series of experiments is a precursor to a full design of experiments, in which the interaction between parameters can be identified and quantified.

References

1. Ren Y, Harvey A. and Zakaria R. (2010), “Biorefining based on Biodiesel Production: Chemical and physical Characterisation of Reactively Extracted Rapeseed,” *Journal of Biobased Materials and Bioenergy*, 4(1), pp. 79-86.

2. Qian J., Wang F., Liu S. and Yun Z (2008) “In Situ Alkaline Transesterification of Cottonseed Oil for Production of Biodiesel and Nontoxic Cottonseed Meal,” *Bioresource Technology*, 99(18), pp. 9009-9012.

3. Haas MJ., Scott KM., Marmer WN. and Foglia TA. (2004), “In Situ Alkaline Transesterification: An Effective Method for the Production of Fatty Acid Esters from Vegetable Oils”, *Journal of the American Oil Chemists' Society*, 81(1), pp. 83-89.

4. Gerpen JV., Shanks B., Pruszko R., Clements D. and Knothe G. (2004), “Biodiesel Production Technology”. *National Renewable Energy Laboratory (NREL) Report*, Colorado.

5. Ma F. and Hanna MA.(1999), “Biodiesel Production: A Review,” *Bioresource Technology*, 70(1), pp. 1-15

6. Meher LC., Dharmagadda VSS. and Naik SN., “Optimization of Alkali-catalyzed Transesterification of Pongamia Pinnata Oil for Production of Biodiesel,” *Bioresource Technology*, 97(12), pp. 1392-1397

7. Meher LC., Vidya Sagar D. and Naik SN., “Technical Aspects of Biodiesel Production by Transesterification - a Review,” *Renewable and Sustainable Energy Reviews*, 10(3), pp. 248-268

(467d) A Novel Approach for Carboxylic Acid Recovery From Fermentation Broths

Alvaro Orjuela, Chemical Engineering & Material Science, Michigan State University & Universidad Nacional de Colombia, East Lansing, MI, Abraham J. Yanez, Chemical Engineering & Material Science, Michigan State University, East Lansing, MI, Carl T. Lira, Chemical Engineering and Material Science, Michigan State University, East Lansing, MI and Dennis Miller, Michigan State University, East Lansing, MI

Biobased fermentation processes uniformly produce dilute aqueous solutions of the carboxylic acid as the primary products. Many of those fermentation systems yield a mixture of acids and other products, depending on the intended metabolic pathways and conditions within the bioreactor. The recovery and purification of individual species from these fermentor effluents can be expensive, accounting for 50 to 80% of the overall production cost. For succinic acid fermentation, significant amounts of acetic, formic, pyruvic and/or lactic acid can be formed as byproducts and are present in the broth in addition to non-converted sugars, biomass (cell growth), salts, and supplements. Because fermentation is most efficient at intermediate pH (5-7), base (NaOH, Ca(OH)2, ammonia) is used to control pH that naturally declines as the acid accumulates. The carboxylic acids produced are thus typically obtained as salts of the base used. Several alternatives have been proposed to recover succinic acid from the acid salt broths; these can be classified according to the approach as direct precipitation (as salt or acid), extraction (solvents, amines), ion exchange, membrane recovery, or esterification. In the traditional process, the aqueous broth containing succinate salt is acidified with H2SO4 at a pH below the pKa1 of succinic acid (pH < 4). With calcium as the cation, calcium sulfate is precipitated; the sulfate is removed by filtration and the SA is subjected to several crystallization steps. Because succinic esters are useful as platform species to produce a variety of value-added commodity and specialty derivatives, we have developed and present here a novel approach to directly recover succinic acid esters from fermentation. Following partial water removal, the process performs acidification with sulfuric acid in ethanol. Acidification of salts in EtOH releases succinic acid which simultaneously reacts to form mono- and di-ethyl succinate, while inorganic salts (sodium or calcium sulfate) precipitate out of the solvent. Other carboxylic acids removed in traditional processes are also recovered as ethyl esters which can be separated easily than original mixed acids. Experiments performed using pure and mixed succinate and acetate salts and actual fermentation products show high recovery efficiency and rapid conversion rates.

See more of this Session: Innovations in Bio-Refinery Processes
See more of this Group/Topical: Process Development Division

A TECHNO-ECONOMIC SIMULATION OF ALTERNATIVE FEEDSTOCK STRATEGIES FOR AN INTEGRATED FOREST BIOREFINERY
AIChE 2010 ANNUAL MEETING

Anna Kalliola, VTT Technical Research Centre of Finland, Espoo, Finland
Juha Hakala, VTT Technical Research Centre of Finland, Espoo, Finland
Pertti Koukkari, VTT Technical Research Centre of Finland, Espoo, Finland

Abstract

A simulation based feasibility study of a paper mill producing simultaneously printing paper and mechanically pre-treated biorefinery feedstock was performed. A process model of a modern paper mill served as a reference for the new process alternatives, in which a TMP reject fraction (feedstock) was conducted out of the papermaking process. The share of the removed pulp fraction in the paper was compensated either by using other raw materials (*e.g.* recycled pulp) or alternative process/paper machine operation practices. Manufacturing costs for each new concept were estimated by computing the variable costs. The cost effects were allocated both for tonne of paper produced and for tonne of TMP reject pulp produced. Thus, a unit cost for the biorefinery feedstock was obtained for each alternative. The unit cost varied from 60 to 360 €/ad-t feedstock.

The simulation method developed can be utilized for the early evaluation of alternative concepts of process renewal.

Keywords: Integrated Forest Biorefinery, Mechanical pulping, Production economics, Simulation Based engineering

Introduction

Economic pressures and more stringent legislation on environmental issues have enforced the renewal of the forest industry. The pulp and paper making (P&P) companies are pursuing new revenues from the emerging lignocellulosic biorefining industry by producing biofuels alongside with their conventional products. The development of sustainable wood-based non-fuel products is also widely investigated. However, the multitude of process and product options, combined with the raw material base in the mill region, makes the industry renewal a difficult task. The current integrated biorefineries are mainly utilizing the P&P side streams or by-products as feedstock. However, it is anticipated that in the future multiple volumes of feedstock will be needed to meet the increasing supply of bioproducts.

The use of mechanical pulp is an established part of current papermaking. In year 2009 the volume of all mechanical and semi-chemical pulp produced in Finland was 3.2 Mt being 37 % of total pulp produced [1]. In thermomechanical pulping (TMP) wood chips are refined at elevated temperatures and under pressure into fibrous pulp. A significant part (25-40%) of the pulp produced at the main refining stage belongs to a sc. reject fraction which needs to be further refined at a reject refining stage to reach the desired fineness for the final paper products. Reject refining is energy intensive. With a successful choice of new uses for this raw material fraction, the process could be simplified and the profitability of the production site could be improved.

The average production of the reject fraction in a modern TMP department is 100-200 t/day. This can be considered as a high volume source of mechanically pre-treated feedstock for new bioproducts. A paper mill producing simultaneously its conventional product *i.e.* paper and 'TMP feedstock' for biorefining then offers a pertinent industrial case for a study with a concept model. In this paper five different process concepts without the reject refining stage were compared for evaluating the techno-economic feasibility of the simultaneous production of printing paper and biorefinery feedstock. The processes were simulated with the steady state process simulator Balas (a software product of VTT) [2] to assess material and energy balances for each concept. Their manufacturing costs were compared by computing the variable costs. The potential bioproducts refined from the TMP reject fraction were yet excluded from this study.

Approach for techno-economic process simulation

A process model of modern paper mill [3] consisting of TMP department, recycled fiber deinking process, paper machine, power plant, and effluent treatment served as a reference (Figure 1) in this comparative study. The reference mill was producing 1335 tons of light weight coated (LWC) printing paper on daily bases. The main raw materials used in the paper production were mechanical fiber (TMP), chemical fiber (bale kraft), filler and coating pigments. Recycled fiber (RCF) was not a paper component in the reference case (RCF line was not operational). However, the RCF line was part of the model to enable the simulation of alternative process concepts.

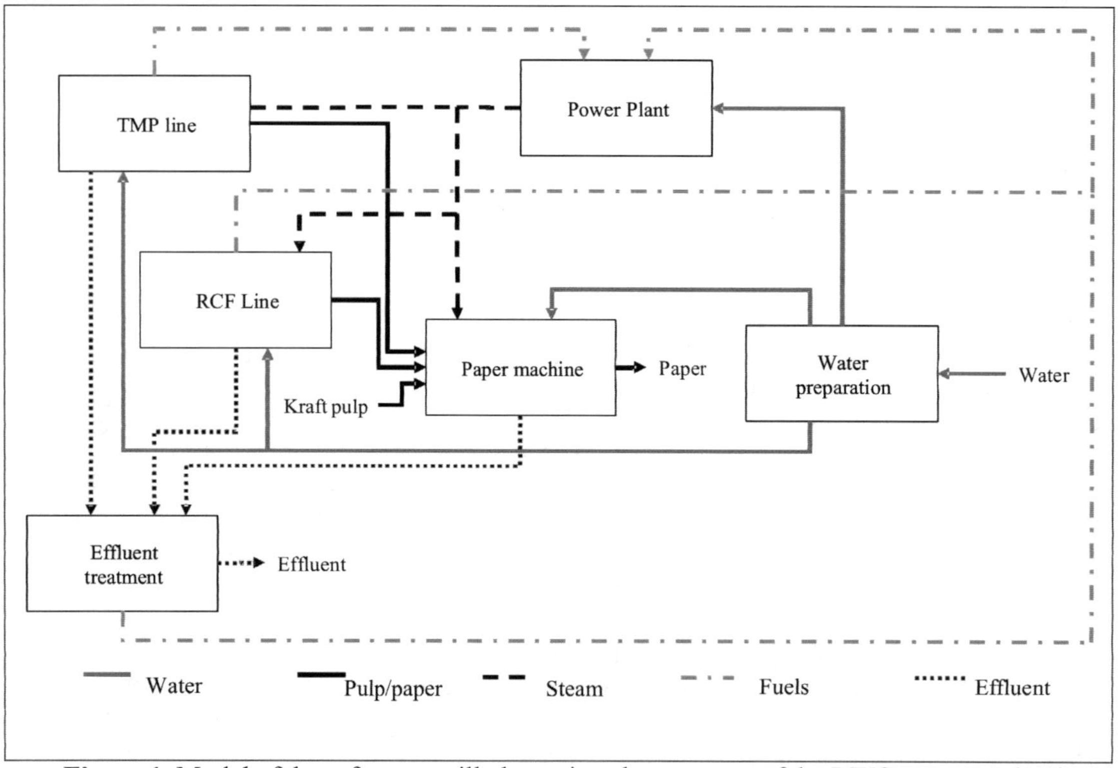

Figure 1. Model of the reference mill: the main sub-processes of the LWC paper production.

A general problem of process analyses, particularly when evaluating emerging technologies, is the confidentiality of process and price data. Yet, by using generic key data from available sources this problem may be circumvented and at least a consistent comparison of alternatives can be produced. For considering the economic aspect in this study, unit prices for the variable costs (raw materials, electricity, oil etc.) were obtained from public databases such as Foex and Eurostat (Table 1), which were applied on the simulated process values by using MS Excel interface. An estimated average price level of January 2009 was then adapted for all simulation alternatives.

Table 2. Unit prices for the variable costs considered in the simulations, based on average price level on January 2009 [4,5,6,7,8,9].

Cost component	Cost	Unit	Notes
Northern bleached kraft pulp (softwood, long fibre)	454	[€/ ad-t]	90% d-c
Bleached kraft pulp (hardwood, short fibre)	418	[€/ ad-t]	90% d-c
Wood (softwood) at the factory	33.7	[€/ t]	42% d-c
Sorted graphic paper for deinking	80.4	[€/ad- t]	
Filler	110	[€/ dry-t]	
Water	0.05	[€/ m3]	
Electricity	50.0	[€/ MWh]	
Natural gas	23.4	[€/ MWh]	
Oil (heavy fuel oil)	0.36	[€/ kg]	
CO2 emission trading	15	[€/ ton CO2]	
Ash treatment cost	0.002	[€/t]	
Effluent treatment costs:			
COD in	0	[€/t COD]	
Flow	0.30	[€/m3]	
COD released	0	[€/t COD]	

Process alternatives

The share of the reject fraction of TMP fibers is 35.9 % and the consistency 35 % (after reject screw press). Because there is no pulp entering the reject refining in the new concept, this stage can be uninstalled. The process model of the TMP line is presented in figure 2. Specific energy consumption of the main and reject refining as well as the option for the new feedstock concept are shown.

The fiber loss in the paper product caused by the removal of TMP reject fraction corresponds to 19 % of total fiber loss. This loss can be compensated using other raw material components or could be ignored (hypothetical and causing deterioration in the strength properties of the paper without new improved paper forming technology).

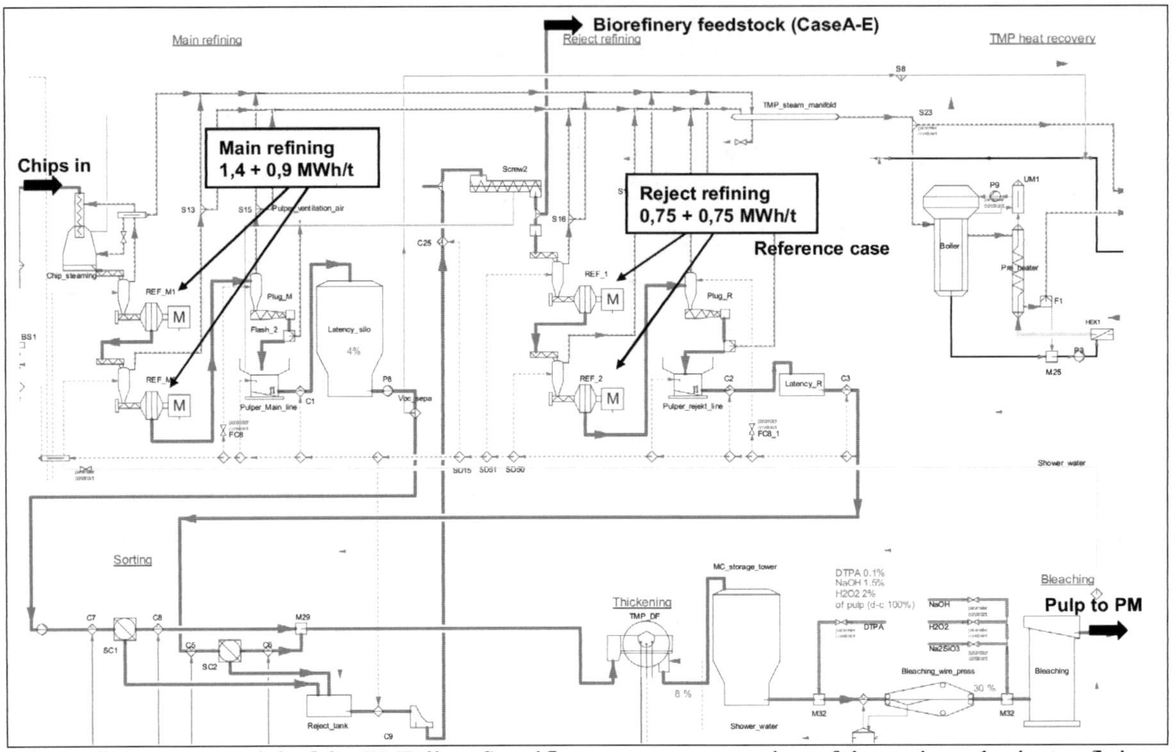

Figure 2. Process model of the TMP line. Specific energy consumption of the main and reject refining, and the option for the new feedstock concept are shown in the figure.

Altogether five different process concepts without the reject refining stage (cases A-E) were simulated as listed below. The TMP reject pulp was conducted out of the process in each alternative. In three cases (B, D, E) the per tonne based paper production was kept constant (the same as with the reference) and the pulp removed was compensated either with increased wood intake or with recycled or chemical kraft pulp. In two cases (A, C) the paper production was assumed lower than in the reference (either thinner paper or less paper) to adjust with the decreased share of TMP pulp.

The net changes in the cost structures (reference mill as comparison basis) were computed for each alternative as cost effects. They were allocated both for tonne of paper produced and for tonne of TMP reject pulp produced. Thus, a unit cost for the biorefinery feedstock was obtained for each process option. For the reference LWC paper mill the costs considered here covered ca. 75% of manufacturing costs even though labor and chemicals were not included.

Also, a process concept (case F) with the acceptance of reject fraction as such for the paper production was simulated. In this alternative the reject refining was not needed and the mill integrate was producing only paper without the feedstock production. The case F is rather hypothetical and describes a condition with lower quality requirements of the paper product.

Reference mill

- Intake of spruce chips to TMP line ca. 1360 t/day (at dry content of 42 %)
- 35,9% of TMP is refined at reject refining stage
- Specific energy consumption (SEC) at reject refining 0.75+0.75MWh/t
- Daily based paper production 1335 t, grammage 70 g/m^2

Case A: Less paper

- Daily based paper production (tons) is decreased to adjust with the decreased share of TMP pulp (reduction of PM speed, paper grammage 70 g/m^2)
- The share of the other paper making raw materials such as kraft, filler and coating pigments is decreased to adjust the decreased paper production, and *e.g.* the flow rate of PM shower waters is connected the production

Case B: More wood

- Increased wood intake to compensate the share of TMP reject fraction (feedstock)
- The share of the other paper making raw materials were kept constant
- No changes in paper machine (same speed and production as with the reference)

Case C: Thinner paper

- Paper grammage decreases to correspond the removed TMP share
- No change in PM speed
- Wood, bleached kraft pulp and filler intake remains constant in tonnage basis, but due to reduced TMP pulp share in the paper, their specific intake values increases

Case D: More kraft

- Chemical fiber (bale kraft) usage is increased to correspond the removed TMP share

Case E: Added recycled paper

- Recycled fiber is used to correspond the removed TMP share
- RCF process line is a part of the mill integrate

Case F: TMP reject to PM

- TMP reject pulp is accepted to the paper production
- No simultaneous biorefining feedstock production

Results and discussion

A unit cost for the TMP reject pulp (feedstock), corresponding the per tonne income needed from the biorefinery products (here unspecified) was obtained for each process alternative (case A-E). Figure 3 indicates the amount of feedstock available in each case - grey squares, while the cost effect to be covered with the new products are shown in black. The unit cost varied from 60 to 360 €/ad-t feedstock. True feedstock production cost in the value chain is higher due to other possible manufacturing costs (direct labor, manufacturing overheads) and fixed costs (*e.g.* investments). Also, feedstock selling price needs to cover the requested profit margin.

Figure 4 presents the paper production in each case, and the cost effect (variable costs) of the reject pulp computed per tonne of paper produced. The distribution of the variable costs per tonne of paper is illustrated in Figure 5. Main cost factors in each alternative are discussed below.

Case A: Less paper

The specific purchased electricity per tonne of paper increased since less paper was produced while the main line refining consumed as much electricity as before (in the reference case). By-passing the reject refining did not cover this effect. Wood purchasing did not change, which led to increase in wood specific consumption per tonne of paper since less paper was produced. Usage of the other raw materials per tonne of paper followed the paper production rate, and did not cause changes in their specific consumptions.

Case B: More wood

Paper production rate was kept constant, but the wood intake was increased. The greater flow through the main line refining was needed to produce the same amount of paper. The specific electricity consumption was thus increased. By-passing the reject refining was not enough to cover this effect. Also, more process steam was generated and it was utilized at the paper machine, so that the need of low pressure steam produced at the power plant was diminished. Thus, in the power plant more steam was conducted to a condensing turbine causing little compensation on the electricity purchase. As in case *Less paper*, there were no changes in the specific consumptions of other raw materials.

Case C: Thinner paper

Production of thinner paper saves raw materials and utilities. Case *Thinner paper* was evaluated both on tonnage and on surface area basis. The latter considers the functionality of paper surface area.

Tonnage basis: Specific electricity savings were obtained since the share of pulp and filler in the paper recipe were increased, replacing the energy intensive TMP pulp.

Surface area basis (not presented in Figure 4): There were no significant changes in the amount of raw materials per surface area of paper, since the produced surface area remained almost constant (19.15 km^2/day). The paper was just thinner. Remarkable savings were obtained in purchased electricity since the same surface area of paper was produced with less energy intensive paper recipe (less TMP pulp) and without the reject refining. Savings in refining electricity reduced the process steam generation, increasing the need of low pressure steam from power plant, and thus also increased the oil usage (as auxiliary fuel).

An important factor is the pricing of the thinner paper. If the price can be kept at the same level as for the original reference paper, assuming that the quality is maintained, savings can be reached despite the fact that biorefinery feedstock is produced simultaneously from the same wood intake as with the reference mill. However, reduction in paper price should be expected, due to the quality decrease. If pricing remains on tonnage basis, price increase is required since more functional area is provided per tonne of paper.

Case D: More kraft

The share of TMP reject pulp was compensated with the bleached kraft pulp to maintain the paper production constant. Purchase of kraft pulp is clearly seen in Figure 5. Specific electricity consumption decreased due to the increased share of bale kraft pulp in the paper recipe, thus providing savings in electricity. Again, process steam generation decreased along with the decreased electricity consumption in refining, however finally increasing the additional usage of auxiliary fuel (oil). Cost effect was quite negative mainly due to purchase of kraft pulp.

Case E: Added recycled paper

The share of TMP reject pulp was compensated with recycled pulp produced from sorted graphic paper at RCF plant. Purchased electricity and auxiliary fuel consumption behaved similarly as in the case D. Main cost effect is caused by purchase of sorted graphic paper. Total cost effect was competitive when compared with other alternatives.

Case F: TMP reject to PM

TMP reject pulp was accepted to paper production. Purchased electricity reduced due to by-passing the reject refining. This effect was comparable with alternatives *More kraft* and *Added RCF.*

Reduced electricity consumption caused additional usage of auxiliary fuel. Total cost effect was favorable.

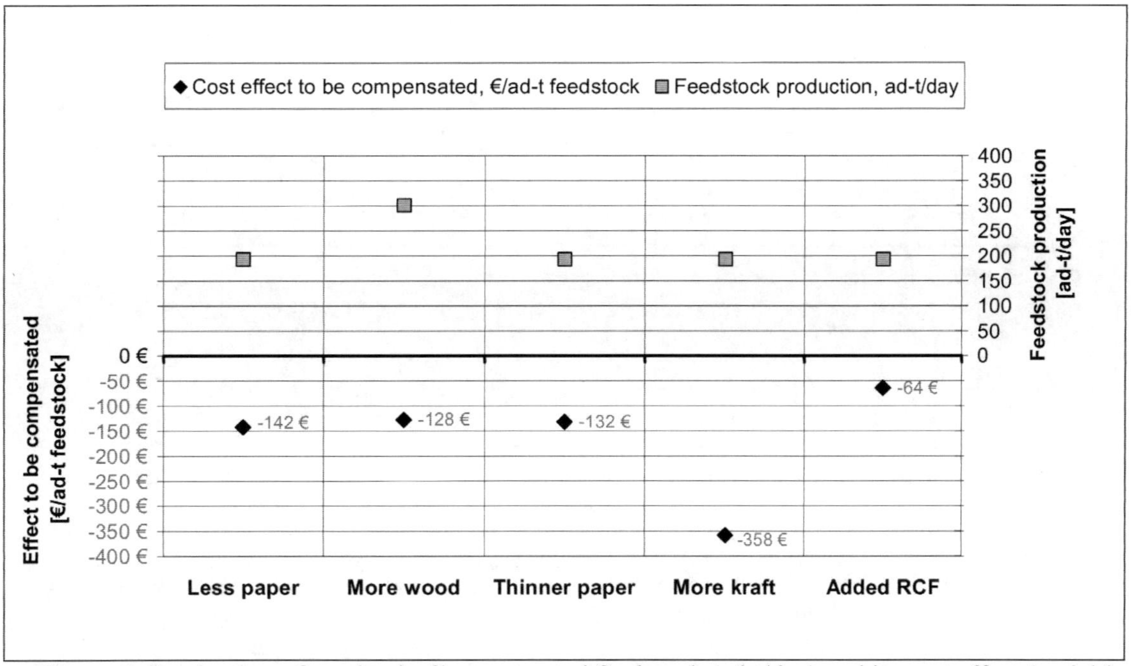

Figure 3. Production of mechanically pre-treated feedstock (ad-t/day) and its cost effect (variable costs, - sign means cost increase) to be covered (€/ad-t feedstock) in different process alternatives.

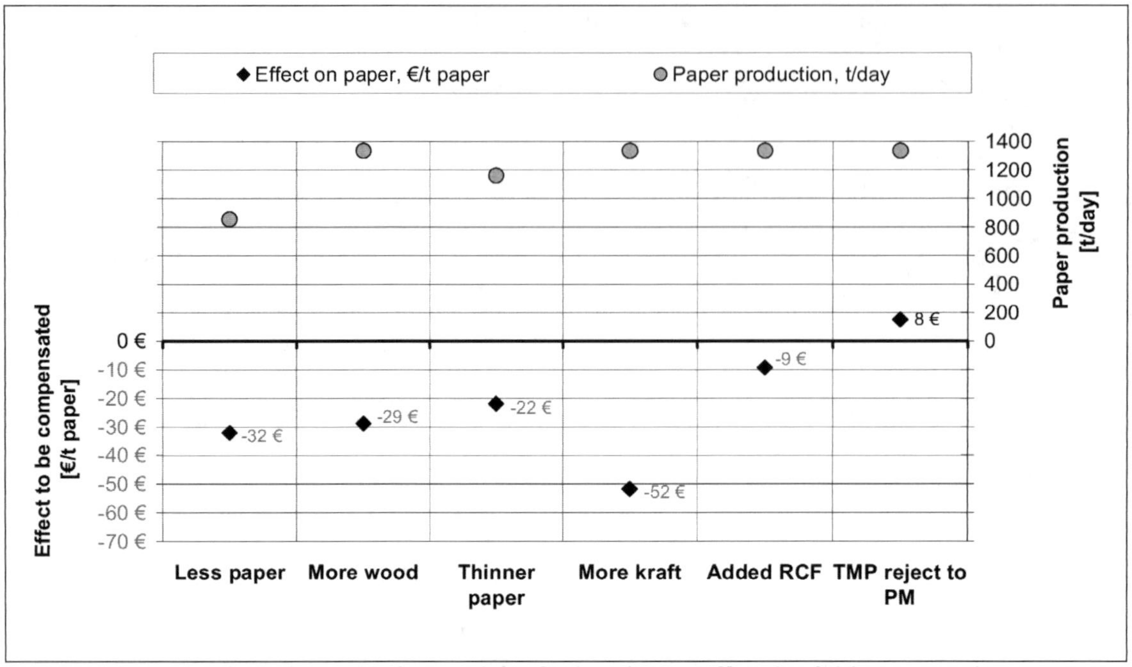

Figure 4. Production of paper (t/day) and the feedstock cost effect (variable costs, - sign means cost increase) to be covered expressed as per tonne of paper in different process alternatives.

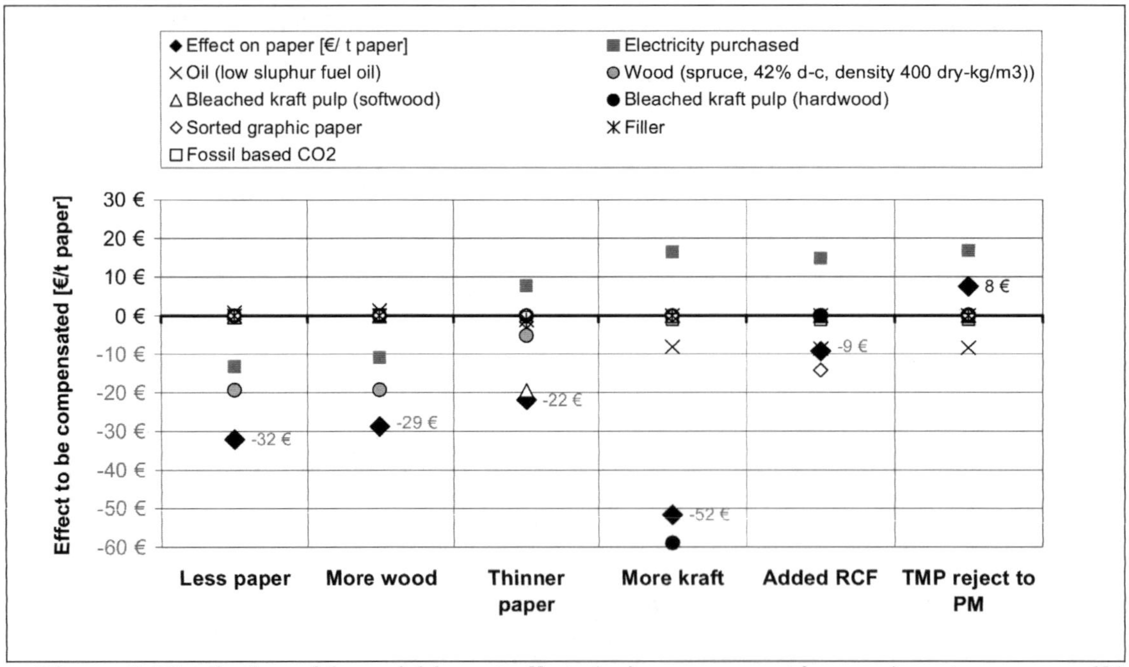

Figure 5. Distribution of the variable cost effects (- sign means cost increase) per tonne paper (€/t paper) in different process alternatives.

Conclusions

A simulation based techno-economic feasibility study of a paper mill producing simultaneously printing paper and mechanically pre-treated feedstock for new bioproducts was performed. Removing the reject refining stage allows for a major feedstock stream from the thermomechanical fraction of papermaking pulp for optional biorefining. The fraction removed may be compensated by reducing paper grammage, its production volume or by increasing other feedstock fractions. The viability of the process renewal alternatives depends on their technical feasibility, price levels and market conditions. The simulation suggests that addition of feedstock, either as wood or recycled fiber is necessary to compensate the lack of the TMP reject in paper production under the prevailing market conditions. With this compensation the combined effect of cost changes and biorefinery feedstock sales appear clearly positive and provide resources for additional investments. Yet, with a changing market situation the production of less or thinner paper together with new bioproducts may emerge as interesting alternatives in the future. For such cases, further studies on the aspects of both paper - and bioproduct qualities will be necessary.

References

1. Finnish Forest Industries Federation. 2009. Metsäteollisuustuotteiden tuotantomäärät Suomessa (Forest industry production rates in Finland) Metsäteollisuuden vuosi- ja nelj.vuositilastot (annual and quarter statistics). (http://www.metsateollisuus.fi) (http://www.forestindustries.fi)
2. www.balas.fi
3. Hakala, J., Manninen, J. Ruiz, J., Kamml. G, Lundström, A., Samuelsson, Å. 2005. D6.1.1 Description of the reference paper mills for ECOTARGET project. EU project Ecotarget deliverables 500345. (http://www.ecotarget.eu/news/ECOTARGET%20Technical%20deliverable%20D6.1.1.pdf)
4. Ekorosk Oy. 2009. Käsittelymaksut 1.1.2008 alkaen -ash (handling fees starting 1.1.2008 -ash (http://www.ekorosk.fi/en/default.html)
5. Eurostat. 2009. Natural gas prices for large industrial standard consumers. (http://epp.eurostat.ec.europa.eu/portal/page/portal/eurostat/home)
6. Eurostat. 2007. Electricity prices for large industrial standard consumers (http://epp.eurostat.ec.europa.eu/portal/page/portal/eurostat/home)
7. Finnish Oil and Gas Federation. 2009. Öljytuotteiden kuluttajahintaseuranta (Oil based products consumer price statistics) (http://www.oil-gas.fi/eng/)
8. Central Union of Agricultural Producers and Forest Owners (MTK). 2009. Kuitupuun hankintahintakehitys, viimeiset 52 viikkoa (pulpwood purchasing price development, the latest 52 weeks). (http://www.mtk.fi/en_GB/)
9. FOEX Indexes Ltd. 2009. (http://www.foex.fi)
   PIX recovered paper indexes Europe
   THE PIX pulp Europe benchmark indexes
   THE PIX paper Europe benchmark indexes

(467f) Integrated Production of Succinic Acid with the Separative Bioreactor

Yupo J. Lin, Cynthia S. Millard, Saurav Datta, Seth W. Snyder and Edward J. St.Martin, Argonne National Laboratory, Lemont, IL

Ion-exchange resins (IXR) have been widely used by industry for catalysis as well as an absorbent for removal of dilute salts from process streams. Electrodeionization (EDI), a combination of ion-exchange resin technology and electrodialysis membrane separation, has further enhanced the utilization of IXR and can significantly reduce capital and operating costs compared to conventional ion-exchange column. Currently, commercial EDI processes are used primarily to produce ultra-pure water for the semiconductor and pharmaceutical industries. ANL researchers have successfully improved on conventional EDI technology by developing "IXR wafers" to immobilize the IX resins into a porous substrate. The novel resin wafer (RW) enables application of EDI to a wider range of applications compared to conventional loose resin EDI processes. The resin wafer is a porous matrix of resin beads immobilized by polymer. Using this technology, ANL has developed several process implementations that take advantage of EDI for bioprocessing as well as bioconversion. The chemical industry is actively seeking bio-catalytic replacements for traditional chemical catalysis reactions. In most cases, separation represents a significant fraction of the overall cost of the process. Therefore, bio-catalytic process designs that focus on maximizing product concentration, limiting impurities, and reducing downstream processing steps are needed. To help meet these needs we developed the EDI Separative Bioreactor to integrate up-stream bioreactor and down-stream product separation into one unit. This membrane-based Separative Bioreactor enables simultaneous bio-catalytic production (fermentations or enzymatic reactions) of organic acids and immediate product separation. Direct capture of organic acid products from bio-transformations offers two immediate benefits: prevention of reactor acidification and avoidance of product inhibition. Argonne also observed that rapid organic acid capture during production offered a new way to control reactor pH without the addition of buffers for pH control. In this presentation, we will discuss an integrated separative bioreactor designed for pure succinic acid production. This system consist a CO2 capture reactor, a succinic acid fermentation and in-situ pure succinic acid recovery using resin-wafer electrodeionization (RW-EDI). The fermentation is setup in a temperature-controlled, stirred bioreactor using a succinic acid producing E.coli mutant developed at Argonne which converts glucose to succinic acid using one mole CO2 per mole of glucose to produce one mole of succinic acid and other products. CO2 is provided from the CO2 capture reactor in the form of carbonate which is formed by bubbling CO2 gas through a basic solution. This solution is used to neutralize the succinic acid fermentator and provide the CO2 for the reaction. The fermentation broth was constantly sent to the RW-EDI device to separate succinic acid from the solution and recycle the base to the CO2 capture reactor. This work was supported by DOE Office of Biomass Program.

See more of this Session: Innovations in Bio-Refinery Processes
See more of this Group/Topical: Process Development Division

(505a) BioProcess Intensification in the Production of Starch-Degrading Enzyme α - Amylase

Dzun Noraini Jimat[1], C. Harwood[2] and G. Akay[1], (1)School of Chemical Engineering and Advanced Materials, Newcastle University, Newcastle upon Tyne, United Kingdom, (2)Institute for Cell and Molecular Biosciences, Newcastle University, Newcastle upon Tyne, United Kingdom

In this study we report the development of a novel microbioreactor which is able to promote the viability, proliferation, differentiation of organisms and enhance productivity by ca. 10 fold compared with the existing techniques. In this study we use *Bacillus subtilis* for the production of α -amylase. Initially, materials for the microbioreactor were developed and their efficiency in the production of α -amylase was determined. The influence of the medium composition on the α -amylase productivity of immobilized cells was also evaluated. The microbioreactor was then connected to the medium reservoir that continuously aerated and under controlled temperature of 37 degree Celsius. After 24h cultivation in continuous culture, the immobilized dormant morphological forms, *B.subtilis* spores were germinated and produced a desired enzyme known as α-amylase. Morphology and distribution of *B.subtilis* cells in the reactor were evaluated using Environmental Scanning Electron Microscopy as well as Transmission Electron Microscopy. The productivity was evaluated as a function of microbioreactor properties and compared with the existing technology. We conclude that the microbioreactors developed in this study promote the viability, proliferation, differentiation of the cells as well as metabolic activity. Detailed results will be presented.

See more of this Session: Advances in Process Intensification III
See more of this Group/Topical: Process Development Division

(505b) BioProcess Intensification In the Production of Bioethanol Using *Saccharomyces Cerevisiae* W303

Wan Salwanis Wan Md Zain, School of Chemical Engineering and Advanced Materials, School of Chemical Engineering and Advanced Materials, Newcastle University, Newcastle, United Kingdom, Janet Quinn, Institute for Cell and Molecular Biosciences, Newcastle University, Newcastle upon Tyne, United Kingdom and Galip Akay, Chemical Engineering and Advanced Materials, Newcastle University, Newcastle upon Tyne, United Kingdom

Fermentation is a very well established and very important technology, especially due to the emerging new applications in bioethanol production from lignocellulosic biomass. Bioethanol production from biomass through fermentation can be achieved in small scale distributed plants provided that the process is intensified so that the burden of 'economies of scale' is not present. We recently developed a BioProcess Intensification technique for fermentation and applied to the conversion of glucose to ethanol using *Saccharomyces cerevisiae* W303. The technique is based on the development of materials for the construction of novel continuous bioreactors with the aim of achieving 10-100 fold enhancement. We investigated the effects of materials' biochemical structure, reactor design and processing conditions such as flow rates, substrate and nutrient concentrations. Productivities were compared with the reported batch processes. The results will be presented in detailed.

See more of this Session: Advances in Process Intensification III
See more of this Group/Topical: Process Development Division

Comparison of Advanced Oxidation Technologies

Muhammad Suleman Tahir[1, 2], Shahid Raza Malik[2], Mahmood Saleem[3], P. Letonja[1], Matthaus Siebenhofer[1]
[1] Department of Chemical Engineering and Environmental Technology, Graz University of Technology, Austria
[2]NFC Institute of Engineering & Fertilizer Research, Faisalabad-Pakistan
[3] Institute of Chem. Eng. & Technology, University of the Punjab, Lahore-Pakistan.
m.s.tahir@student.tugraz.at

Advanced oxidation processes AOPs cover a variety of technologies and combinations of. The basic principles of technologies are of physical/mechanical nature such as UV-irradiation, sonication or corona discharge and chemical nature such as ozone treatment, hydrogen peroxide application or oxygen admixture. The oxidizing power is achieved by admixture of oxidizers to the effluent and activation by either chemical or physical measures such as the Fenton processes or several UV-radiation processes or combination of chemical and physical measure as provided by corona discharge.

The major obstacle for successful industrial application of advanced oxidation processes is caused by the composition of effluents. Except fresh water preparation effluents from any production source are rarely specified as contaminated by a specific constituent only. Overwhelmingly effluents are composed of a diffuse organic and/or inorganic basic load and dilute target constituents. Latter actually need to be addressed in application of advanced oxidation processes. As a consequence treatment may be very costly or even too expensive because cost of treatment is determined by the effluent matrix and not by the target constituent.

With these limitations in mind a selection of advanced oxidation processes was investigated and compared, based on the treatment of synthetic effluents spiked with test constituents. The experimental part of technology evaluation considered the Fenton reaction, the Photo-Fenton Reaction, UV-H_2O_2 oxidation, selected photocatalytic processes, anodic oxidation and corona discharge/ozone treatment. Comparison of technologies was carried out with phenol, acetone and EDTA. Investigations considered degradation as well as TOC-depletion. Comparison was carried out by determining the specific energy demand per mass unit of constituent and peripheral process specifications such as fouling.

Beside cost evaluation, technology selection for practical application of AOPs also has to consider maintenance of equipment which is governed by fouling control. Anodic oxidation processes for example are fouling sensitive. Fouling prevention mainly depends on the setup of equipment. Fouling prevention of photochemical reactors needs separation of the radiation source from the effluent, limiting the quantum yield due to scattering. Several Fenton reactions are sensitive to process poisoning by adsorption, precipitation or formation of stable complexes or overdosing of Fe^{2+}. Commercially available methods for effluent treatment may need costly chemicals and infrastructure which increases operational costs as well as the chemical impact on the environment. Degradation of organic pollutants with chemical processes / advanced oxidations has been extensively investigated by various authors [1-5]. Major photochemical advanced oxidation processes even with combination of some chemicals are discussed in Table 1.

Table 1: Representative advanced oxidation processes.

Combination and source of process	**Process name**
UV	photolysis
UV/H2O2	UV-peroxide process
UV/O3	Ozonolysis
UV/O3/ H2O2	PEROXON-process/ technology
Fe2+/ H2O2	FENTON-Reaction/Process
UV/Fe2+/ H2O2	Photo-Fenton process
UV/TiO2	photo catalysis

Improvement of technologies for wastewater and off-gas treatment processes is of high priority. Development of new technologies for the efficient removal of hazardous pollutants from effluents as well as off-gas is a major task. Corona discharge, in principle a gas phase process, enables separation of the power source from the effluent on account of the operation principle. When applicable, corona discharge treatment may therefore be a serious alternative or extension to radiation based processes. While state of the art processes are well explained corona discharge reactors may need some attention in getting the mode of operation explained. Therefore the specific characteristics will be discussed.

Experimental set up

Figure 1 Shows the experimental set up of the corona test facility

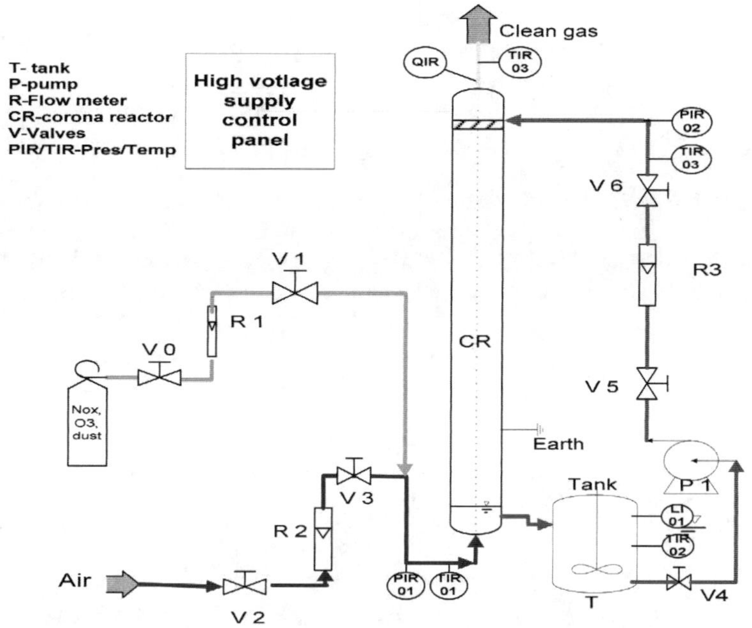

Figure 1: Experimental set up of the test facility; Counter electrode: stain less steel with water film; tube diameter: 66 mm; tube length: 1550mm; Discharge electrode: stainless steel (brush type electrode length: 700mm, brush diameter 8mm and brush wire diameter 0.15mm).

According to Fig.1 air is fed at the bottom and rinsing water laden with constituents is continuously fed on top of the reactor and a required water film thickness is developed. For analytical purposes water samples are taken at the bottom of the reactor or circulation tank (T). To quantify the change in TOC, the samples are analyzed with "TOC-5000A SHIMADZU". 0.5g/Liter of acetone or EDTA was mixed with sodium sulphate for conductivity purposes and with deionized water. Water circulation was set at 50 litres per hour with flow meter (R3) and the air flow rate was set at 0.5 m3/h with flow meter(R2). Air exits at top of the reactor and air constituents are also analyzed.
Corona discharge was investigated at ambient conditions.

Results

Figure 2 shows the shape and design of brush-type discharge electrodes used for investigations of current voltage characteristics

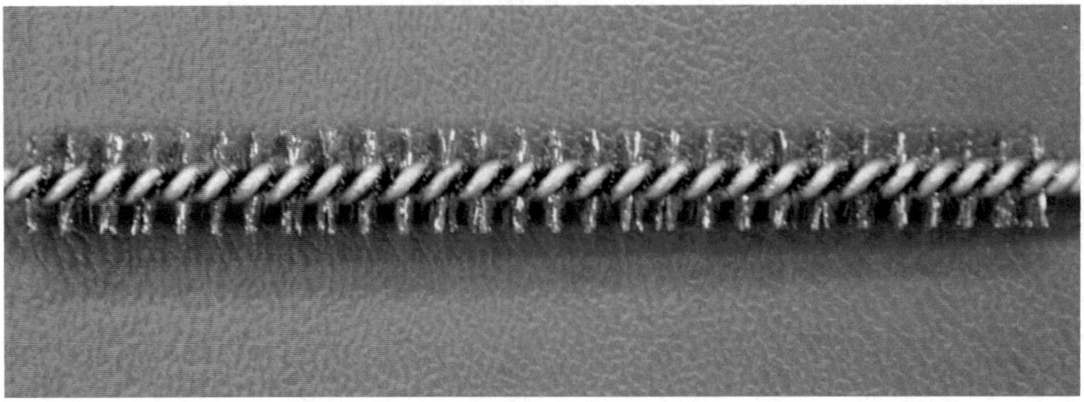

Figure 2: Brush-type discharge electrode (diameter=8mm, Total Length=800mm, active length =700mm, brushing wire diameter=0.15mm, central wire diameter=2.7mm); material of construction: stainless steel

Figure 3 shows the image of the discharge characteristics of the investigated electrodes.

Figure 3: Corona discharge for brush type discharge electrode, the image corresponds to 21kV operation voltage at ambient conditions.

Figure 4 shows the experimentally obtained current/voltage curves under various gas flow conditions. For given operation voltage the corona current significantly depends on gas moisture and slightly depends on the gas velocity.

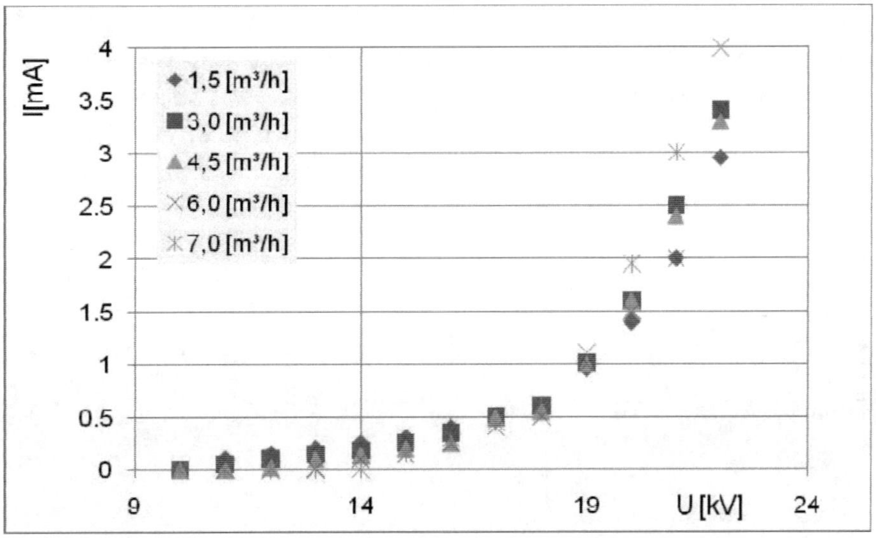

Figure 4: Comparison of current/voltage curves for brush type discharge electrode for different air flow conditions

Figure 5 shows ozone generation with brush type discharge electrode for stagnant air and gas flow.

Ozone formation rapidly increases at low corona current. It passes a maximum and deplets at elevated corona current because ozone undergoes intermolecular destruction.

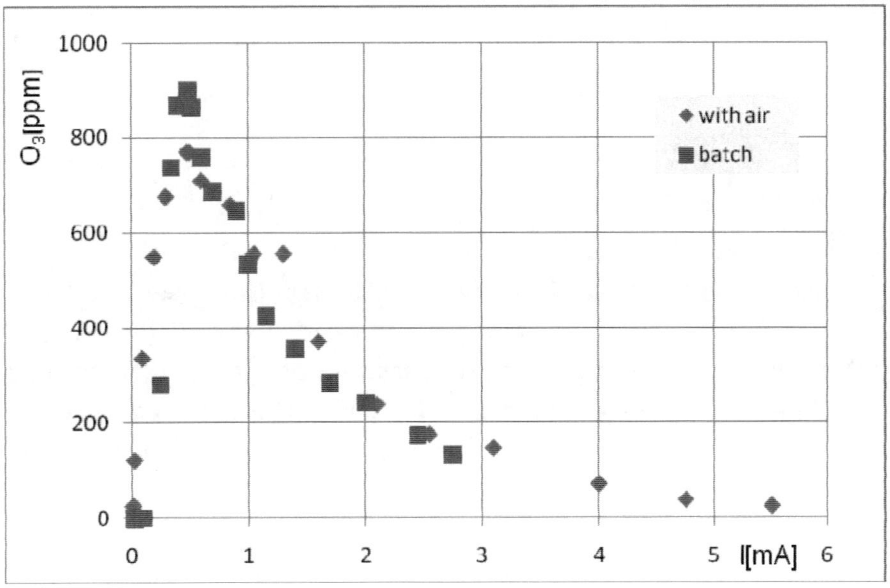

Figure 5: Comparison of the ozone generation with a brush-type discharge electrode for stagnant air and air flow rate of 1 [m^3/h] at ambient conditions (T, P: ambient)

Ozone may either attack constituents by forming ozonides or it may form OH-radicals according to the processes mentioned in Table 1. To demonstrate the oxidation power of ozone the circulation water of the corona discharge reactor was spiked with acetone. The TOC-value was determined from samples of the circulation tank. **Figure 6** shows the comparison of acetone degradation in terms of TOC depletion with corona discharge and UV-irradiation.

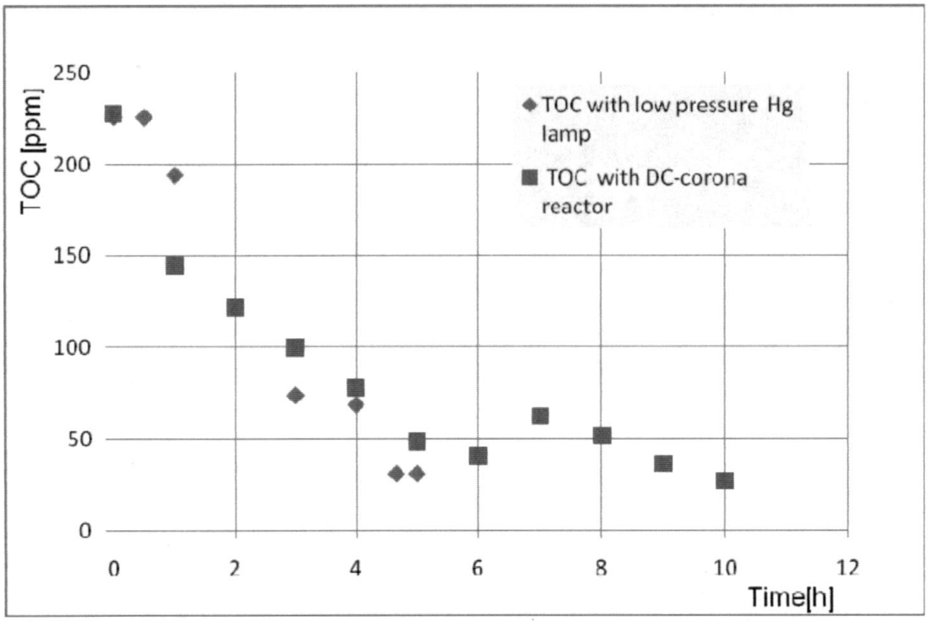

Figure 6: TOC depletion for acetone oxidation in the rinsing water of the corona discharge reactor. Depletion is compared with UV-irradiation of oxygen saturated effluent.

Summary

The experimentally obtained corona current with brush-type discharge electrodes is significantly higher than observed for state of art electrical discharge in air with wire-type electrodes [6] and it produces significant amounts of ozone. Compared with ozone generation according to state of the art processes the specific energy demand is similar. Opposite to operation of ozone generators the gas phase does not need any pretreatment. The mechanism of ozone generation was investigated. Ozone generation was quantified over a wide rang of corona current. Ozone formation passes a maximum and will deplete with increasing corona current because of intermolecular interaction. Actually ozone formation and degradation of generated ozone is a simultaneous reaction occurring in the corona reactor. At low corona current ozone formation is the dominating reaction. High energy discharge directly attacks the generated ozone to disintegrate it into molecular oxygen plus dissociated or reactive oxygen. As a result decreased ozone concentration is observed at increased corona current.

Experimental investigations were conducted to investigate the effect of stagnant air and air flow on corona current and ozone formation over a specific range of applied voltage.

Experiments were conducted to investigate the degradation of various organic constituents with a corona reactor. The effect of corona discharge on TOC depletion is comparable with the effect of low pressure Hg Lamp UV irradiation. The specific energy consumption of corona discharge reactors is favorable over irradiation processes because of the incomparably higher current efficiency.

References

[1] Thomas Oppenlander, Photochemical purification of water and air, page 5-6

[2] Taek-Soo Kim, Removal of organic impurities in an aqueous solution using uv irradiation produced from high and low pressure hg lamps. Industrial & Engineering Chemistry Research, 46(14):4799–4810, July 2007.

[3] Ranajit K. Talukdar, Reaction of hydroxyl radical with acetone. 2. Products and reaction mechanism. The Journal of Physical Chemistry A, 107(25):5021–5032, June 2003

[4] Carolina Baeza1. Degradation of EDTA in a total chlorine free cellulose pulp bleaching effluent by uv/h2o2 treatment. Journal of the Chilean Chemical Society version On-line ISSN 0717-9707, 1(52), 2007.

[5] S. Chitra. Ultrasonic treatment of liquid waste containing EDTA. Journal of Cleaner Production, 12(4):429–435, May 2004.

[6] M. S. Tahir. Optimization of Electrostatic precipitations with brush type discharge electrode, 6th Minisymposium Verfahrenstechnik , 2010

(505d) Effect of Relative Volatility Ranking On the Design and Operation of Batch Reactive Distillation Systems

Yu-Lung Kao and Jeffrey D. Ward, Department of Chemical Engineering, National Taiwan University, Taipei, Taiwan

Batch reactive distillation (BREAD) is an attractive process alternative which combines the advantages of reactive distillation and the flexibility of batch processes. There are three basic batch distillation column configurations: (1) Conventional batch distillation column (CBD): Feeds are charged at the bottom and products are taken out at the top. (2) Inverted batch distillation column (IBD): Feeds are charged at the top and products are taken out at the bottom. (3) Middle-vessel column (MVC): Feeds are charged in a middle-vessel of the column and products are withdrawn at the top and the bottom.

Tung et al. [1] used an ideal system to study the effect of relative volatility ranking on the design of column configuration for continuous RD systems. A reversible reaction A+B <=> C+D is considered and constitutes a quaternary system which has 24(4!) possible rankings according to the relativity volatility among reactants and products. They further grouped these 24 possibilities into 6 (24/2!/2!) distinct categories since the two reactants and two products are interchangeable. In this work, these 6 distinct volatility rankings are applied to the 3 basic BREAD column configurations to study the effects of relative volatility ranking on BREAD process design. The process design focuses on the choice of column configuration and the optimal operating policy including when and where to collect products and off-cuts and the corresponding reflux profile. The designs are optimized based on the batch capacity (CAP), defined as the total quantity of products meeting the purity specification produced divided by the total batch time [2]. The results indicate that if one of the reaction products is the lightest key, CBD or MVC performs the best. On the other hand, if one of the reaction products is the heaviest key, IBD or MVC performs the best.

Furthermore, we investigate a new column configuration: a modified MVC in which the location of the reaction vessel is not exactly in the middle of the column. We consider the general case where the reaction vessel can be connected to the column at any point. If the reaction vessel is connected all the way at the bottom, we recover the CBD process whereas if it is connected all the way at the top, we recover the IBD. Finally, for each relative volatility ranking, we propose the most suitable column configuration and identify the operating policy with the highest CAP.

References

1. Tung, S. T. and C. C. Yu (2007). Effects of relative volatility ranking to the design of reactive distillation. *AIChE Journal* **53**(5): 1278-1297.

2. Luyben, W. L. (1971). Some Practical Aspects of Optimal Batch Distillation Design. *Industrial & Engineering Chemistry Process Design and Development* **10**(1): 54-59.

See more of this Session: Advances in Process Intensification III
See more of this Group/Topical: Process Development Division

(505e) Stripping Methanol From a High Viscosity System by a Rotating Packed Bed

Chia-Ying Chiang and Hwai-Shen Liu, Chemical Engineering, National Taiwan University, Taipei, Taiwan

With the trait of high mass transfer coefficient in an rotaing packed bed (RPB), the bottle neck of stripping methanol from viscous glycerol solution could be improved. The gas flow rate, liquid flow rate, rotor speed, liquid viscosity, temperature, and liquid concentration were taken into consideration as the operating variables to pursue better stripping ratio and mass transfer coefficient in the system. According to the experimental results of 2% and 8% (methanol/glycerol) data sets, as the rotor speed increased from 1200 to 2400 rpm, gas flow rate increased from 40 to 50 L/min, and temperature increased from 30 to 60°C, the stripping ratio could be raised to 3.4 folds. As considering the ratio of (gas flow rate/liquid flow rate) in 0.5% data, we could find that while the liquid flow rate decreased from 200 to 100 mL/min, there would be a higher ratio of (gas flow rate/liquid flow rate) and a higher capacity to remove methanol from the liquid phase to gas phase. Thus the stripping efficiency could be found to be doubled. On the other hand, equilibrium constant became triple by increasing temperature from 30 to 60°C, and 2~3 times of stripping efficiency was obtained. In this research, a correlation applicable to both aqueous and viscous systems is presented.

See more of this Session: Advances in Process Intensification III
See more of this Group/Topical: Process Development Division

(505f) Electrodeposition of Aluminum in Ionic Liquid Under High Gravity Field

Lianbin Xu, Pengyuan Zhang, Guangtao Tang, Jiezhen Cui and Jian-Feng Chen, Research Center of the Ministry of Education for High Gravity Engineering and Technology, Beijing University of Chemical Technology, Beijing, China

Aluminum was electrochemically deposited on the stainless steel electrode in $AlCl_3$-BMIC ionic liquid under high gravity field at 80℃. The effects of high gravity field on the aluminum deposition rate and current efficiency were studied, and the aluminum deposition layers were characterized by scanning electron microscopy (SEM) and energy-dispersive X-ray spectroscopy (EDS). The results revealed that the current efficiency ranged from 70-90% when the electrodeposition current density was in the range of 15-30 mA/cm^2. The deposition rate of aluminum increased with the enhancement of high gravity coefficient, *G*, and when $G = 162$, the deposition rate was 1.5-2.5 times of that of normal gravity. Under *G* values of 80 and 104, with a current density of 18 mA/cm^2, the current efficiency for aluminum deposition was 83.2% and 84.5% respectively. The SEM micrographs of aluminum coatings obtained in constant current experiments showed that the dendrite aluminum was avoided for electrodeposition under high gravity field. Homogeneous, uniform, dense, shining and tightly adherent aluminum coatings were obtained under $G = 104$.

See more of this Session: Advances in Process Intensification III
See more of this Group/Topical: Process Development Division

(539a) Reactive Scheduling for Ethylene Cracking Furnace System

Chuanyu Zhao, Chaowei Liu and Qiang Xu, Dan F. Smith Department of Chemical Engineering, Lamar University, Beaumont, TX

Thermal cracking is crucial for ethylene plants because the major product yields of an ethylene plant are mainly determined in this operation. The cracking furnace system follows a certain schedule to process all the coming feeds. In practice, because of the uncertainties hidden in the feedstock market and/or during the shipment, the lead time and the incoming quantity may not follow the expectation very precisely. Thus, it is better for the furnace scheduling conducted in a reactive way, such that various feeds from new shipments and existing inventories can be smartly allocated to different furnaces for processing to obtain the maximum production profit per unit time. The main challenges for this reactive scheduling study include: i) thermal pyrolysis process is a performance decay semi-continuous process, and thus a furnace has to be shut down after some processing time for decoking; ii) non-simultaneous decoking policy usually has to be ensured for mitigating the productivity upsets from the cracking furnace system; iii) the cracking product of ethane has to be recycled for a secondary cracking and usually needs a specific furnace for such purpose; iv) reactive scheduling has to seamlessly connect the previous schedule and the being determined schedule for taking care of different feeds from new shipments and existing inventories. Facing these challenges, this paper developed a novel reactive scheduling strategy, which can dynamically generate a reschedule based on the new coming feeds and the leftover feeds from the current schedule. A new MINLP model is also developed to maximize the profit per time under various stringent constraints. The efficacy of the study and its significant economic impact are demonstrated by case studies.

See more of this Session: Industrial Innovation in Process Design and Operations
See more of this Group/Topical: Computing and Systems Technology Division

(539b) Optimizing Design Specification for LNG Receiving Terminal Using Dynamic Simulation

Chul-Jin Lee[1], Youngsub Lim[2] and Chonghun Han[2], (1)Chemical & Biological Engineering, Seoul National University, Seoul, South Korea, (2)Chemical and Biological Engineering, Seoul National University, Seoul, South Korea

Since the demand for natural gas has been continuously increased due to its cleanness, the construction of LNG receiving terminal which imports and stores LNG in on-shore area also grows up over the world. Although the design specification for LNG receiving terminal needs to be adjusted to the individual characteristics of the terminals such as the whole length of the pipeline, the number of the LNG storage tanks and atmospheric temperature, the systematic approach to determine a design specification to reduce the operating cost seems to be absent. Such various characteristics can make effect on terminal operation, which leads to the change in operating cost including pump and compressor. Because an inadequate operation such as excessive recirculation flow rate or compressor movement brings about increase in operation cost, we need to precisely examine the design factors effecting on operating costs and optimize them for each terminal. In this work, we suggest the systematic methodology for optimal design specification to cope with the various characteristics of LNG receiving terminals. Proposed method contributes to minimize the operating cost in LNG receiving terminal.

See more of this Session: Industrial Innovation in Process Design and Operations
See more of this Group/Topical: Computing and Systems Technology Division

(539c) Recent Advances in Reactive Distillation

Philip Lutze, Department of Chemical & Biochemical Engineering, Denmark Technical University, Lyngby, Denmark, Rafiqul Gani, CAPEC, Department of Chemical and Biochemical Engineering, Denmark Technical University, Lyngby, Denmark, John M. Woodley, Department of Chemical & Biochemical Engineering, Technical University of Denmark, Lyngby, Denmark and Emmanuel A. Dada, ChemProcess Technologies, LLC, League City, TX

The integration of tasks in one unit has attracted considerable academic and industrial interest using synergistic effects of this integration as a potential means of process improvement. The most studied and applied operation in that field is Reactive Distillation (RD) comprising simultaneously chemical reaction and vapor (gas)-liquid phase separation in one column. RD has the ability to separate difficult separation mixtures in one column such as azeotropic or close boiling points mixtures as well as increasing the conversion of the reaction beyond its equilibrium and suppressing side-reactions through product removal into the non-reactive phase. This enables a number of potential process improvements such as lower capital and operational costs, a simplified flowsheet by replacing a series of separation steps, higher energy efficiency through usage of exothermic heat of reaction for evaporation, higher raw material efficiency, higher selectivity as well as improving safety due to usage of heat of reaction for evaporation.

Even though many applications are reported, design decisions have been made mostly case-based, so one may conclude that RD has not reached its full potential yet. Therefore, in order to support a wider implementation of RD in industry, a systematic framework for design/synthesis of reactive distillation systems is valuable.

This contribution presents recent industrial as well as academic research in application of reactive distillation and discusses the design decisions reported through application of a systematic design framework considering operational design and synthesis as well as potential selection of different (catalytic) column devices. First of all, a set of criteria/ rules is used in order to establish for which systems RD might be potentially advantageous. Next, a synthesis/ design framework is applied to generate alternative RD-designs from a superstructure; create models for each alternative; analyze/ evaluate the most promising alternatives to identify the best; and finally design experiments for verification of design.

The application of the model-based synthesis design framework will be illustrated through examples of industrial interest.

See more of this Session: Industrial Innovation in Process Design and Operations
See more of this Group/Topical: Computing and Systems Technology Division

Optimization for Energy Conservation of Crude Oil Atmospheric Distillation Columns by Modifying Vapour Feed Inlet Tray from Pre-Fractionation Train

[1]*Mehdi Arjmand, Graduate School of Chemical Engineering, Royal Institute of Technology (KTH), Stockholm, Sweden,* arjmand@kth.se

[2] *Luis Moreno, Department of Chemical Engineering and Technology, Royal Institute of Technology (KTH), Stockholm, Sweden,* lm@kth.se

[3] Longcheng Liu, Department of Chemical Engineering and Technology, Royal Institute of Technology (KTH), Stockholm, Sweden, lliu@kth.se

ABSTRACT

Optimization of a typical crude oil atmospheric distillation unit and reduction of energy conservation were carried out through modifying the implementation and change in the flash zone of the tower. Conventional procedure in such units involves combination of liquid and vapour product of the pre-fractionation train surge drum upon insertion to the tower. However it is theoretically illustrated and represented by simulation means that introducing the vapour feed into the upper stages of the distillation column separately can lead to an energy saving of 12.6% in the condenser duty, increased liquid to gas flow (L/G) at certain points of the column and hence a reduction in diameter and investment costs of new tower designs of approximately 0.7×10^6 \$/y. The proposal can be put into practice without the need of additional equipments or additional cost of difficult re-routing the streams. An industrial case study of a steady-state crude oil distillation unit is given by simulation provision of AspenHysys™.

Keywords: Optimization, Crude Oil, Atomspheric Distillation Column, Flash Zone, Simulation

I. INTRODUCTION

Crude Distillation Units (CDUs) are still the first major processing units in refineries used to separate the crude oils by fractionating the feed according to boiling point so that each of the processing units following will have feedstocks that meet their particular specifications [1, 2]. CDUs exhibit such a vital role that they are addressed as the heart of any petroleum refinery and one of the most complicated operations among separation processes [3]. However, the drawback in spite of their indispensable existence is the large energy consumption. Hence, new design methodologies for increasing overall energy efficiency are all aimed to reduce the energy requirements. Furthermore, the cost of a CDU comprises of the capital cost of the column, determined largely by the number of trays and diameter of the tower, and the operating costs, determined by the steam and cooling water requirements [4]. Since over half of a plant's energy demands for either process heat or cooling used in full operational distillation columns [5], optimizing the process with maximum energy recovery through more engineering time given to the design phase and retrofitting older operating plants to save or recover even minuscule amount of energy used can significantly affect the economic efficiency of a plant.

Despite the fact that the art of crude oil distillation has been practiced for many years [6,7], studies still continue to determine the best design procedures with great effort and scrutinised research put to place on optimization [8,9] which as a result has enormously improved our understanding and design for enhanced tower capacity and efficiency. To date, problems of determination of the minimum number of stages, optimum operating conditions and optimal locations of feed stage and side stream withdraws, are all of basics that have already been thoroughly addressed [10-13]. However, the problem encountered by all professionals with optimization of crude oil distillation is the proverbial complexity of the feed composition ranging from different type of hydrocarbons to inorganic compounds and also the multitude of products derived from the unit. Nevertheless, processes for treatment of such respect leads to complex equipments which further increases the considerable factors involved in optimization of crude oil distillation process. To this extent, the traditional procedure for optimization of crude oil distillation columns is carried out based on past experiences and case studies from steady-state simulation. Accordingly, several design alternatives have been proposed for more energy efficient considerations of operating distillation towers as some require additional capital investment like making use of feed/products in exchangers for heat recovery, column revisions, better insulation or column control. In contrast to these changes requiring small capital investment, there are other methods requiring higher expenditure as addition of vapour recompression, heat pump or side heaters/coolers and condensers [14].

Therefore, the search for an optimum solution has been on quest, and heuristic methods as feed splitting [15], introducing small changes in feed composition [3] and optimizing preheat train network of exchangers [8] have recently gained attention amongst researches. Although these proposals yield significant energy saving, not only majority of them require additional equipments which increase the investment costs, but also due to operational complexity of the crude oil, their application is limited only to few basic binary feedstocks or even require fundamental changes which may not be possible in current existing full operational towers. It is to this extent that a significant breakthrough in practice of optimization of crude oil distillation demands not only approval of a cost-benefit trade-off, but also provision of the optimization goal at minimum infrastructural changes.

Nevertheless, what has been failed to recall during all past efforts is the probable revision of the outline, topology and simple re-streaming of the process without causing a cardinal change in the process. The approach adopted in this work is the first one focusing on the possibility of an improved energy efficient design of new towers and also optimization of existing units considering minimum expenditure. These objectives are met through re-streaming of the process and making use of light components from the pre-fractionation unit in conventional crude oil distillation topologies to conserve energy in the column.

II. CRUDE DISTILLATION UNIT (CDU)

A. Process Description

In a simplified conventional crude distillation unit, energy is added to the reboiler at the base of the column while it is partly recovered in the condenser at the top. The basic equipments required for a standard continuous industrial crude oil distillation are simply shown in Fig.1. It contains a feed stream F, a top product stream or distillate (D) and a bottom-residue stream (B). Side products can also be withdrawn from the intermediate trays where products of intermediate composition and properties not obtainable merely by mixing distillate or bottoms with feed are desired [16]. Design of the column is based on the theoretical consideration that heat and mass transfer from stage to stage (theoretical) are in equilibrium [17] and as vapour and liquid are brought into contact on trays (Stages), vapour flows up the column while the liquid counter-currently streams downwards. Other equipments, a part of the tray section, are a condenser (with its associated heat duty, Q_c), where part of the condensate is returned to the top of the column to provide liquid flow above the feed point (to maintain reflux), and a reboiler (with its associated heat duty, Q_r), where part of the liquid from the base of the column is vaporized and returned to provide vapour flow.

The separation of liquid mixtures depends on the differences in volatility between the components, initiated from the difference in vapour pressures [6]. The larger the difference, the easier a separation is achieved; hence light hydrocarbons play a vital role in increasing the volatility and promoting a better fractionation, causing their existence to be undoubtedly essential.

B. Process Topology

The conventional procedure for reducing CDU's energy demand is based on increasing the enthalpy of the feed. Separation can begin by preheating the crude oil with hot overhead and product side streams of the crude unit. After desalting the feed is introduced to a flash drum (surge drum), where light components are vapourized and discharged off from the top and by gravity force the liquid settles at the bottom of the vessel, where it is withdrawn. The liquid is further preheated with the hot bottom stream and side products in an array of several heat exchangers followed by heating using an external heater (a furnace) to increase the feed enthalpy as much as possible and achieve a perfect separation in the tower [18]. The vapour product of the flash drum will meet up again with the preheated liquid upon entrance to the tower where they are inserted as one single stream. Products as gasoline, naphtha, kerosene, gas oils and lubricants are obtained as distillate, side products and column residue each of which are mixtures of hundreds of hydrocarbons.

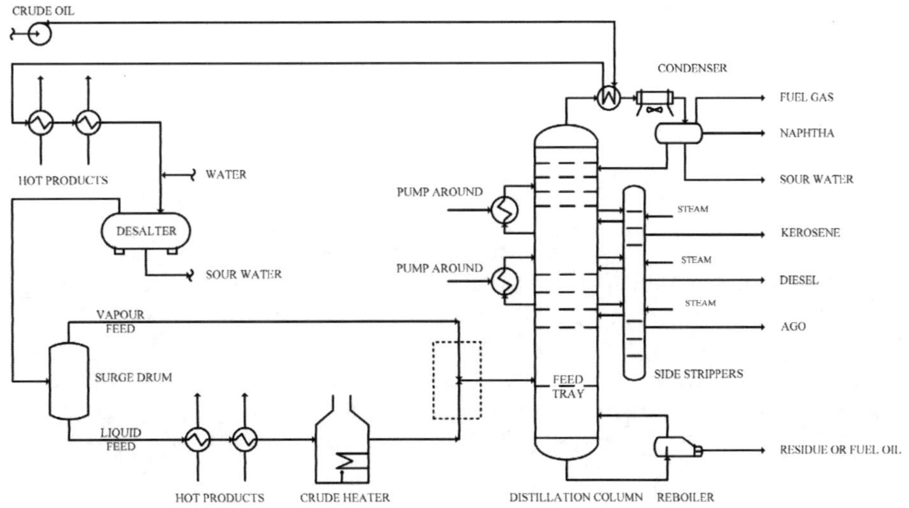

Figure 1. Conventional topology of a typical Crude Distillation Unit (CDU) with the liquid and vapour product of the surge drum inserted at the same stage to the tower

In this case, in order to avoid decomposition of sensitive components, the heat required can be provided by admission of steam directly to the bottom of the tower, hence the reboiler is then dispensed with [4,16].

III. PROPOSED SOLUTION

What will be demonstrated here is modifying the basis of common CDU topology with a detailed example in order to reduce the investment costs and column heat duty, using minimum change in the existing topologies and thereby disclosing the proposal to be more beneficial in terms of process economy than the conventional approach.

Current topology of insertion of both feeds together at the same stage causes a notable problem. The feed mixture will become more volatile, increasing the probability of vapour losses upon introducing the mixture to the tower. This is while the vapour product of the flash drum contains light end hydrocarbons which correlate more to the top of the column components and products, rather than the middle or lower section (where feed is usually inserted). The conventional topology will actually result in a higher tower heat duty as it requires additional elevation of light components to the higher stage. The solution offered here is to insert the vapour feed of the flash drum into upper stages of the column separately, rather than same stage insertion with the liquid feed from the bottom of the flash drum, which may be followed according to Fig. 2.

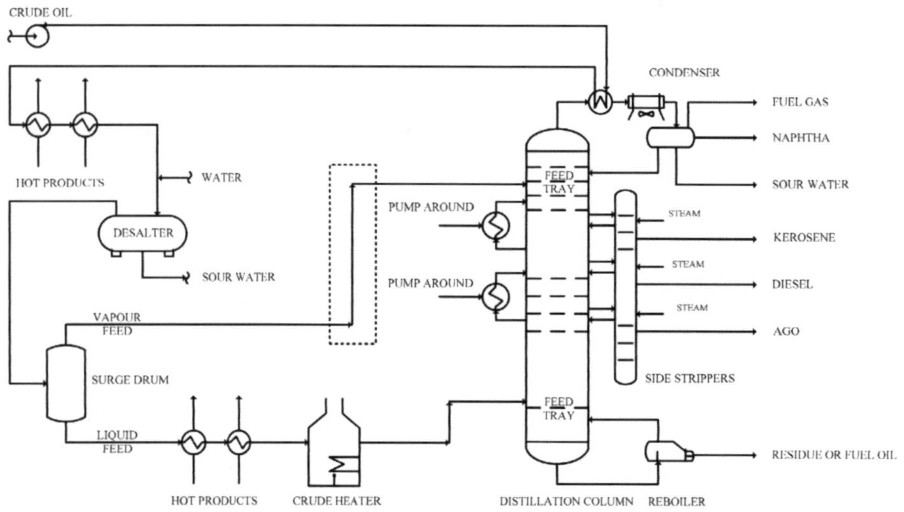

Figure 2. Proposed solution: inserting the vapour feed of the surge drum in upper stages of the column to reduce the condenser duty

With a composition of mainly light hydrocarbons corresponding more to higher stages of the column, inserting this feed to upper trays results in an increase of purity of side stream products or column's main products (Distillate and Bottom). This is due to less contact of the vapour entering and the liquid falling down from upper trays. Eventually the rising vapour is less likely to experience condensation and be detected in the side stream products. Therefore the tower heat duty would decrease as no further boiling of light components would be required and consequently can maintain the same purity of products with a lower duty. In other words, the condenser requires less work to pull up the light hydrocarbons as they are closer to the top of the column in the proposed theory.

The principal factor that determines a column diameter is the vapour flow-rate [19], hence an additional superiority of this solution would be an increase of Liquid to Gas flows (L/G)

between the two feed stages. The ultimate result is a decrease of column diameter between the two feeds as shown in Fig. 3, and further decreasing the investment costs.

This can also be expressed as

$$L_m/V_m > L_r/V_r \qquad (1)$$

Therefore

$$d_m < d_r \qquad (2)$$

However, if one also desires same pervious energy utilization in reboiler (or steam) and condenser can be considered, whereas in this case, higher purity of products will be achievable in compare to the basic common topology case. The first advantage of reducing reboiler energy consumption can be in an operating column while the second benefit of reducing investment costs can be made use of in the design phase of new distillation processes.

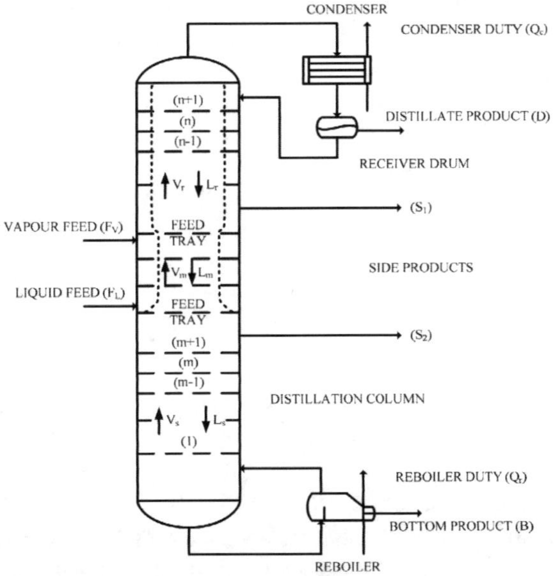

Figure 3. Detailed scheme of proposed topography; Insertion of the vapour product of the surge drum to upper stages of the tower reduces the mid-upper diameter

IV. FORMULATION

The optimization problem consists of an objective function of recovered heat to be maximized and constraints from design and operation conditions. Constraints are usually enforced as physical limitations of the process (e.g. product rates of naphtha, kerosene, LGO, and HGO).

A. Objective Function

Operating costs of industrial distillation towers are usually dependent on the condenser and reboiler heat duty, hence significant strive exists to reduce this amount. In formulation of optimum economic design various complicated objective functions can be defined, however the use of some simple function usually serves the purpose of optimizing a distillation column design. It would be sensible to choose an energy recovery in the column as the objective function. A global energy balance of the system can be written as follow based on Fig. 3:

$$F_V.h_V + F_L.h_L + Q_r = Q_c + B.h_B + D.h_D + \Sigma_{i=1}^{N} S_i h_i \tag{3}$$

For the purpose of optimization, with rearranging, the objective function of recovered heat from the process Q_{rec} can be defined in terms of heat duties of the reboiler and condenser and also feed and product enthalpies as follow

$$Q_{rec} = F_V.h_V + F_L.h_L + Q_r - Q_c - B.h_B - D.h_D - \Sigma_{i=1}^{N} S_i h_i \tag{4}$$

Considering the reboiler eliminated in crude oil distillation the objective function would simplify to only minimizing the condenser duty. Since the change imposed to the process as shown in the proposed solution is only observed in the tower diameter and condenser duty, change in Total Annual Cost (TAC), will be considered to reasonably study the effect of maximizing the objective function. This measure incorporates both energy and capital costs which are estimated [20,21] as the sum of the costs of the column and condenser duty. Annual cost of capital is obtained by dividing the cost of capital by a suitable payback period.

$$C_{col} = 17{,}640[Dia.(m)]^{1.066}[Len.(m)]^{0.802} \tag{5}$$

$$\textit{Energy Cost} = 4.7\$/10^6\,kj \tag{6}$$

$$C_r = \frac{(1+i)^k}{(1+i)^k - 1} \tag{7}$$

where, interest rate of return, *i*, is assumed to be 9.5% and a project life, *k*, of 25 years.

B. Constraints

Stage location of liquid feed and side strippers draw and return are considered to be constant. Pumparound and side strippers constraints as flow rate and duties are considered. In addition other variables as products rates, pressure profile, constant number of tower and side stripper stages are imposed to give the physical meaning of the process.

V. CASE STUDY

A. Process Description

Since simulation and optimization requires a comprehension of the process, a brief description of the unit is necessary and in order to give proof of the proposed solution, an industrial example of a Crude Distillation Unit (CDU) will be considered as in Fig. 4. The crude oil is processed in a fractionation facility to produce naphtha, kerosene, diesel, atmospheric gas oil (AGO), and atmospheric residue products. Preheated crude from an upstream preheat train and after preliminary heat exchange with several of the products from the plant is fed to the flash drum, where due to sudden change in volume, a portion of the feed is vaporized while the rest exits the bottom following by passing through the tubes of a gas-fired furnace. By acquiring necessary heat and elevating to a temperature of 343.3 °C and formation of yet another vapour phase, the feed exits the heater. While the flash vapours bypass the furnace, they are each separately introduced to the atmospheric distillation column. The column consists of a refluxed absorber with three side strippers and three cooled pump around circuits. Overall operating conditions of the process and the pre-fractionation train composition analysis are also given in Table 1 and 2.

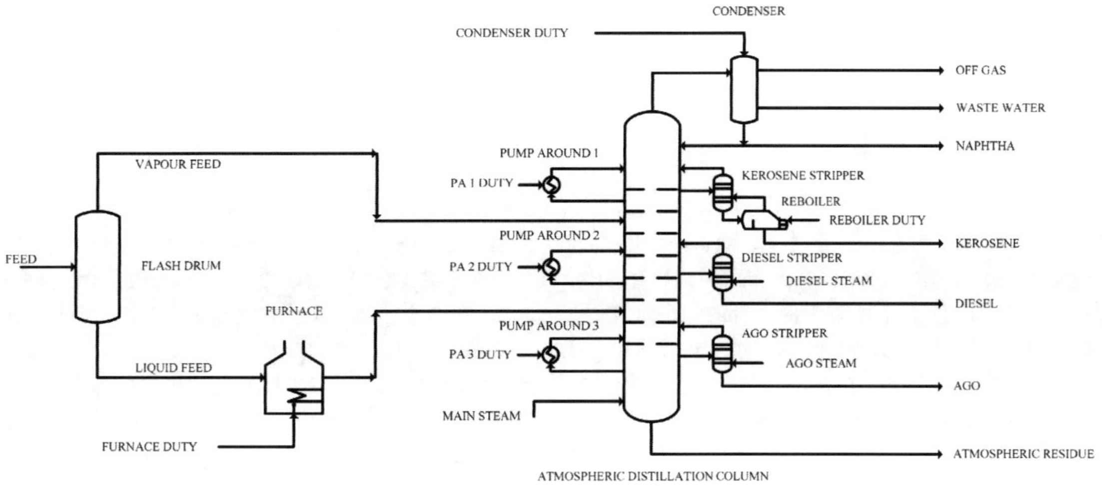

Figure 4. Case study crude oil distillation unit in a plant with feeds inserted separately as in proposed topology

Note from Table 2 that approximately 14.4 % vol. of the feed is vaporized, and therefore will greatly influence the outcome. The Column contains 29 trays, however, a total of 40 theoretical stages including condenser, 9 for the side strippers and 1 for the reboiler of the kerosene side stripper can be considered. Superheated steam is inserted to the bottom of the column at the very last stage to strip the remaining traces of volatile substances from the residue products and passing up the tower, where it lowers the effective pressure and temperature of the distillation. The hot crude feed is inserted at stage 28 and light hydrocarbon vapours escalade to the above and the liquids are shed on the lower stages. Details of the column operating conditions and configuration are given in Tables 3 and 4.

Table 1. Flash Drum vapour analysis

Component	Mole Fraction (%)
Methane	0.1
Ethane	0.2
Propane	2.7
i-Butane	1.6
n-Butane	5.7
H_2O	0.0
NBP 49	10.1
NBP 79	11.6
NBP 111	10.7
NBP 144	9.7
NBP 176	9.4

NBP 208	8.6
NBP 240	7.2
NBP 272	5.8
NBP 304	4.5
NBP 336	3.3
NBP 368	2.4
NBP 400	1.7
NBP 433	1.3
NBP 464	1.0
Total	97.6

Distillate stream is cooled down using a partial condenser and returned to maintain reflux. The steam and most volatile substances leaving the top tray are condensed and waste water, off gas and naphtha are the products obtained at the three-phase point. Several trays down from the top of the tower, short auxiliary three-stage side strippers are used to yield straight run products. Kerosene is produced from the reboiled side stripper, while diesel and AGO are produced from the steam-stripped side strippers. Considering marketable petroleum fuels and the fact that motive fuels are the ones that are the most desired [18], off gas production would be minimized (no production) and considered as one of the determined specification in this case study.

From the fact that it is not the presence of specific substances in such fuel products that is of importance but rather properties [18], specifications can be made as to boiling ranges (blend cuts), specific gravity, viscosity and the like.

Table 2. Unit overall operating conditions

Process Variables	Values
Raw Crude Oil MW	300
Feed API Gravity	48.75
Feed Flow Rate (barrel/day)	100,000
Feed Temperature (^oC)	232.2
Feed Pressure (bar)	5.17
Hot Crude Oil Temperature (^oC)	343.3
Vapour Feed Flow Rate (barrel/day)	14,410
Liquid Feed Flow Rate (barrel/day)	85,590
Column Top Pressure (bar)	1.36
Column Bottom Pressure (bar)	2.25
Condenser Pressure Drop (bar)	0.62
Furnace Pressure Drop (bar)	0.69
Off Gas Flow Rate (barrel/day)	0.00
Naphtha Flow Rate (barrel/day)	23,000
Kerosene Flow Rate (barrel/day)	9,300
Diesel Flow Rate (barrel/day)	19,250
AGO Flow Rate (barrel/day)	4,500

Table 3. Column side operating conditions

Stripping Steam	
Main Column	
Temperature [^{o}C]	190.6
Pressure [bar]	10.34
Flow Rate [kmole/hr]	3402
Diesel Side Stripper	
Temperature [^{o}C]	148.9
Pressure [bar]	3.45
Flow Rate [kmole/hr]	75.54
AGO Side Stripper	
Temperature [^{o}C]	148.9
Pressure [bar]	3.45
Flow Rate [kmole/hr]	62.95
Kerosene Reboiler	
Flow Rate [kmole/hr]	327.5
Duty [MJ/hr]	7,913
Pumparound	
PA 1	
Flow Rate [kmole/hr]	2,183
Duty [MJ/hr]	58,030
PA 2	
Flow Rate [kmole/hr]	830.1
Duty [MJ/hr]	36,930
PA 3	
Flow Rate [kmole/hr]	648.8
Duty [MJ/hr]	36,930

Table 4. Column streams configuration

Number of Trays	
Main Column	29
Kerosene Side Stripper	3
Diesel Side Stripper	3
AGO Side Stripper	3
Locations	
Main Column Liquid and Vapour Feed	28
Main Column Steam Feed	29
Kerosene SS Draw and Return Stage	9,8
Diesel SS Draw and Return Stage	17,16
AGO SS Draw and Return Stage	22,21
PA 1 Draw and Return Stage	2,1
PA 2 Draw and Return Stage	17,16
PA 3 Draw and Return Stage	22,21

Other determined specifications for this separation are the products flow rate, pump around duties and rates, kerosene reboiler duty and an overflash rate (usually 3-5% of the total column feed [20,24]), all of which are applied to maintain the required properties which are held constant throughout the simulation. In addition, the well-known Peng-Robinson equation of state is employed to compute the thermodynamics properties. Pressure drops are considered in the furnace, condenser and the column to further approach realistic conditions and make the case more practical.

The ultimate goal would be on how to take full advantage of the flash drum vapour product (mainly consisting of light hydrocarbons) and how to maximize its effects. The proposal presented in this work is to insert the mentioned vapour stream on a stage corresponding to similar in-tower composition, which results in minimum condenser duty. Another objective of the vapour fraction would be to increase L/G in order to decrease the diameter and therefore reduce the capital cost of a tower. There shall exist an optimum tray which corresponds to above mentioned advantages.

B. Optimizing Vapour Feed Stage

Firstly, in-column composition profile of applied mixture will be given to obtain an initial estimation of the composition corresponding stage to that of the vapour feed exiting the flash drum (Fig. 5). Vapour stream composition of the flash drum from Table I is shown for NBP 49 to NBP 304 (accounting for approximately 78% of the total composition) on the respective stages corresponding with the light hydrocarbons mixture composition in the column (dashed line). Light compounds are located on the very early stages of 2 and 3 while heavier hydrocarbons broaden the span to stage 11. Therefore, an early standpoint would be to insert the vapour feed between these stages.

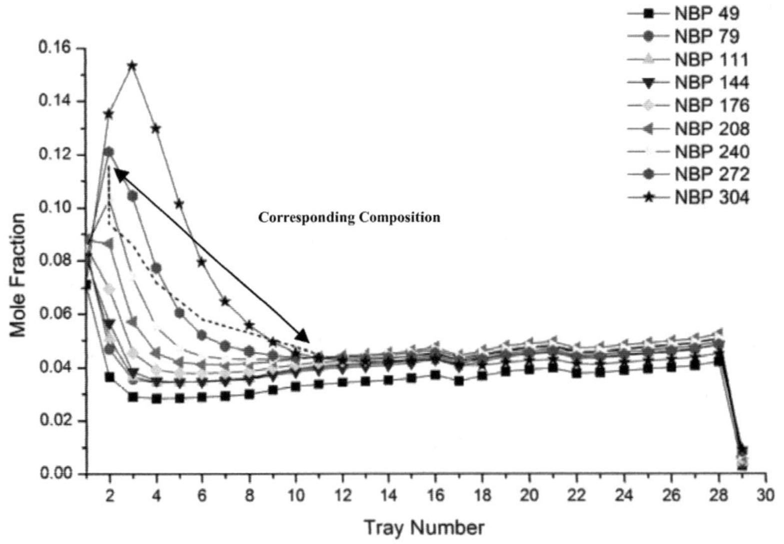

Figure 5. Correspondency of light hydrocarbons composition profile along the tower with the light fraction composition of the vapour feed from the surge drum

However relying only on composition similarity would not be adequate for any act of modifying the topology, hence other means for further confirmation and finding the minimum condenser duty is essential. Consequently, results form a computer simulation can be plotted to determine the optimum feed stage and examine the precision of above approximation. Simulation runs are performed at several different feed points, keeping the material balance and total number of stages constant.

For this case, predetermined specifications as mentioned in the case study process description will be held as constants for the simulation to observe the variations in the condenser duty through change in the feed stage. AspenHysys™ [20,21,25], a steady state simulator is used to find such an optimum tray. Now for finding the desired tray, several vapour feed inlet stage runs is examined and results have been gathered for condenser duties shown graphically in Fig. 6.

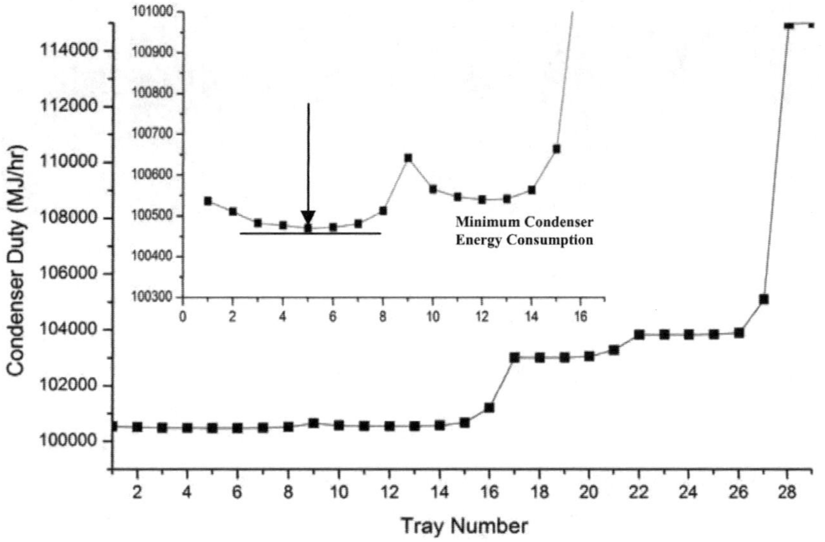

Figure 6. Results of optimization for the condenser duty through different runs of the vapour feed inlet stage

The optimum stage for this case yielding the minimum condenser duty is found on stage 5. The simulated results having both feeds inserted at one stage (Tray 28), will be considered as the base case, to compare to the alternative proposal of separate insertion of the feeds. Table V summarizes the result of such optimization. Furthermore, the result is in agreement with the initial estimation as in Fig. 5 that at proximity of stage 5, minimum condenser duty was estimated to be obtained. By inserting the vapour feed at the upper section of the column, it is possible to decrease the energy demand of the condenser down by 14,530 MJ/hr. Considering this optimum feed stage, it is possible to save 12.6% of condenser duty with respect to the base case. Needless to say, that this energy conservation is achieved at no additional cost of implementing any sort of additional equipment, difficult and expensive rerouting of the streams or unlikely possible fundamental changes.

Finally, due to the fact that L/G will increase at few sections of the column by insertion of vapour feed to upper stages, one can expect to design new columns having lower associated section dimension than for using the ordinary topology, thus reducing capital or investment costs of new designs. For the example shown, the L/G parameter at different stages of the column is given in Fig. 7 for comparing both cases.

As shown, the number of times that the L/G of the proposed solution is higher than of the base case exceeds the times that it is lower. With a more precise search, the increase of L/Gs have actually occurred at side products withdrawal and return stages (Table IV) as expected which then confirms the configuration given in Fig. 3. Therefore, at these points, same process conditions and product purities can be achieved using smaller in diameter and lower in cost towers.

Table 5. Results of the comparative study

Case study name	Condenser Duty (MJ/hr)
Base case (Fig.1)	115,000
Proposed Solution (Fig. 2)	100,470

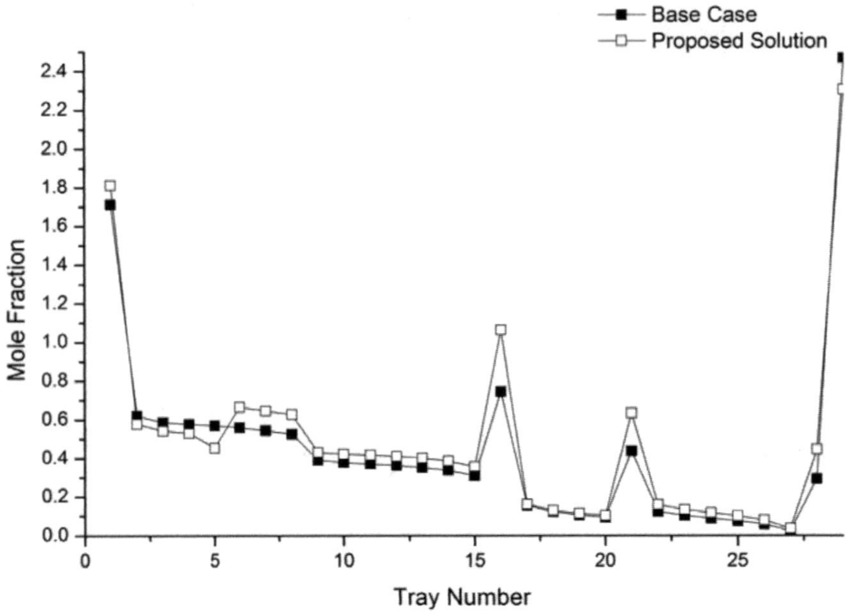

Figure 7. Comparison of the Liquid to Gas flow ratios (L/G) at different stages of the column for the base case and proposed topography

Table VI shows the result of optimization to minimize the annual costs. Compared with the base case, a decrease of approximately 0.7 million US$/y can be achieved with the proposed topology for new designs or around 0.6 for modification of online towers. In other words, one can conserve approximately 9% in the annual cost through modification of the current topologies.

If one also desires, simulation can be performed by use of a reboiler for heat generation instead of steam with proposed topography to vividly see the advantages and also observe the energy conservation in both reboiler and condenser. Also, behold that inserting this feed to the upper trays results in an increase of products purity with the same condenser or reboiler duties. It must be kept in mind that such proposal will yield different energy savings for different mixtures as have also been examined (not shown in this work). While it is found out this conservation is very dependent on the design and operating conditions, the number of products which varies considerably from one refinery to another can also influence the outcome. Should other optimization proposals be integrated with the solution given here, accurate simulations must be carried out to be ascertain of the results. Although, one must not forget that an increase in capital and operating cost caused by any sort of improvements can be compensated by better sells due to the product quality improvement, however the determining key factor to the reinstate in these improvements is a comprehensive knowledge of the capital cost of the column, operating cost associated with utility requirement and cost of products for different quality, which by then optimal distillation column designs may be obtained.

Table 6. Results of annual cost optimization

Design Variable	Values
Base case (Fig. 1) ave. diameter (m)	5.334
Proposed Solution (Fig. 2) ave. diameter (m)	4.8768
ΔC_{col} (\$)	95,837
Δ*Energy Cost (\$/y)*	598,299
Δ*Total Annual Cost (TAC)* (10^6 *\$/y)*	0.705

VI. CONCLUSIONS

Optimization of crude oil fractionation has been considered herein with the use of change in the basis of general outline of the current topologies. The approach presented is simple and practical, and it is attained through slight re-routing of the vapour feed product of the surge drum on the insertion stage. With respect to the conventional topology of the same feed stage insertion as the liquid feed, inserting the vapour feed from the pre-fractionation train to upper stages of the column on a separate stage, provides the following results:

- Splitting the feed into two feeds of liquid and vapor upon vaporization in the flash drum maintains a simpler control of the process.
- Decrease the energy demand of the condenser and reboiler (if available) to an extent that finding an optimized stage can achieve an energy conservation of 12.6% in condenser duty for a typical fractionation of crude oil.
- Having L/G increased, same purity of products can be achieved with a smaller tower diameter. In fact, the fixed (capital) cost of a tower is reduced.
- With L/G increased, a decrease of approximately 9% in new tower design or in current units with modification can be attained.
- Since the flash drum vapour product contains component corresponding more to the upper stages of the distillation column, inserting this feed to the upper trays results in an increase of products purity with the same condenser or steam/reboiler energy consumption.
- Such topology requires negligible change in existing units for the introduction of vapour feed to an upper tray.
- Further runs of simulation is required if the same approach is to be used for other mixtures or in combination with other methods.

ACKNOWLEDGMENT

Support from the Royal Institute of Technology (KTH) and also Mr. Yohannes Kiros, for his technical support and help, are gratefully acknowledged. The authors also wish to acknowledge support from Shazand-Arak refinery complex.

NOTATIONS

L	Liquid molar flow rate
G	Gas molar flow rate
D	Distillate product molar flow rate
B	Bottom product molar flow rate
S	Side stream product molar flow rate
Q_c, Q_r	Condenser duty, Reboiler duty
L_m, V_m	Liquid, gas molar flow rate between the feeds
L_s, V_s	Liquid, gas molar flow rate in stripping section
L_r, V_r	Liquid, gas molar flow rate in rectifying section
d_m	Column diameter at between feed section
d_r	Column diameter at the rectifying section
F_V, F_L	Vapour, liquid molar flow rate
h_v, h_L	Enthalpy of the vapour, liquid feed
h_B, h_D	Enthalpy of the bottom, distillate product
h_i	Enthalpy of the *i*th side stream product
C	Cost
TAC	Total Annual Cost
i	Interest rate of return
k	project life
LGO	Ligh GasOil
HGO	Heavy GasOil

REFERENCES

[1] J. H. Gary and G. E. Handwerk, Petroleum Refining–Technology and Economics, 4th Ed., Marcel Dekker Inc., New York , 2001, pp. 1-66.

[2] Wiley Critical Content: Petroleum Technology, vol. 2, New Jersey : Wiley & Sons Inc., 2007, pp. 315-326.

[3] Dave, D. J. and M. Z. Dabhiya, "Online tuning of a steady state crude distillation unit model for real time applications", J. of Process Cont. vol. 13, April 2003, pp. 267–282, doi: 10.1016/S0959-1524(02)00026-4

[4] J. F. Richardson and J. H. Harker, Coulson and Richardson's Chemical Engineering, 5th Ed., vol. 2, Oxford : Elsevier Butterworth-Heinemann Ltd. 2002, pp. 575-576.

[5] J. Kunesh, H. Kister, M. Lockett, J. Fair, "Distillation: still towering over other options", Chem. Eng. Progr. vol. 91, 1995, pp. 43–54, doi: 0360-7275

[6] C. D. Holland, Fundamentals of Multicomponent Distillation. McGraw-Hill Inc., New York, 1981, pp. 299-300

[7] H. Z. Kister, Distillation Operation, McGraw-Hill Inc., New York, 1990, pp. 21-30.

[8] J. W. Seo, M. Oh and T. H. Lee, "Design Optimization of Crude Oil Distillation", Chem. Eng. Technol. vol. 23, 2000, pp. 157-164, doi: 0930-7516/00/0202-0157.

[9] A. Lucia, A. Amale and R. Taylor, 'Distillation pinch points and more", Comp. & Chem. Eng., vol. 32, June 2008, pp. 1342-1364, doi: 10.1016/j.compchemeng.2007.06.019.

[10] L. M. Naphtali and D. P. Sandholm, "Multicomponent separation calculations by linearization", J. of AIChE, vol. 17, Jan. 1971, pp. 148-153, doi: 10.1002/aic.690170130.

[11] N. L. Ricker and E. A. Grens, “A calculational procedure for design problems in multicomponent distillation”, J. of AIChE, vol. 20, Jun.1974, pp. 238-244, doi: 10.1002/aic.690200205.

[12] R. W. H. Sargent, K. Gaminibandara and L. C. W. Dixon, “Optimum design of plate distillation columns, optimization in action”, Ind. Eng. Chem. Res., vol. 39, Oct. 2000, pp. 4326–4335, doi: 10.1021/ie0001974.

[13] J. M. Srygley and C. D. Holland, “Optimization design of conventional and complex columns”, J. of AIChE, vol. 11, Jun. 1965, pp. 695-701, doi: 10.1002/aic.690110424.

[14] Engineering Data Book: Gas Processors Suppliers Association, Sec. 19th, 11th Ed., Oklahoma, 1998, pp. 28-29.

[15] G. Soave and J. A . Feliu, “Saving energy in distillation towers by feed splitting”, App. Thermal Eng., vol. 22, Jun. 2002, pp. 889–896, doi: 10.1016/S1359-4311(02)00006-6.

[16] R. E. Treybal, Mass-Transfer Operations, 3rd Ed.,McGraw-Hill Inc., 1980, pp. 394-397.

[17] Applied Process Designs for Chemical and Petrochemical Plants, 3rd Ed., vol. 2, Gulf Professional Publishing, Houston, 1997, pp. 1-4.

[18] D. S. J. Jones and P. R. Pujado, Handbook of Petroleum Processing, Springer, Dordrecht, 2006, pp. 114-118.

[19] R. K. Sinnott, Coulson and Richardson’s Chemical Engineering Design, 4th Ed., vol. 6, Elsevier Butterworth-Heinemann Ltd., 2005, pp. 557-558.

[20] Aspen HYSYS – Tutorials and Applications, Aspen Technology Inc., Cambridge, 2006.

[21] Aspen HYSYS – User’s Guide, Aspen Technology Inc., Cambridge, 2006.

[22] W. L. Luyben, Distillation Design and Control Using ASPEN Simulation, John Wiley & Sons Inc., New Jersey, 2006.

[23] M. S. Peters and K. D. Timmerhaus, Plant Design and Economics for Chemical Engineers, 4th Ed., McGraw-Hill Inc., New York , 1991, pp. 235-236

[24] J. R. Couper, W. R. Penny, J. R. Fair and S. M. Walas, Chemical Process Equipment, 2nd Ed.,Butterworth-Heinemann, Massachusetts, 1990, pp. 550-557.

[25] *Aspen HYSYS – Simulation Basis,* Aspen Technology Inc., Cambridge, 2006.

Applying Best Practices in Procedural Operations in a Continuous Process

vigilantplant.®

The clear path to operational excellence

Marcus Tennant
Principal Systems Architect
Yokogawa Corporation

November 2010

YOKOGAWA

Agenda

- Issues around Continuous Process Industries
- Procedural Operations
  - Definitions
  - Examples
- The Impact and Benefit of ISA-88 on standardization
- Modularizing Procedural Operations
- Manufacturing Benefits
- A New Procedural Automation Standard
- Conclusion

Safety and the Environment

vigilantplant®

- The operator workforce is diminishing but the workload is increasing
- Operators under pressure can make mistakes or not notice a developing situation
- These can lead to safety and/or environmental incidents
- Sometimes with disastrous consequences

November 2010

YOKOGAWA

Average Loss Per Major Incident

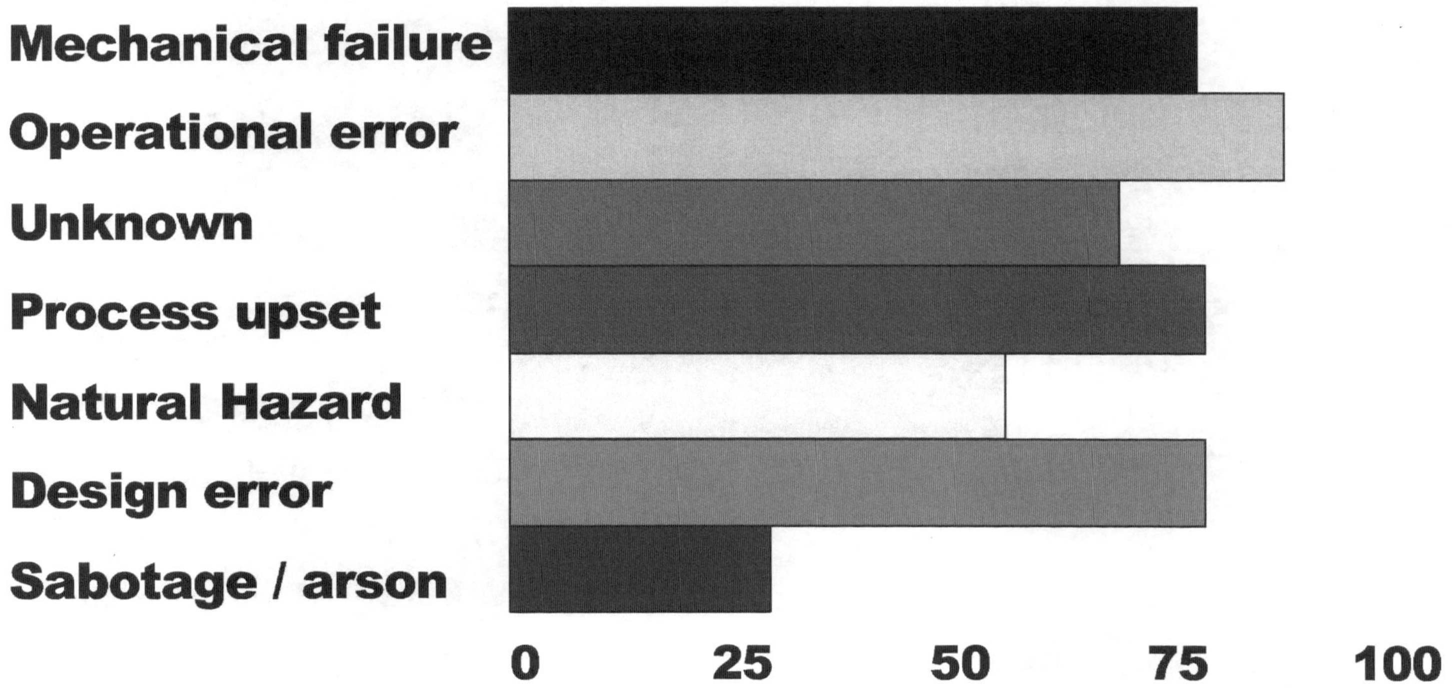

Average Dollar Loss Per Major Incident by Cause
Millions of Dollars
Source: J & H Marsh & McLennan, Inc.

At an average cost of $90 Million in losses per major incident, operational error is the most expensive cause of error in production facilities

Key Skills Leaving the Workforce

- Major Refining Company
  - Lost 2500 years of experience in 2006 when 100 operators retired at one site, each with an average of 25 years experience

- Major Chemical Company
  - Analyzed plant demographics
  - Found one of their largest plants was going to lose 75 percent of its operating staff to retirement within 3 years

Availability and Reliability

vigilantplant®

- Industry in the "Developed World" is moving from expansion to reliability
- With tighter margins any unplanned downtime MUST be avoided
- Includes the process equipment AND the control system and network infrastructure
- In other words a move from Capex to Opex and the need for Improved Operational Excellence

Capturing Procedural Knowledge

vigilantplant®

- Most of these issues facing the process industries involve procedures in one form or another, many carried out by skilled operators
- But these operators will be retiring over the next 5-10 years
- One manufacturer had to bring back retired operators to restart a unit after shutdown as the current operators had never been through a start-up
- Automated procedures can capture the knowledge of the best operator on his/her best day…every day
  - Remove shift to shift inconsistencies
  - Ensure that a procedural operation is being conducted safely

Procedural Operations

vigilantplant®

- Process plants are designed and run according to operational procedures
  - Every process unit has a SOP (Standard Operating Procedure)
- Procedures consist of a set of tasks that are conducted in a set way to achieve a certain goal
  - Starting up
  - Shutting down
  - Transitioning
- Three types of procedure
  - Manual
  - Prompted
  - Automated
- Procedures vary by industry/company/plant, but they require consistent execution

Examples of Procedural Operations

- Bulk Chemicals
  - Startup
  - Shutdown
  - Grade Changes
  - Line Switchover
  - Cleaning
  - Decoking
  - Transition Management
- Refining
  - Crude Switchover
  - Regeneration
  - Decoking
  - Transition Management
- Polymers
  - Startup
  - Shutdown
  - Grade change
  - Switchover
  - Transition Management
- Power Generation
  - Startup
  - Shutdown
  - Load Management
  - Grade Changes
  - Cleaning
  - Decoking

The Need for Standardization – ISA-88

vigilantplant®

- Production facilities with traditional batch or sequenced operations adopted a standardized approach to automating procedures several years ago: ISA-88 – The Batch Automation Standard
  - Revolutionized batch control
  - Operations and Process Units were modularized
  - Suppliers and manufacturers worked from the same principles
- Significant savings were realized by the introduction of the standard
  - Better safety
  - Fewer environmental incidents
  - Improved throughput
  - Repeatable operations
  - Reduced engineering costs

ISA™

Procedural Operations - Modularized

vigilantplant®

- ISA-88 was successful in moving batch applications from large monolithic sequences to modular, re-usable sets of procedural logic
- Many process industries currently use procedures (even in SFC format) that are very complex and potentially difficult to maintain
  - A large refining company made a presentation at the 2008 NPRA Technology Forum showing scheduling sequences that consisted of many SFC steps
  - As with batch prior to ISA-88 these sequences are custom coded and difficult to maintain by anyone but the person who designed and implemented them
- To make procedures easier to design, understand and implement it is necessary to use the same techniques of modularization from the ISA-88 standard in ALL procedural operations

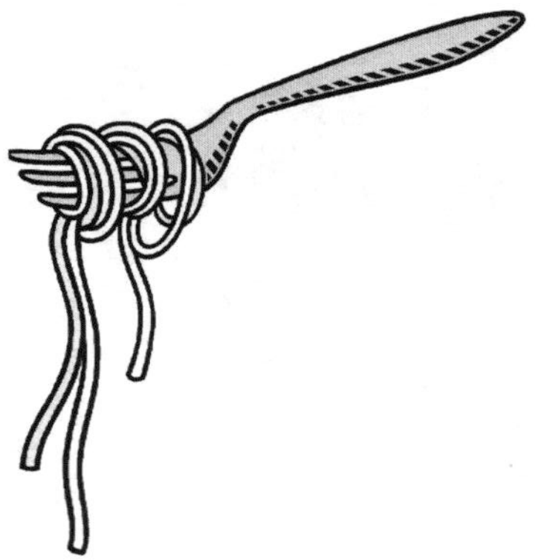

Modular Procedural Automation (MPA)

vigilantplant®

- Modular Procedural Automation (MPA) is a methodology for handling procedural operations in continuous processes
- It is modular to standardize implementation within sites and across companies
  - improves acceptance and understanding by operations
  - reduces engineering and maintenance costs
- Procedural automation
  - increases safety and reliability
  - provides consistent, repeatable and verifiable procedural operations in an era of skills depletion

Benefits of Automated Procedures

vigilantplant.®

- Capture and implement operating know-how of experienced operators
- Improve Plant Operating Efficiency
  - Increase productivity by reducing the time that operators take during procedural operations
  - Reduce costs by reducing usage of energy, utilities, and materials during procedural operation
  - Reduce alarms and events to allow operator to concentrate on other activities.
- Improve Safety of Plant Operations
  - Operates plant exactly as defined by the stored operating procedures
  - Reduce operational errors and omissions

Capturing Best Practices

vigilantplant.®

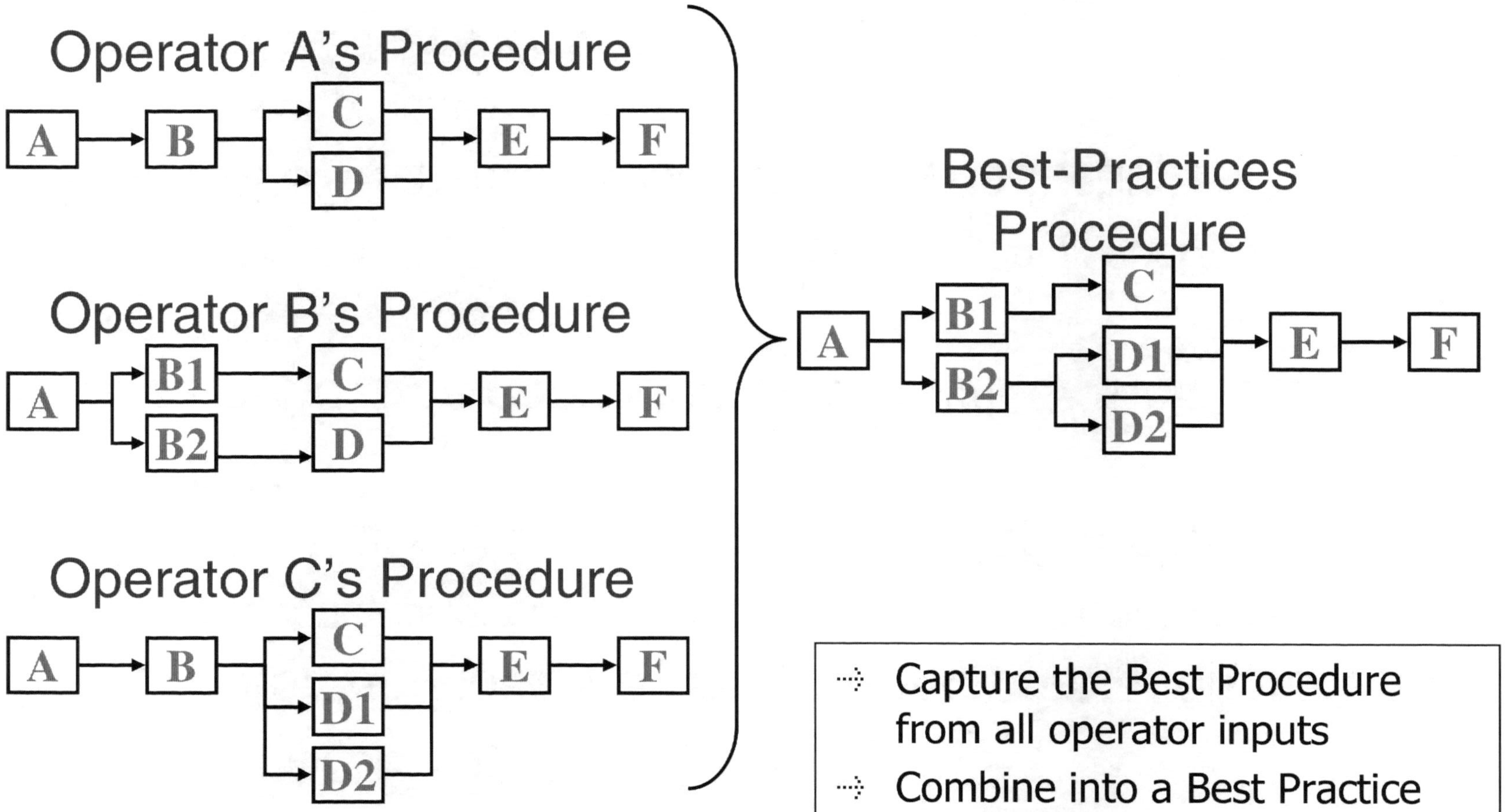

- Capture the Best Procedure from all operator inputs
- Combine into a Best Practice Procedure

Automation of Crude Switch Operation

Driving Force for Automation

- Operator workload was higher than normal operation
  - Frequency of grade change was two or three times per week
- There were 130 DCS manipulations/operation on average
- Operating time depended on the skill of the assigned operator
  - More than eight hours in the case of a junior operator
- Operational errors caused loss of productivity
  - Loss of product quality
  - Increased off-spec product
  - Waste of utilities (electricity, steam, compressed air, etc.)

Automation of Crude Switch Operation

Benefits

- Operators' workload was drastically decreased
  - From 130 manipulations to 10 for one crude switch operation
  - Monitoring work was fully automated
- 2000 process alarms were avoided during crude switching
  - By automatic alarm optimization according to process conditions
- Operating time was minimized
  - From 8 hours to 2.5 hours
- Operator errors and omissions were prevented
  - Productivity was increased
- Operational procedure was standardized
  - Productivity was not dependent on operator skills
- Operation know-how was shared
  - Education time for new & junior operators was reduced

Manual Transitions

Most control operations are automated using DCS functions. However, knowledge based (transition) operations are generally conducted manually by operators. Operations such as:

- PP/PE grade change
- Crude switch-over
- Lubricant switch-over
- Reactor switch-over
- Boiler load change
- Olefin reactor purge
- Silo change
- Decoking operation
- Plant start up/shut down
- Trouble shooting, etc.

> Operation time, product quality, quantity of transient products, and safety depends on the skill of the assigned operator

- Time loss due to the difference in operator skills
- Product quality loss due to the difference in operator skills
- Inventory loss due to difference in operator efficiency
- Safety loss caused by operational error and omissions

Need for a New Standard

vigilantplant®

- As with batch automation prior to ISA-88 there is no consistency in the documentation, design and implementation of procedural operations
  - This can lead to unplanned shutdowns at best or an unsafe situation at worst
- Manufacturers often have their own, in-house, methodology for working with procedures
  - These usually rely on long, serially coded sequential function charts
  - Often also rely on the original designer for modifications
  - Could require a shutdown to modify them
- There is no consistency in the supplier community either
- Hence – we need a new industry standard
  - To bring manufacturers and suppliers together
  - To improve the safety of procedural operations

ISA
IEC
ANSI

November 2010

YOKOGAWA

Procedural Automation Standard ISA-106

vigilantplant®

- Standards and Practices committee was formed June 2010
- Targeted at continuous processes
- Define standards for automating procedural operations – Topics considered.
  - Models and terminology
  - Modularization of procedural steps to foster re-use and lower TCO
  - State models for procedural logic
  - Process unit orientation with operational perspective
  - Exception handling process abnormal situations
  - Implementation of start up, shutdown, abnormal situations, hold states, and transition logic
  - Recommended target platform (i.e.. control system vs. safety system) for different types of procedures
  - Training and certification best practices
- Has input and support from 39 Manufacturers and Suppliers

Conclusion

- Process industries are facing a skills and knowledge shortage leading to potential production and safety issues
- Industry standards such as ISA-88 have helped the batch industries to alleviate this problem and save millions of dollars
- The same standards concepts currently being developed in ISA 106 can also help with the design, implementation, documentation and operation of procedures in the process industries

(539f) Investigation Based On Electric Field to Optimize and Design An Energy-Saving Magnesium Electrolysis Cell

Yun Zhao[1], Ze Sun[2], Guimin Lu[1] and Jianguo Yu[2], (1)National Engineering Research Center for Integrated Utilization of Salt Lake Resources, East China University of Science and Technology, Shanghai, China, (2)State Key Lab of Chemical Engineering, East China University of Science and Technology, Shanghai, China

Fused salt electrolytic technology for light metal production is complex process, which includes six important physics fields affecting performance of electrolysis cell, such as electric field, magnetic field, temperature field, flow velocity field, concentration field and stress field. To obtain an advanced electrolysis cell by optimizing the behaviors of these physics fields, Conventional experimental methods have to take a lot of money, time and labor requirements, and have to bear a risk of failure. However, with a great increase of computer power in the last ten years, a novel numerical simulation (FEM) has become an effective tool in the design or modeling of a physical phenomenon in various engineering disciplines with the help of ANSYS®. This approach makes it possible to study the effect of each physics field in isolation and evaluate the electrolysis cell responses to variation of the process and design parameters, relatively successful applications were made in design advanced aluminum reduction cells by optimizing electric field, magnetic field, and temperature field. As is known to us, the cell performance mainly depends on current efficiency and cell voltage including decomposition voltage, excess voltage and resistance voltage, which decide the energy consumption. In order to develop the energy-saving magnesium electrolysis cell by FEM, this paper aimed at optimizing electric field that is energy foundation of operation and source of the other physics fields, mainly focused on investigating how to reducing resistance voltage (Ohmic voltage) by changing cell structure parameters, such as relative positions of anodes and cathodes, et al. Firstly, an investigation based on electric field of magnesium electrolysis cell was employed to optimize a 120 kA cell, where the cell with the dimension 2.91×1.87×1.40 m, 8 anodes with dimension 2.20×1.00×0.15 m, 9 cathodes with dimension 0.950×2.235×0.050 m, and the anode and cathode distance as 0.07 m. The major work was to look for the best cell structure by adjusting the extended distance of cathode relative to anode in the direction toward cell bottom, the extended distance between cathode and back wall, the shorten distance between cathode and cell bottom, the shorten distance between cathode top and electrolyte surface and the extended distance between cathode and partition. In theory, effective work is done in the region between anode and cathode, the least bypass current going into other regions is expected. Comparing figure 1 and figure 2, the voltage potential distributions in regions II and IV are close to desired state. Thus, simulation results indicate that suitable modification of cathode and cell profiles could obtain good resistance voltage distribution. Furthermore, an orthogonal test was conducted to verify the optimum 120 kA cell, and the optimization criterion was applied to the amplification of 250 kA cell. The computed resistance voltage of the optimized 250 kA is also the minimum voltage 1.631312 V among provided solutions, and the potential distributions in region B and D is more perfect than that in regions A and C, which are shown in figure 3 and figure 4. If decomposition and over voltages are 2.735 V, external voltage is assumed 0.25 V, and then theoretical cell voltage could attain 4.616312 V. Comparing with the magnesium electrolyzers in commercial use, this cell is more competitive in energy consumption. Hence, the developed model and simulated results will be useful for optimization and design of the magnesium electrolyzers.

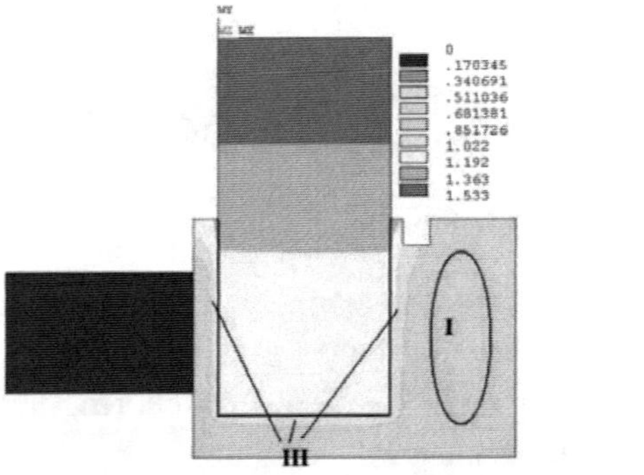

Figure 1 contours of potential in original 120 kA cell

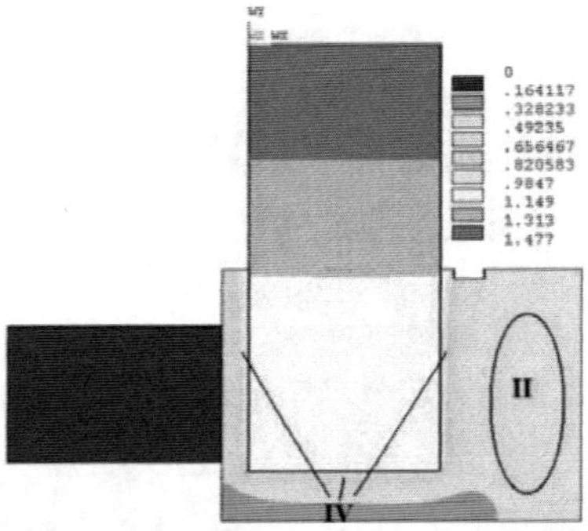

Figure 2 contours of potential in optimized 120 kA cell

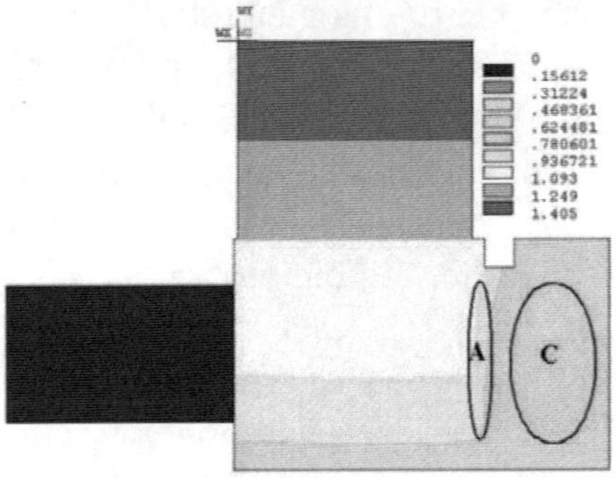

Figure 3 contours of potential in original 250 kA cell

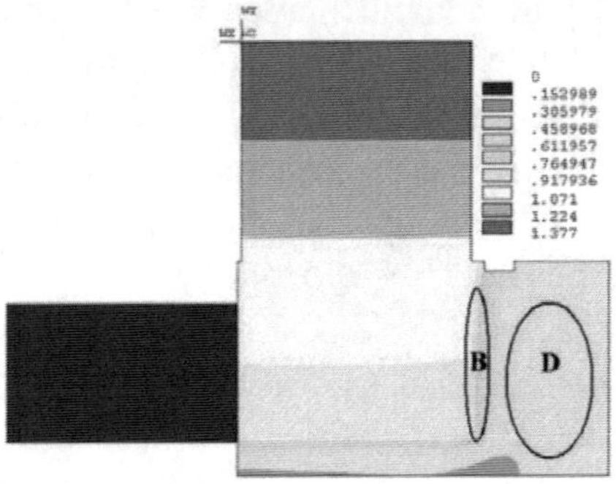

Figure 4 contours of potential in optimized 250 kA cell

See more of this Session: Industrial Innovation in Process Design and Operations
See more of this Group/Topical: Computing and Systems Technology Division

(555a) Photocatalytic Oxidation of Rhodamine B Over P-Doped Titania Nanotubes

R. Asapu[1], **Daniel H. Chen**[2], Bin Wang[3], Rakesh Sadu[1] and Zhanhu Guo[2], (1)Chemical Engineering, Lamar University, Beaumont, TX, (2)Dan F. Smith Department of Chemical Engineering, Lamar University, Beaumont, TX, (3)Chemistry, Lamar University, Beaumont, TX

Pure titania nanotubes and nanotubes doped with different metals like Cu, Zn, Cr, Fe etc. and non-metals like N, S, C have gained much attention in recent years. In this work, titania nanotubes (TNTs) were first synthesized with a hydrothermal method, then doped with phosphorous using dimethyl phosphite in ethanol solution and calcined at 350oC for 1 hour. TEM image of the prepared P-TNTs synthesized at 180oC show tube formation with 25-50nm OD and 400-1500 nm length. XRD analysis revealed peaks characteristic of monoclinic sodium titanate oxide (Na2Ti3O7 and Na2Ti6O13) phases, which could correspond to the H2Ti3O7 monoclinic unit cell structure or the orthorhombic protonic lepidocrocite phase. Surface areas of TNT (@180C) and P-TNT are 100 and 105 m2/g, respectively, while TNTs (@130C) have a surface area of 306.65 m2/g. The photocatalytic activities of the P-doped TNT (P-TNT) were tested with the degradation of Rhodamine B (C28H31ClN2O3), a dye commonly used as a tracer and a fluorescence agent, in water. Liquid samples were analyzed with a CARY 50 UV-VIS spectrometer. The optimum phosphorous content was found to be 0.75 wt %. P-TNTs (synthesized at 130oC, with 0.75wt% Phosphorous) has the highest activity when compared with undoped TNTs and Degussa P-25 nanoparticles (NP). A UV filter was used to cut off the UV light below 400nm from a 9W fluorescent lamp (Philips PL-S) and the results confirm the P-doped TNTs have better visible-light activities than P-25 TiO2 NP.

See more of this Session: Process and Product Development for Sustainability
See more of this Group/Topical: Process Development Division

(555b) Adsorptive Product Recovery From a Fermentation Broth Using a Nitrogen Source for Regeneration

Paul J. Th. Bussmann and Ronald C. Vroon, Food and Biotechnology Innovations, TNO Quality of Life, Zeist, Netherlands

Continuous product removal can increase the productivity and economic validity of fermentations when the productivity is limited by product degradation, inhibitory or toxicity effects of the product. Furthermore, in state-of-the-art fermentations for the production of high purity acids or bases, excessive large volumes of salty waste streams are produced. Integration of these fermentation processes with the appropriate product removal technique during fermentation can increase the productivity and overcome the production of salty waste streams. In this study a process based on adsorptive monoethanolamine (MEA) removal from a fermentation broth using the nitrogen source (NH3) for regeneration and thereby preventing the production of salty waste streams was developed. MEA is an intermediate in the manufacturing of many chemicals which are applied in the fields of gas purification, surfactants, cement additives and textile chemicals. Currently MEA is produced chemically, but the rising price of the raw material naphtha is an important incentive to study biotechnology routes. Furthermore, the relatively low concentration levels in a fermentation broth necessitate efficient product removal.

See more of this Session: Process and Product Development for Sustainability
See more of this Group/Topical: Process Development Division

(555c) An Alternative Process for Oxidative Methane Coupling for Ethylene, Electricity and Formaldehyde Production

Daniel Salerno, Harvey Arellano-Garcia and Günter Wozny, Chair of Process Dynamics and Operation, Berlin Institute of Technology, Berlin, Germany

Oxidative coupling of methane (OCM) is a promising route for the production of ethylene by utilizing methane, the main constituent of natural gas. OCM is a reaction of two methyl radicals coupled after the abstraction of one hydrogen atom from the methane molecule. This reaction is highly exothermic. The conventional processes are endothermic, energy intensive and low in C2 selectivity, depending on the choice of the feedstock. However, the development of the industrial process still remains in its very early stage with the main efforts focused on the search for a sufficiently selective, active, and stable catalyst. The motivating force of this work is to design an alternative process for ethylene production via OCM and the co-generation of electricity and formaldehyde that can make the process economically attractive and designed so as to be industrially implemented. The total project investment, based on total equipment cost (TEC), as well as variable and fixed operating costs, was developed based on mass and energy balance information taken from ASPEN ICARUS® simulation results. With these costs, a discounted cash flow analysis was used to determine the production cost of ethylene, electricity and formaldehyde, when the net present value of the project is zero. The feasibility was evaluated in terms of energy savings, CO2-emission reductions, and costs, in comparison to the separate production of ethylene with conventional OCM technology alone. In addition, the process was screened to look for possible future technology developments. The economics have been reviewed and compared to the classical OCM process. Capital requirements and the potential to enhance cash margins are the primary focus of the analysis. The basis of the analysis is a world-scale conventional OCM plant that converts 2750 Mm3 per day of methane into 320 metric tons per day of ethylene, 17 MW of electricity and 77 metric tons per day of formaldehyde. Capital and operating costs are for an arbitrary remote location where natural gas is available at 246.9 € per 1000 m3. Payout time for this process, with OCM plant, is less than ten years. This payout time is generally much higher, which indicates that the OCM process alone currently is not candidate for commercialization without consideration of special incentives, such as power generation and formaldehyde production. The analysis also includes an evaluation of the effects of process yields on the economics for two potential technologies: oxidative coupling to ethylene and co-generation of electricity as well as formaldehyde production. This analysis suggests areas for research focus that might improve the profitability of natural gas conversion.

Acknowledgment The authors acknowledge support from the Cluster of Excellency "Unifying Concepts in Catalysis" coordinated by the Berlin Institute of Technology and funded by the German Research Foundation – Deutsche Forschungsgemeinschaft.

(555d) Developing E3 Crude Oil Blending Strategy for Petroleum Refineries

Vibhor Mittal, Jian Zhang and Qiang Xu, Dan F. Smith Department of Chemical Engineering, Lamar University, Beaumont, TX

Abstract

Crude oil blending is a very common practice in petroleum refineries, where the main focus is to minimize the total purchasing cost of the crude oils under specified blending oil properties. Crude-oil blending also has a significant impact on the energy consumption during the crude-oil processing, because the product qualities of crude-oil processing to the downstream process usually should be kept as constant as possible. Conceivably, the energy consumption will heavily influence the emission of greenhouse gas (GHG) such as CO2. In this paper, an energy and emission considered economic crude-oil blending (E3) strategy is presented. The E3 strategy couples crude blending model rigorous crude processing model and is implemented in Aspen Hysys simulation package, such that it can simultaneously take into account concerns of production profitability, energy saving, and emission reduction. The development will provide technical support for optimal crude oil blending and processing in petroleum refineries. The efficacy of the methodology is demonstrated by a real case study.

See more of this Session: Process and Product Development for Sustainability
See more of this Group/Topical: Process Development Division

(555e) The Sustainable and Scalable Synthesis of Piperylene Sulfone: A "Volatile" and Recyclable DMSO Substitute

Gregory Marus[1], Eduardo Vyhmeister[1], Pamela Pollet[2], Megan Donaldson[1], Veronica Llopis Mestre[1], Sean Faltermeier[1], Renee Roesel[1], Leslie Gelbaum[2], Charles L. Liotta[2] and Charles A. Eckert[3], (1) School of Chemical & Biomolecular Engineering, Georgia Institute of Technology, Atlanta, GA, (2) Chemistry and Biochemistry, Georgia Institute of Technology, Atlanta, GA, (3)School of Chemical & Biomolecular Engineering and Specialty Separations Center, Georgia Institute of Technology, Atlanta, GA

We have established a roadmap to a sustainable and scalable synthesis of piperylene sulfone (PS), a potentially important new dipolar aprotic solvent. PS is a fully recyclable solvent with solvatochromic properties similar to dimethyl sulfoxide (DMSO). Although DMSO is an important and powerful polar, aprotic solvent for conducting chemical reactions, the separation of the reaction products from the solvent is difficult and expensive. In contrast, PS can be decomposed by undergoing a reversible retro-cheletropic reaction at 110ºC, permitting facile solvent removal and recycle. Because PS is not commercially available, we synthesized laboratory quantities using a method not optimized to scale-up due to expensive chemicals and significant waste generation. In order to develop and optimize a scalable process, we first determined the kinetic parameters associated with the reaction by employing in-situ proton NMR measurements and then studied the effects of radical inhibitors in reducing unwanted side reactions. In addition, we have now recovered PS from the reaction mixture via a sustainable carbon dioxide separation method, which resulted in a substantial waste reduction. Thus, we have developed a cost-effective, safe, and sustainable scale-up method for PS, a "volatile" and recyclable DMSO substitute.

See more of this Session: Process and Product Development for Sustainability
See more of this Group/Topical: Process Development Division

(555f) A Novel Solution Strategy with the Event-Tree-Based Modeling Approach for Crude Oil Scheduling

Xiao-Rong He[1], Lai-Xi Zou[1], Chufu Li[1] and Chaowei Liu[2], (1)Department of Chemical Engineering, Tsinghua University, Beijing, China, (2)Dan F. Smith Department of Chemical Engineering, Lamar University, Beaumont, TX

Crude oil scheduling is an important sector for the supply chain sustainability. In order to quickly obtain an optimal scheme for the large-scale crude oil scheduling, a novel solution strategy is proposed with the event-tree-based modeling approach. The new modeling approach is dynamically solving the event-tree induced real-time models consisting of all feasible states in the scheduling horizon, which is completely different from mathematical programming techniques. The event triggered rules are used to search for the single node of the event-tree model, where the continuous time presentation is adopted to divide the scheduling horizon. To reduce the complexity of the problem, the entire scheduling problem is decomposed to an unloading sub-problem and a charging sub-problem. Finally, the depth-first search algorithm (DFSA) is employed to search the optimal solution. The major improvement of this work is that it greatly improves the searching efficiency compared with the breadth-first search algorithm (BFSA). Case study results show that the event-tree-based approach with DFSA is an effective and efficient way for dealing with the crude oil scheduling problems.

See more of this Session: Process and Product Development for Sustainability
See more of this Group/Topical: Process Development Division

(613a) A "Green" Pilot Plant Process Cooling Alternative

Frank P. Lee and Michael K. Corrigan, Xerox Research Centre of Canada, Mississauga, ON, Canada

The Xerox Research Centre of Canada (XRCC) Pilot Plant is a multi-purpose specialty chemicals pilot plant. Originally designed and built in early 1980s, the pilot plant has the mission of enabling pre-manufacturing R&D and commercialization of specialty materials, including organic, inorganic, and composites, for use in document printing and other applications.

To satisfy temperature requirements of different types of chemical processes, dedicated heat transfer systems, containing media including steam, water, glycol and hot oil, were installed in the pilot plant to cover temperature ranging from -25oC to 350oC. Common to these heat transfer systems had been domestic water supply, which was the primary source of process cooling. Domestic water in distribution headers was piped to heat transfer systems for removing heat generated by chemical processes and was immediately discharged to the sewer systems. As the plant grew in size and increased in utilization, more systems were connected and more domestic water was consumed. In recent year, the plant had been using significantly more water. The open water cooling system was costly and it warranted a greener and more innovative alternative.

As part of XRCC's long standing commitment to "Green" and "Sustainable" technologies and solutions, a team representing operations, plant support and facilities engineering was commissioned to design a new eco-friendly pilot plant cooling system that would offer better environmental benefits and provide improved operational features.

Based on thermodynamic and heat transfer studies carried out on existing equipment and processes, the team estimated that the plant's peak demand for process cooling would be close to two million BTU/hr. HVAC suppliers contacted suggested that a 200-refrigeration ton Closed-loop Chilled Glycol Circulation System, which consists of a 200-ton chiller, heat exchangers, pumps, a reservoir and dedicated process control system, would be required to handle heat loads of this magnitude.

After review systems designed by HVAC suppliers, the plant's engineering team made the following proposals to further improve the efficiency of the system and to reduce installation and operating costs:

- Incorporate a roof-mounted air cooler for cooling glycol medium with outside air, taking advantage of the plant's location in Southern Ontario, where winters tend to be long and cold, and average daily temperature frequently drops below 8oC. - Use variable speed motors and compressors, and to control temperature and flow rate of the new chilled glycol circulation system with the plant's exiting distributed control system - Tie in the chilled glycol system to the existing building automation system which is programmed weekly to start up and shut down plant utilities based on operation requirements - Purchase components instead of a packaged system and utilize as much as possible, in-house engineering, project management and installation capabilities

The project took about a year to complete from concept and design to commissioning and start up. Sizing and selecting the system's components to work adequately and efficiently under different demand conditions was a major challenge. Installing the system and keeping the plant in operation was another. Once completed, chilled glycol was first fed to major consumers of domestic water, including the plant's hot glycol and hot oil circulation systems. Heat exchangers for reactors, receivers and dryers would soon follow.

The new chilled glycol circulation system has been in operation for over a year. The project has proven successful and has dramatically reduced the plant's consumption of domestic water and sewage discharges. From late Fall to early Spring, the system was able to supply chilled glycol at 5oC to 8oC by circulating glycol to the roof-mounted air cooler without running the 200-ton chiller, a practically "free" cooling for optimum performance and less downtime. This and other built-in energy saving features has made the system more efficient and the project more "Green".

See more of this Session: Pilot Plants for Green and Sustainable Processes I
See more of this Group/Topical: Process Development Division

(613b) A Case Study in Miniplant Scale for the Carbon Dioxide Capture of the Oxidative Coupling of Methane Process

Steffen Stünkel[1], Reinhard Schomäcker[2] and Günter Wozny[1], (1)Chair of Process Dynamics and Operation, Berlin Institute of Technology, Berlin, Germany, (2)Institute of Chemistry, Secr. TC8, Technical University of Berlin, Berlin, Germany

The Oxidative Coupling of Methane (OCM) to ethylene is a promising alternative for the oil based industry. Due to yield limitation is the process uneconomically and not applied in the industry yet. Several alternatives were purposed, but an efficient and economical process could not found in the literature. In this novel approach the whole process, including downstream separations is investigated simultaneously theoretical and experimental in a miniplant scale. The process is divided into three units: the reaction, the purification and the separation units1. This article focuses on the purification part particularly. The design task for the purification unit in this case study is the removal of the acid gas CO2 from 22 mol% down to 1 mol% at an operating pressure of 32 bar after drying. As possible process alternatives chemical absorption, membrane separation and a combination of both to form an amine-membrane separation process are discussed. For the chemical absorption the Monoethanolamine (MEA) is a widely used solvent and is the bases in this study. The regeneration of the loaded amine is the most energy consuming step in the process. The case study was performed using Aspen Plus® and Aspen Custom Modeller® (ACM). For the absorption unit an in-build rate based model with an electrolyte NRTL - model and chemical equilibrium reactions for the water-MEA-CO2 system was used. For a standard absorption Process with 95 % CO2 removal including solvent regeneration and heat integration the thermal energy demand in the stripper of 3.7 MJ/kgCO2 at 50% solvent regeneration of a 40 wt% MEA solution was found. For the membrane unit an ACM model was developed using the permeability-solubility model with the free-volume-theory for the gas permeation including the Joule-Thomson effect as well as the concentration polarisation[2]. For the case study a rubbery Polyethylene oxide (PEO) and a glassy Cellulose Acetate (CA) membrane with experimental determined parameters by the GKSS3 was used. It was found, that the total CO2 removal with a single membrane unit causes product losses in the range of 30 % and is uneconomically for the OCM process. The use of a membrane unit for bulk removal of the CO2 from 22 mol% to 10 mol% causes 9 % ethylene losses. Therefore an integrated separation process was developed as a combination of a membrane and an absorption process for 95 % CO2 capture. In the first part a GKSS envelope-type CA membrane unit with a CO2/C2H4 selectivity of 12 is used for the bulk removal. In the second part the gas stream is purified from 10 mol% down to 1 mol% CO2 by an absorption process using 15 wt% MEA with thermal energy efficiency of 3.15 MJ/kgCO2. The combined membrane-absorption process achieves the design task of 95 % CO2 removal with a thermal energy demand of 1.6 MJ/kgCO2 with acceptable ethylene losses. Reducing these losses from 9% down to 2 % can be achieved with a two step membrane process for the bulk removal with an electrical energy demand of 2.75 MJ/kgCO2. In the next step the simulations of the amine process has to be extended to kinetic reactions and the results have to be validated experimentally for the miniplant scale.

References:

1. Stünkel et. al. Modeling and simulation of a hybrid separation process for the carbon dioxide removal of the oxidative coupling of methane process, Comp. Aided Chem. E., Vol. 26, 2009, p. 117-122 2. Brinkmann et al. in Ohlrogge, Membranen, Wiley-VCH, 2006, pp. 273 3. GKSS – Research Centre Geesthach, Department of Process Engineering, Max-Planck-Straße 1, 21502 Geesthacht, Germany

See more of this Session: Pilot Plants for Green and Sustainable Processes I
See more of this Group/Topical: Process Development Division

Deep Dehydration of Organic Solvents by Pressure Swing Adsorption

Akitoshi Ito[1], Yasushi Fujimura[1] and Hiroshi Umino[2]

[1] JGC Corporation, 2205 Narita, Oarai-machi, Higashi-ibaraki-gun, Ibaraki, 311-1313 Japan
[2] JGC Corporation, 2-3-1, Minato Mirai, Nishi-ku, Yokohama 220-6001, Japan

1. Introduction

Dehydrated ethanol is in high demand for industrial use. It is well known, however, that ethanol forms azeotropes with water at 89.4 mol% (95.6 wt%) ethanol, and it is impossible to separate water from ethanol at concentrations lower than this by common distillation[1)].

For these reasons, ethanol dehydration has been researched. Azeotropic distillation using benzene, toluene, cyclohexane, diethyl ether or pentane as entrainer and extractive distillation using ethylene glycol or glycerol as solvent are examples. Although azeotropic and extractive distillation are proven and widely used processes, both have drawbacks in terms of their complicated process flow, difficulty of operation and high energy consumption.

As global warming becomes a serious problem, many industries are required to reduce energy consumption. One way to reduce energy consumption is to adopt a process which uses less energy. Since distillation, including azeotropic and extractive distillation, is one of the most energy consuming separation processes, the benefits of using an alternative process can be huge.

Vapor Phase Pressure Swing Adsorption (PSA), which JGC had developed with Japan Synthetic Alcohol Co. Ltd., is a solution for these issues. The process can deeply dehydrate ethanol with low energy consumption, easy operation and compact plant size. It is also found that other organic compounds which form azeotropes with water can be treated with this process.

In this paper, the advantages of Vapor Phase PSA are explained and the effects of operation parameters such as cycle time and the height of the adsorption bed are discussed based on experimental results.

2. Advantage of Vapor Phase PSA Technology

2.1 Process Flow of Vapor Phase PSA

Figure 1 shows the Vapor Phase PSA overall flow which is combined with the ethanol distillation process. Aqueous ethanol is supplied to the distillation column at first and near azeotrope product is distilled as overhead vapor. Then, overhead vapor is superheated and sent to the PSA unit. In the PSA unit, dehydrated ethanol is obtained from the top of the adsorption column as a product. Adsorbed water is recovered under reduced pressure using a part of the product as purge gas. Since this desorbed water contains ethanol used as purge gas, it is recycled to the distillation column. So the recovery ratio of the ethanol is almost 100%.

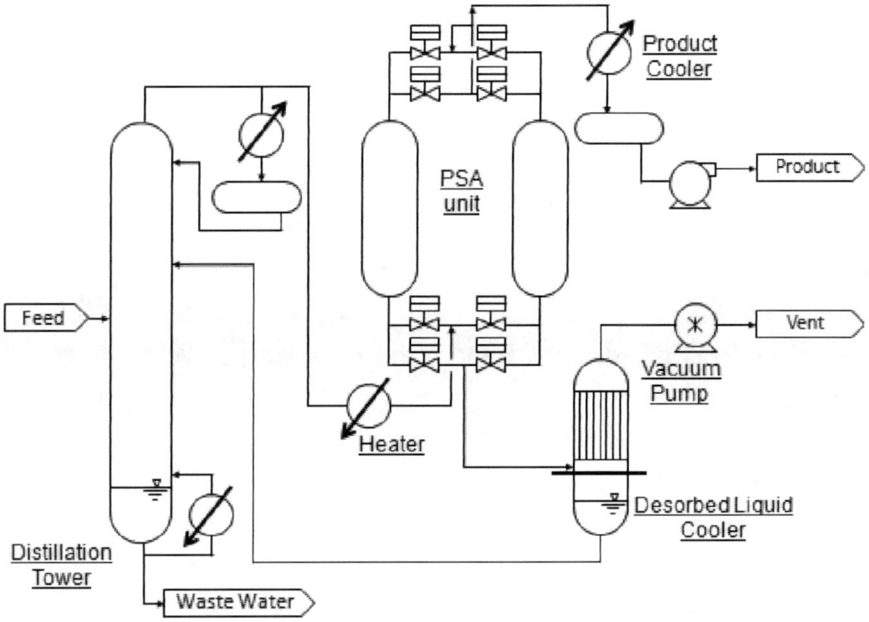

Figure 1. Overall process flow of ethanol dehydration.

Vapor Phase PSA uses two adsorption columns. By switching each column's role, dehydrated ethanol is produced continuously. Figure 2 shows the flow of one cycle. In Figure 2(a), feed vapor is sent to the left column which is in the adsorption-step. Water is adsorbed on adsorbent. In our test, molecular sieve 3A is packed in the column as adsorbent.

While the left column is at the adsorption-step, the right column is at the desorption-step. Here, the right column is regenerated by desorbing water. Regeneration is carried out under reduced pressure using a part of the dehydrated product as purge gas to reduce the partial pressure of the water.

After the desorption-step, the right column proceeds to the re-pressurization-step and pressure is raised to adsorption pressure (Fig. 2(b)). The pressure raised by feeding a portion of the dehydrated product by shutting the outlet valve. Then, the right column proceeds to the adsorption-step and the left column proceeds to the regeneration-step (Fig. 2(c)). By repeating this sequence, continuous production of dehydrated ethanol can be achieved. These operations are run by the opening and closing of 8 valves, placed on the tops and bottoms of the columns. This makes operation very simple and easy.

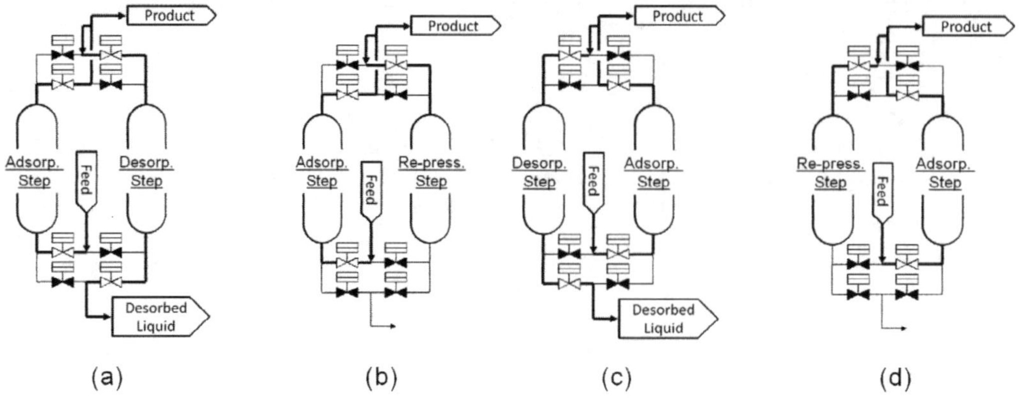

Figure 2. Valves action and process flow on a cycle.

2.2 Energy Consumption of Vapor Phase PSA

Energy consumption of PSA is compared with azeotropic distillation by a heat balance calculation. Energy consumption to distill feed ethanol to near azeotropy was not included. For the PSA unit, the main energy consumption is the additional distillation energy for treating desorbed liquid and heating energy of feed vapor before PSA. Assuming the reflux rate is 4 and 20% of the product is used for purge gas, additional distillation energy consumption is 1306 kJ/kg. The heater uses 67 kJ/kg.

Dehydration by azeotropic distillation consumes 3167 kJ/kg to dehydrated ethanol production using n-pentane as entrainer[2)]. This calculation suggests Vapor Phase PSA can reduce energy consumption by more than half.

2.3 Dehydration Performance

Vapor Phase PSA dehydrates ethanol and other organic compounds to the extent of 100 wtppm water or less. It is not easy for azeotropic or extractive distillation to dehydrate to this extent. For membrane technology, it still needs more improvement to dehydrate lower than 100 wtppm, economically.

3. Experimental Apparatus

Two PSA plants, a pilot plant with 5 m column and a bench plant with 1.5 m column, are installed in the JGC Technology Research Center. Pilot Plant can change adsorption bed height. Figure 3 is pictures of pilot plant and bench plant.

Figure 3. Pictures of Pilot Plant (left) and Bench Plant (right)

Figure 4 shows the process flow of the pilot plant. The flow of the bench plant is basically the same. This plant recycles product and recovered desorbed liquid, and uses it as feed stock in order to operate without make-up of feed. Therefore, continuous operation can be achieved without manual operation except for tuning the condition.

Feed solvent is supplied to the bottom of the adsorption-step column by the feed pump after it is vaporized. Water in the feed vapor is removed and dehydrated product is obtained from the top of the adsorption column. Product is cooled through the product cooler and stored in the product tank. A part of the product is sent to the vaporizer by the regeneration pump to use as purge gas. Purge gas is sent to the top of the desorption-step column. Adsorbed water is desorbed and flows with purge gas from the column, then it is condensed and stored in the desorbed liquid tank.

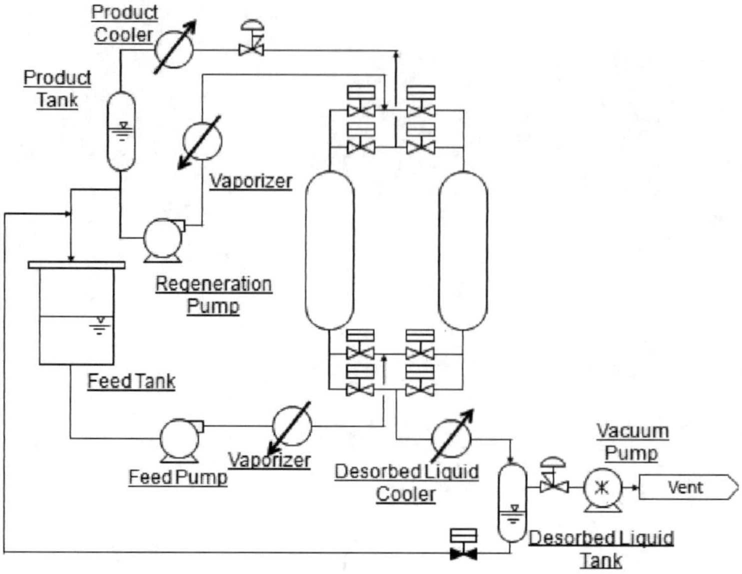

Figure 4. Process flow of 5m-Column Pilot Plant.

4. Result and Discussion

4.1 Performance of Dehydration

Figure 5 shows the relationship between the relative superficial velocity and the water concentration of the product. The superficial velocity is expressed as a relative value, setting the maximum allowable superficial velocity as a base. Hereafter, the maximum allowable superficial velocity is defined as the highest superficial velocity at which the water content in the product does not exceed 100 wtppm. Also, water concentrations below 10 wtppm are expressed as "10 wtppm" on graphs, because this value is the detection limit of water content analysis.

As can be figured out from the graph, by keeping superficial velocity under the maximum, the water concentration of the product is maintained stably at less than 100 wtppm.

Also, this graph shows that the water concentration of the product increases very rapidly when the maximum allowable superficial velocity is exceeded. This result suggests Vapor Phase PSA has achieved the formation of a very sharp adsorption band, which is a characteristic of this technology.

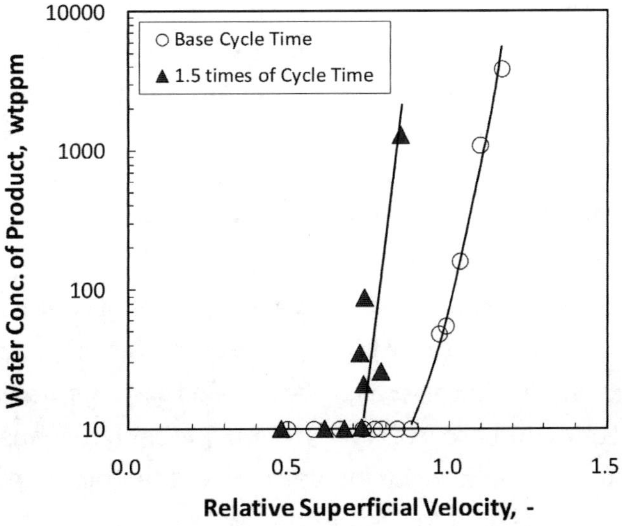

Figure 5. Relation between relative superficial velocity and product water concentration.

Base of relative superficial velocity is set at the maximum allowable superficial velocity for base cycle time.

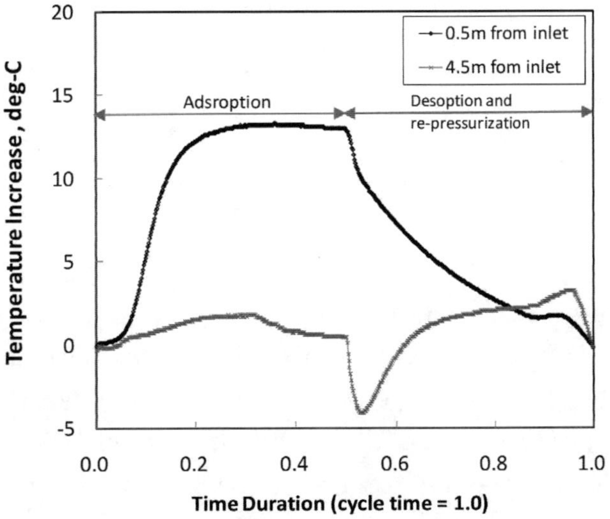

Figure 6. Temperature change of adsorption bed during one cycle

Figure 6 shows temperature change of adsorbent at 0.5m and 4.5m position from bottom of the adsorption bed at 0.7 of relative superficial velocity. In adsorption-step, adsorbent temperature at 0.5m position from bottom increases quickly and reaches to plateau due to adsorption of water. In the desorption-step, desorption of water proceeds by using heat of adsorption. At 4.5m position, temperature is stable during a cycle. This result indicates adsorption band don't reach to the position.

4.2 Effect of Cycle Time

The effect of the cycle time was investigated. It is expected that, as the cycle time gets shorter, the maximum allowable superficial velocity increases. Figure 5 indicates the effect of cycle time on the dehydration performance. By extending the cycle time to 1.5, the maximum allowable superficial velocity decreases by 0.75 times. This is because the total amount of water fed to the adsorption bed in one cycle is 1.5 times compared to the basic condition for the same superficial velocity condition.

As Table 1 shows, the maximum allowable amount of water in one cycle to keep the water content of product below 100 wtppm decreases as the cycle time decreases. This indicates that the maximum allowable superficial velocity in the same operation duration is not inversely proportional to the cycle time. This is probably because the adsorption band gets broader as the superficial velocity increase. The adsorption band gets broader as the superficial velocity increases and the front of the adsorption band, where the vapor water content exceeds 100 wtppm, does not progress forward proportionally to the change in the superficial velocity.

Table 1. Relationship of cycle time and maximum allowable amount of water in one cycle.

Relative Cycle Time	Relative value of Maximum Allowable Amount of Feed Water*
1.0	1.0
1.5	0.75

*Fed water in one cycle is calculated by multiplying the water flow rate at the maximum allowable superficial velocity and the cycle time.

4.3 Effect of Adsorption Bed Height

This investigation is carried out to evaluate the maximum allowable superficial velocity at various heights. In Figure 7, the relative superficial velocity and the water concentration of the product are shown for bed heights of 3, and 5 m. Although higher superficial velocities can be used at higher bed heights, the maximum allowable superficial velocity is not proportional to the bed height. This result also can be explained by the broadened adsorption band caused by higher superficial velocity.

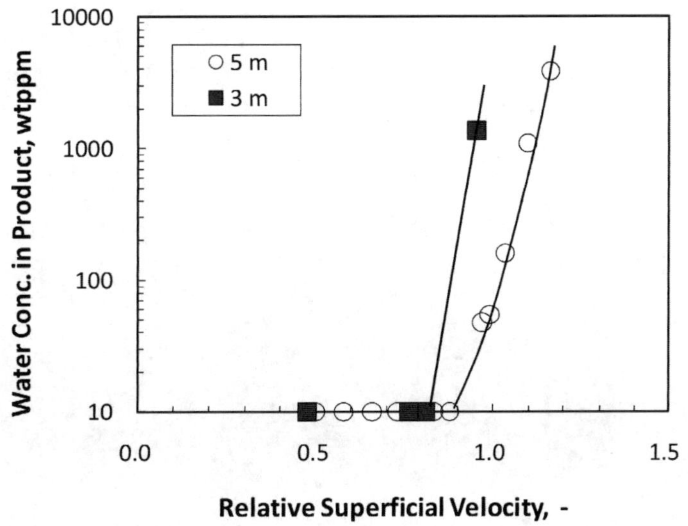

Figure 7. Relationship of superficial velocity and product water concentration for 3 m and 5m columns.

Relative superficial velocity is based on the maximum allowable superficial velocity for a 5m -Column.

4.4 Application to other solvents

Dehydration performance for other solvents than ethanol is also evaluated. Table 2 summarizes the result. These all form azeotropes with water and good deep dehydration performance by Vapor Phase PSA was confirmed. The PSA can be successfully applied.

Table 2. Dehydration performance of Vapor Phase PSA.

Feedstock	Water Content	
	Feed wt%	Product wtppm
Ethanol	6.6	<100
t-Buthyl alcohol	4	<100
i-Propyl alcohol	13	<50
Acetone	5	<100
Tetrahydrofran	5	<200

4.5 Industrial Achievement

JGC has achieved completion of the engineering, procurement, and construction for six PSA commercial plants, one of which is shows below. The oldest one has been operating for 9 years without any major trouble and without any need to change or make-up the adsorbent.

Figure 8. Achieved Vapor Phase PSA commercial plant

5. Conclusion

JGC had been investigating the several operation parameters of Vapor Phase PSA dehydration technology and constructed a number of commercial plants. We will continue investigating the applicability of this dehydration technology to other organic compounds

Reference

1) Santosh Kumar, Neetu Singh and Ram Prasad (2010) "Anhydrous ethanol: A renewable source of energy," Renewable and Sustainable Energy Reviews 14 pp.1830–1844

2) Black. C., (1980) "Distillation modeling of ethanol recovery and Dehydration Processes for ethanol and gasohol," Chem. Eng. Progr., Sept. pp.78-85.

(613d) Esterification of Succinic and Acetic Acid Mixtures with Ethanol by Reactive Distillation

Alvaro Orjuela[1], Andrea del Pilar Orjuela[2], Aspi Kolah[3], Carl T. Lira[4] and Dennis Miller[3], (1) Chemical Engineering & Material Science, Michigan State University & Universidad Nacional de Colombia, East Lansing, MI, (2)Chemical Engineering, Universidad Nacional de Colombia, Bogota, Colombia, (3)Michigan State University, East Lansing, MI, (4)Chemical Engineering and Material Science, Michigan State University, East Lansing, MI

Bio-based succinic acid (SA) obtained by fermentation is considered a major platform molecule for the future chemical industry. It can be used as feedstock and intermediate in production of solvents, surfactants, biodegradable polymers, specialty chemicals and commodities. With some organisms, fermentation to produce SA also produces acetic acid (AcAc) and small amounts of pyruvic, lactic, and/ or formic acid, making difficult to isolate pure SA crystals. Among proposed alternatives to economically separate and recover these carboxylic acids esterification has proven an attractive alternative. Because esterification is equilibrium limited, the ongoing removal of products during reaction is required to achieve complete reaction. In that sense, reactive distillation (RD) columns can be used not only to overcome equilibrium limitations by removing reaction products, but it also can separate succinic esters from other carboxylic acid esters formed. In this work, a pilot scale reactive distillation process is demonstrated for simultaneous succinic acid and acetic acid mixed esterification with ethanol. Experiments in an elevated pressure, pilot-scale RD column demonstrate succinic acid and acetic acid conversions greater than 99% with recovery of diethyl succinate in a purity exceeding 98%. Ethyl acetate is recovered in a mixture of water and excess ethanol, from which it can be isolated as a valuable byproduct. Using experimental phase equilibria and reaction kinetic data, AspenPlus simulations of the reactive distillation column have been performed. An equilibrium stage model with kinetically controlled reaction reasonably predicts the pilot-scale column behavior and can be used to simulate a commercial scale unit. An economic analysis of such a commercial process has been carried out. Experimental and simulation results demonstrate that reactive separations provide the opportunity to economically produce bio-based chemicals produced in the biorefinery.

See more of this Session: Pilot Plants for Green and Sustainable Processes I
See more of this Group/Topical: Process Development Division

Development of Technology for Small-Scale Drying of Fish and its By-products

Emrod Elisante[1], Bernard Chove[2] and Charles Muchunguzi[3]

[1]Department of Chemical & Process Engineering, University of Dar es Salaam

P. O. Box 35131 Dar es Salaam, Tanzania, elisante@udsm.ac.tz, Tel +255-754-865695

[2]Department of Food Science and Technology, Sokoine University of Agriculture

[3]Department of Development Studies, Mbarara University of Science and Technology

ABSTRACT

A technology for small-scale drying of fish has been developed, tested and demonstrated to a group of artisanal fish processors at the Dar es Salaam Ferry fish market. A matrix of factors for choice of technology was developed and used to determine the most suitable technology between saw-dust drying, solar driers, and oil burner. The technology for heavy fuel oil burner with air-flue gas heat exchanger was selected which can be converted to use waste oil 130,000 Btu/gal as energy source. The fish were put on wire mesh trays and stacked in shelves installed inside a tray drier chamber measuring 6 m L x 3 m W x 2.7 m H. Different types of fish like Anchovy- *Stolephorus baganensis*; Ponyfish- *secutor insidiator;* and Herring- *spratelloides delicatulus* were dried. The envisaged drier capacity is 1.0 ton/day accommodating 250 kg of fish per batch each taking 2 hours for drying and one hour for loading and unloading. A desired drying temperature of 70 ^{o}C could not be uniformly attained throughout the chamber due to lack of mechanism to distribute the hot air. The product obtained was packed and subjected to water activity and culinary tests, which showed a water activity of 0.6 and 90% acceptance in favor compared to similar products in the market for aspects like taste, appearance, and softness after cooking. More work need to be carried out prior to dissemination of the technology. This include optimization of the drier parameters like: (i) uniform distribution of hot air inside the chamber; (ii) establishing suitable purging schedules to remove moisture load during operation; (iii) installation of a parallel chamber so that the oven doesn't have to be switched off during unloading and loading; and (iv) establishing comprehensive cost-benefit analysis once sufficient data has been gathered.

Key words: Small-scale fish processing, tray drier, technology development, drying.

INTRODUCTION

Fish give a good balance of protein, vitamins and minerals constituting more than 60% of total protein intake in adults especially in rural areas (Bukola *et al.*, 2008). It is a widely accepted delicacy that cuts across socio-economic, age, religious and educational barriers (Adeleye, O. A., 1992). The processing and preservation of fresh fish and its by-products is of utmost importance because they are highly susceptible to deterioration immediately after harvesting. If fish is not sold fresh then proper preservation measures must be taken to prevent losses and extend its shelf-life (Davies *et al.*, 2009, Okonta and Ekelemu, 2005). Large amount of fish waste and deteriorated whole fish are discarded daily in the fish canning industry (Faid, *et. al.*, 1997) due to microbial activity.

In many developing countries, the processing of fish and its by-products is carried out by subsistent fishermen or artisanal processors using traditional technologies like drying, smoking and salting. The processing is usually undertaken on lake shores under temporary shades using tools like wire-mesh, wood spikes and ropes for hanging fish. Since mid 1990s about ten large-scale Nile perch fillet packaging industries have been installed in East Africa towns of Jinja, Kisumu and Mwanza (Elisante *et al*, 2006). Most of the fillet is exported and by-products like skin, skeletons and heads are sold to artisanal processors, who employ various techniques (Figure 1-4) like: (i) de-fating the skin using boiling water; (ii) sun-drying the fish skeletons on wooden rafts; (iii) salting the skeletons for preservation; (iv) heat drying of heads over wire mesh using burning saw-dust; and (v) deep frying the fish heads in cooking oil. The dried skeletons are used by the local milling industry as a source of calcium in animal feeds and the fried fish heads are sold as protein for human food.

Figure 1: De-fating Fish Skins by Boiling Employing Used Tyres as Fuel at Buhongwa, Mwanza- Tanzania

Figure 2: Sun-drying of Nile Perch Heads and Skeletons at Mwaloni Market in Mwanza-Tanzania

Figure 3: Drying Fish-Skeletons at Obunga Site, Kisumu-Kenya

Figure 4: Drying Fish-Heads by Saw-dust, Kanyama Site, Mwanza-Tanzania

The de-fated skin is dyed and pressed for making leather products like belts and shoes. Despite these advantages, the processing of by-products using the above techniques has several environmental and technical drawbacks, which include:

1. Discomfort to the general public due to fouling smell and smoke pollution;
2. Energy inefficient and unhygienic processing, as shown in Figs 3 - 4, which result into poor product quality, sub-optimal utilization of resource, and environmental degradation;
3. Poor safety of operators characterized by hazardous working conditions; and
4. Lack of control in process parameters like, drying temperature, moisture content, and throughput.

Health and safety problems have been cited in areas where fish drying using firewood is practiced especially lung related ailments caused by smoke inhalation (Peters, *et. al.*, 1999). Furthermore, lack of proper premises cause wastage and difficulties in processing during cloudy or rainy days in the case of sun-drying. This research aims at developing a technology for processing fish or its by-products in a manner that improves indigenous technologies to tackle the above mentioned shortfalls.

METHODOLOGY

Choice of Technology

The choice and development of technology for large-scale industry is relatively easy compared to small and medium enterprises (SMEs) which are constrained by many factors like: affordability, attitudes, social acceptance, entrepreneurial and technical skills. In a survey conducted in 2006 (Elisante *et al.*, 2006) it was found that the fish processors in Mwanza, Jinja and Kisumu use three types of technology to process fish and its by-products, namely, (i) sun- or saw-dust drying (ii) de-fating and drying of skins; (iii) deep frying of heads and skeletons; and (iv) salting. Out of these techniques, only drying was practiced in all the three areas because the dried fish or by-products had multiple uses as animal feed and human consumption.

Since it was not possible to develop a technology that caters for different processing needs, it was decided to focus on drying technology only because it was practiced in all areas surveyed. The main challenge was to keep an energy source that is accessible and familiar with the local processors while weighing on other factors like affordability, reliability, and the environment. The following energy sources were therefore proposed: (i) solar drying; (ii) firewood/saw-dust drying; and (iii) waste oil burner technology. The matrix shown in Table 1 was used to choose a technology that would suit the processing needs taking into account factors like productivity, income generation, and availability of fish or its by-products in the surveyed areas.

Table 1: A Matrix for Choosing a Drying Technology Based on the Source of Energy

Attribute	Energy Source						
	Weigh (w)	Solar tunnel driers	Score (S/10)	Saw-dust drying	Score (S/10)	Waste oil cabinet driers	Score (S/10)
Economic Capital costs *(Land building, machines and utilities).*	0.3	About US$ 2,000 for a five stage visquen covered cabinets.	6	Requires open drying racks, wire mesh, and tools costing roughly US$ 500.	9	Equipment cost roughly US$ 9,000. Requires built-up space.	7
Operating costs *(labor, fuel, raw-materials)*	0.15	Labor intensive to load, unload and turn..	7	Labor intensive attending firing, cost of firewood.	8	Labor only for loading and unloading. Utilities- fuel, water, and electricity	6
Cost-benefit: value addition *(price, quality, appearance, shelf life ,consumer taste).*	0.15	Regarded as post-harvest facility. Product has moderate price.	6	Low product cost sold to low income bracket. Short pay-back on investment.	5	High quality due to automatic temperature control. Product has long shelf life hence expensive.	9
Social Employment generation, productivity, and gender distribution.	0.1	Low productivity ~160 kg/day per person. Equitable gender mix.	6	Subsistence occupation practiced mostly by women.	4	High productivity. Gender skewed towards men as machine operators..	8
Technical Spare parts, skills, maintenance, availability and reliability.	0.2	Medium throughput. Used during day-time only. Unavailable in rainy or cloudy days. Low maintenance cost. Low skills required.	5	Low throughput. No special skills required. Can be done under a shade.	6	High throughput. Operable 24 hrs a day. Machine may break down. Prone to power outages. Requires skilled manpower to do maintenance.	7
Environmental Pollution, sanitation, degradation, sustainability.	0.1	Risk of exposure to dust, insects and harmful micro-organisms and maggots.	4	Unhygienic handling. Cause deforestation. Exposure to smoke and dust.	5	Operated under sanitary conditions- protected drying cabinets. But cause CO_2 emissions from flue gases.	6
Total	**1.0**		**5.75**		**6.6**		**7.15**

In Table 1 the main attributes were assigned weights commensurate to their importance then an 'expert opinion' was employed to score points between 0 and 10 for each attribute based on the indicated criteria. The attribute score is obtained by multiplying the weight by scored points then the total is obtained by adding all scores for a particular energy source. As shown in Table 1, the waste oil-cabinet drier had highest aggregate score of 7.15 so it was selected for development.

Technology Development

Burner design: After choosing the waste oil burner technology, it was found convenient to first design a simple inexpensive low-pressure burner powered by a single phase 1.5 Hp electric motor. Flame ignition is done manually by the operator who also controls fuel inlet through a valve positioned on the fuel line. An assembly drawing of the burner unit and detail of the firing nozzle are shown in Figure 5. Atomization is achieved by forcing high velocity air blown by the motor fan rotating at 2600 rpm in a narrow spiraled neck that imparts a swirling motion to atomize the fuel as it comes out at the center of the nozzle. An adjustable air door (3) is provided to help maintain the required air-fuel ratio. The estimated power of the unit was 100 kW when burning 2.4 L/ hour of heavy furnace oil of viscosity 125 cSt and energy content of 130,000 Btu per gallon.

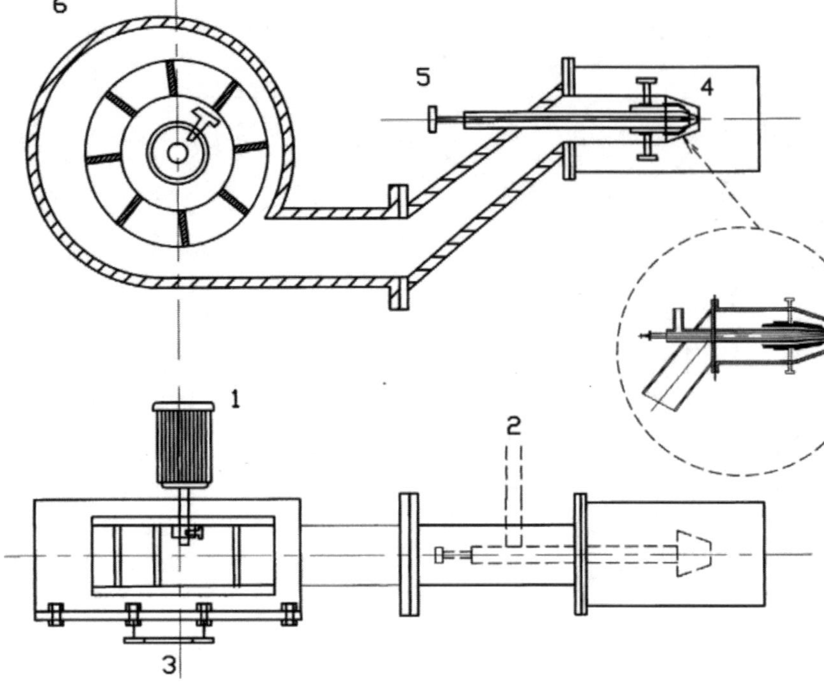

1-Motor; 2- Fuel inlet; 3- Adjustable air inlet; 4- Nozzle; 5- Fuel adjusting valve; 6-Blower

Figure 5: Details of Burner and Fuel Nozzle Design

Heat Exchanger Design: Based on the field survey (Elisante *et al.*, 2006), the daily total average of fish waste in Mwanza, Kisumu and Jinja was 10 tons so the drier capacity was set at 3 tons per day assuming that not all waste is dried, and multiple driers can be used per site. In order to obtain high productivity and throughput, a batch size of 500 kg per 2 hours of drying was proposed but as shown later, the load was reduced to cut down batch time. The desired heat transfer area A was obtained from energy balance in the burner chamber and the fish drying chamber using a coupled lumped parameter model:

$$m_b C_{ps} \frac{dT_b}{dt} = F_a \rho(\lambda_{in} - \lambda_{out}) + Q - UA(T_b - T_f) \qquad 1$$

$$m_f C_{pf} \frac{dT_f}{dt} = UA(T_b - T_f) - Q_w \qquad 2$$

Where: T_b, T_f = temperatures of the burner and fish drying chambers respectively, m_f = mass of fish load batch; m_b = mass of steel for burner and heat exchanger; F_a = volumetric flow rate of air through the burner; Q = power dissipated by burner; λ_i = enthalpy of inlet air and flue gases; C_{ps}, C_{pf} = heat capacity of steel and fish respectively; and U = overall heat transfer coefficient for flue gases inside tubes forced air outside (Coulson and Richardson, 1999). Equations 1 and 2 were simulated using the MATLAB software and gave a heat transfer area of 6.2 m^2, which was provided by a cylindrical steel chamber Ø600 x 1500 mm and twenty nine steel tubes Ø2 inch x 1500 mm arranged in a triangular pitch as shown in Figure 6. The burner at position (2) fires hot flames into the hollow cylinder (4), the flue gases reverse direction and pass through the tube bank (3) before exiting through the chimney. A photo of the fabricated heat exchanger is shown in Figure 7.

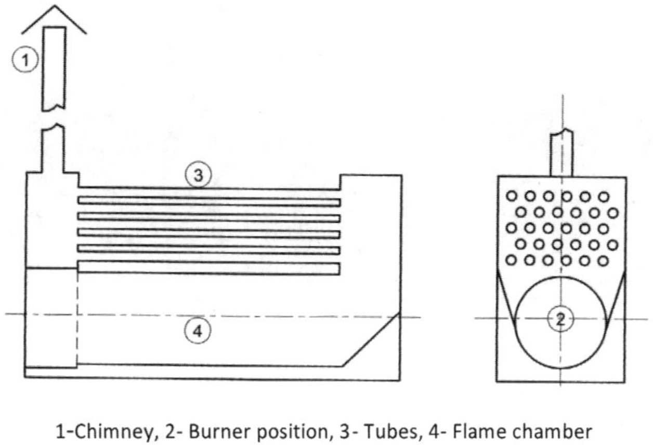

Figure 6: Heat Exchanger Tube Arrangement

Figure 7: The Fabricated Heat Exchanger

Experimental Layout: A schematic layout showing the process and instrumentation (P&I) diagram is shown in Figure 8 and Figure 9 shows a photo of the experimental drying chamber measuring 6 m L x 3 m W x 2.7 m H that was constructed at the Dar es Salaam Ferry fish market. The burner E4 represented in the mid-section is supplied with furnace oil fuel 125 cSt from tank E2. The flue gases exchange heat with process air before exiting from the chimney (1). During operation, a 5 Hp centrifugal fan blows the process air at a rate 3 m^3/min on the outside surface of the heat exchanger tubes into the drying chamber. The hot process air circulates through the materials to be dried, which are loaded in coffee mesh trays 18inch x 36 inch and stacked in shelves as shown in Figure 10. As the air cools down, naturally it sinks to the bottom, where it is drawn once again by forced draught into the heat exchanger and re-heated. Thermocouple sensors type K are positioned at various locations as shown in Figure 8 to read temperature via a ten channel Comark programmable microprocessor thermometer Model ZC6600.

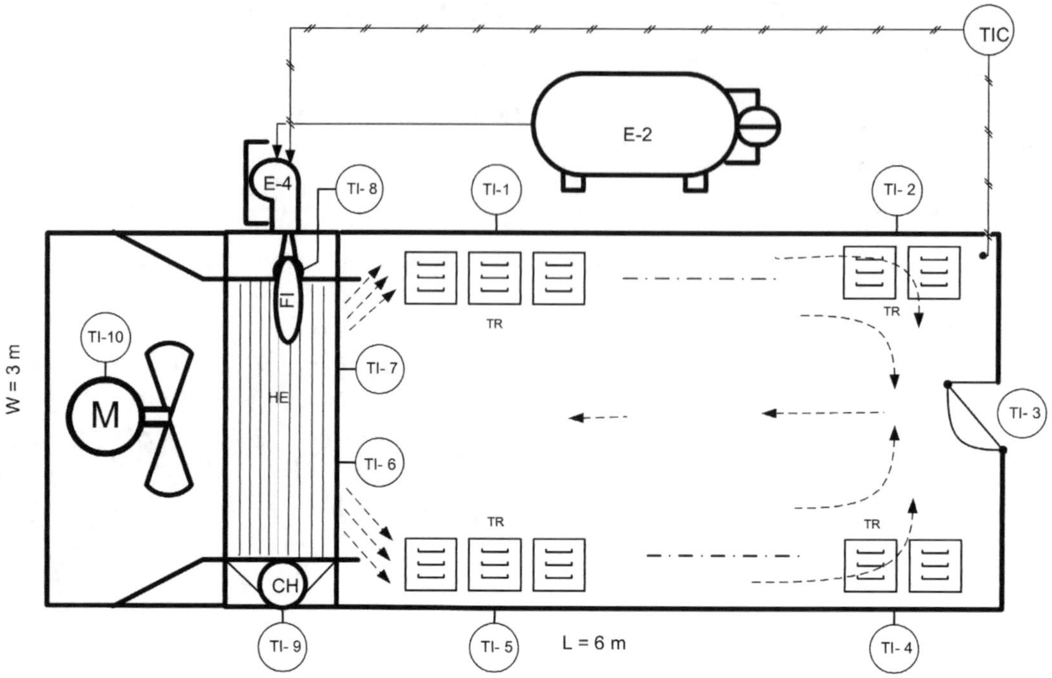

E2- Fuel tank, E4- Oil burner, HE- Heat exchanger, FI-Flue gases, M- Blower; CH- Chimney, TR- Trays, TIC- Temperature controller, TI- Temperature Sensor

Figure 8: Schematic Process and Instrumentation Diagram

Figure 9 Drying Chamber at Ferry Fish Market

Figure 10: A Tray with Fish

The burner and blower are isolated from the hot drying chamber using fiber glass insulation while the walls and roof are insulated by an air pocket created by two inch soft wood planks placed between walls and 18 gage aluminum sheets. The drying chamber is sealed everywhere except the front door which is used for loading and unloading.

RESULTS AND DISCUSSIONS

The equipment was tested for drying fish obtained from local fishermen fishing overnight using small boats in the Indian Ocean. As part of technology dissemination, the tests were done in conjunction to a group of 73 fish mongers called *Umoja wa Waanika Dagaa na Uuzaji wa Samaki Wakavu* (UWADUSWA) at the ferry fish market in Dar es Salaam. The fish to be dried was thoroughly washed in perforated plastic baskets then spread evenly about 1.5 kg per tray before inserting them in drying shelves. In total the drier shelves can accommodate up to 180 trays hence about 270 kg can be dried per batch. The different fish species used for testing are shown in Figure 11 and their Swahili names as identified by local fishermen are given in Table 2.

Table 2: Local and Foreign Names of the Tested Fish

Swahili Name	English Name	Species Name
Dagaa Mchele	Anchovy	*Stolephorus baganensis*
Dagaa Papa	White sardinella	*Sardinellar albella*
Dalidali	Delicate herring	*Spratelloides delicatulus*
Kibua	Mackerel	*Rastrelliger Kanagurta*
Kolekole	Kingfish	*Carangoides fulvoguttatus*
Palawe	Pugnose ponyfish	*secutor insidiator*
Saladini	Blue stripe herring	*Herklotsichthys quadirmaculatus*

1- Herring; 2- Ponyfish; 3- Anchovy; 4- Sardinella

Figure 11: Different Species and Sizes Used for Test Drying

Figure 12 shows no-load open-loop temperature transients for simulated and measured values of the burner flue gases, cold air inlet to the heat exchanger, and hot air exit from the heat exchanger to the drying chamber. It shows that the measured flue gas temperature reaches a steady-state value of 576 °C within 15 minutes while the heat exchanger outlet and inlet temperature reach 200 °C and 145 °C respectively. The simulated temperature values for burner chamber matches the measured values beyond 25 minutes but for the heat exchanger outlet, the simulated temperatures do not match with the measured values. This could be attributed to either: poor choice of model parameters; excessive heat absorbed by the brick walls and timber shelves; or heat lost through the roof due to poor insulation.

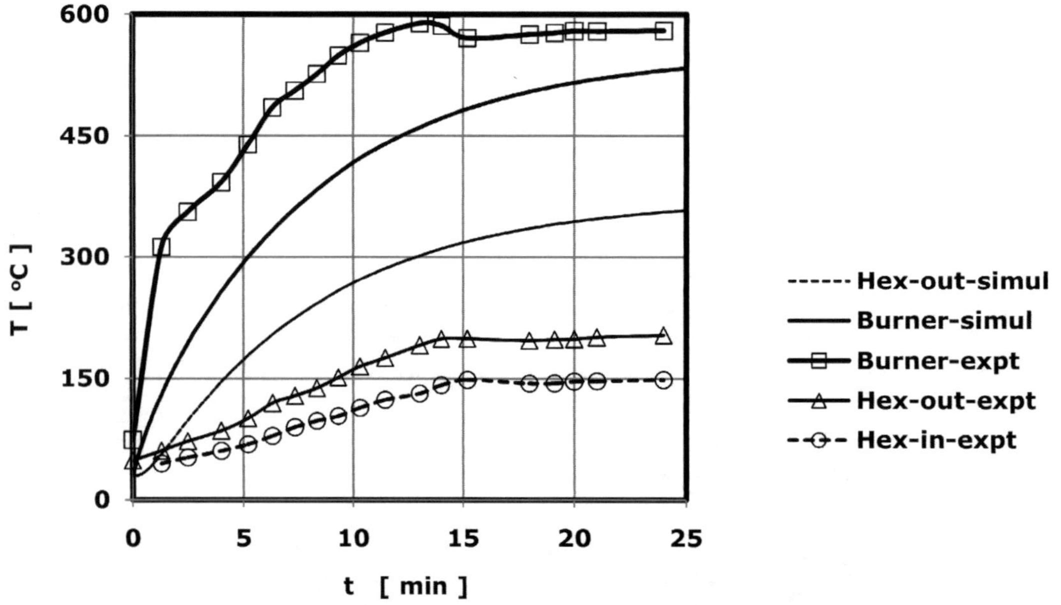

Figure 12: Open-loop Temperature Response Under No-load Conditions

Figure 13 shows temperature response at different sensor locations (see also Figure 8) when the drying chamber was loaded with 90 kg of Anchovy fish *Stolephorus baganensis*. The test was done under closed-loop feedback condition with a desired drying temperature of 70 °C at sensor TI-2. Compared to the empty chamber, the laden chamber temperature response is sluggish taking 30 to 60 minutes to achieve steady-state temperature at various sensor locations. Furthermore, due to load, the steady-state temperature at the chamber inlet TI-7 and outlet TI-6 dropped from 200 °C and 145 °C respectively to 127 °C and 64 °C. Due to uneven air mixing, the desired steady-state drying temperature of 70 °C could not be achieved uniformly inside the chamber. The average obtained was 87.5 °C with the lowest of 42 °C at sensor TI-4 and the highest of 109 °C at TI-3 near the door, directly facing the hot gas pipe from the heat exchanger. The non-uniform temperature distribution caused the following drawbacks:

1. Delayed drying for fish placed in the low temperature zones and fast drying rates at high temperature zones. This necessitated interchanging of trays from time-to-time so as to compensate the difference; and

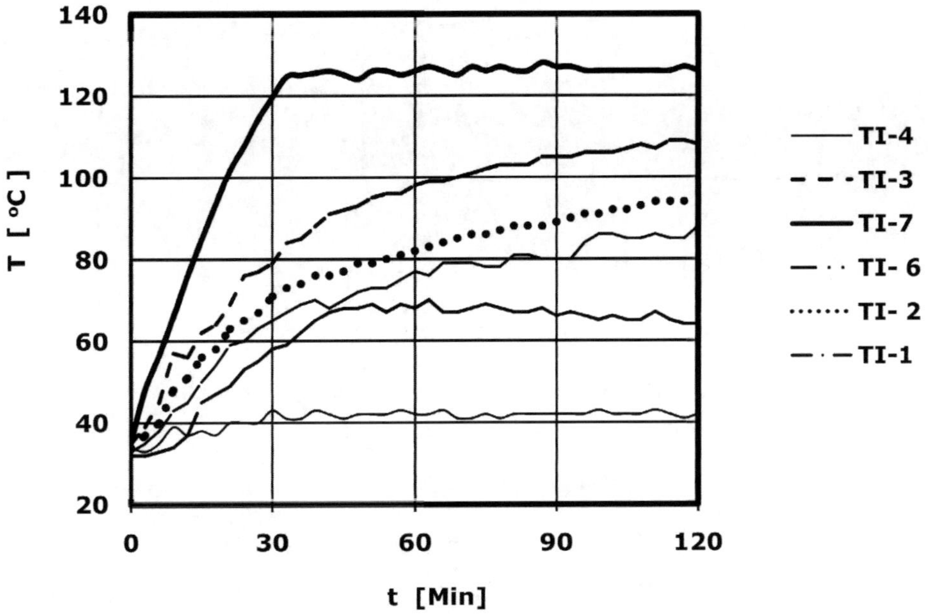

Figure 13: Temperature Response at Different Sensor Location

2. Sensor location problems whereby positioning a feedback sensor at a low temperature zone cause the thermostat to maintain burner firing when the high zone has already exceeded the desired temperature.

In order to ensure uniformity it will be necessary to install a distribution mechanism like a fan near the hot air inlet pipe. Figure 14 shows results of moisture content determination where 10 g fresh fish samples were dried in a laboratory oven until a constant weight was attained. Except for Anchovy the average water content of the species was about 70% however in order to prevent microbial activity the recommended moisture content for dried fish must be below 15% (w/w) and water activity (a_w) of below 0.85 (Eriahu *et al.*, 2006, Eboh, *et al.*, 2006).

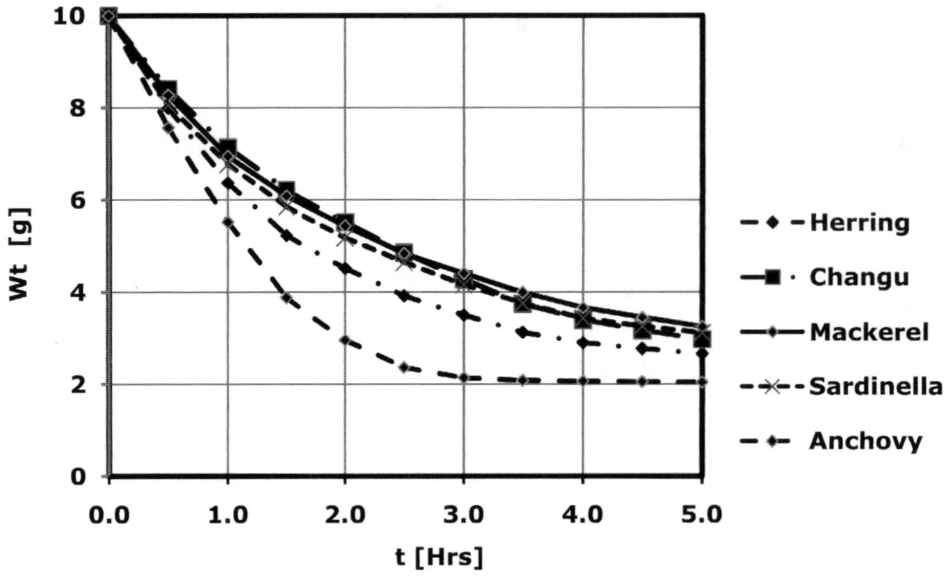

Figure 14: Moisture Content Determination for Various Fish Species

Figure 15 shows results of drying different species whereby to reach the acceptable moisture content the original fish weight had to be reduced by two-thirds. From cold start the oven took 2 - 3 hours to dry the fish, after which it must be switched off and cooled for 15 minutes before unloading. Since the batch size is 270 kg, it means that a total of 1.08 tons/day of fish can be dried in 4 batches-allowing 2 hours for drying and one hour for loading and unloading. Due to their size and weight, Anchovy fish with average weight of 10 g per piece and measuring 5 cm L x 1.5 cm W dried faster compared to Herring and Mackerel weighing 40 g measuring 12 cm L x 4 cm. When a bigger fish were sliced open to expose its inner flesh, the drying rate was similar to small size fish as shown for a 700 g Red sniper fish in Figure 15. This is consistent with drying theory (Coulson and Richardson, 1999) that rate is dependent on area exposed rather than weight of material being dried. Hence the drier can be used for drying different products like fruits, herbs, or fish skeleton as long as sufficient area is exposed and appropriate drying duration is prescribed.

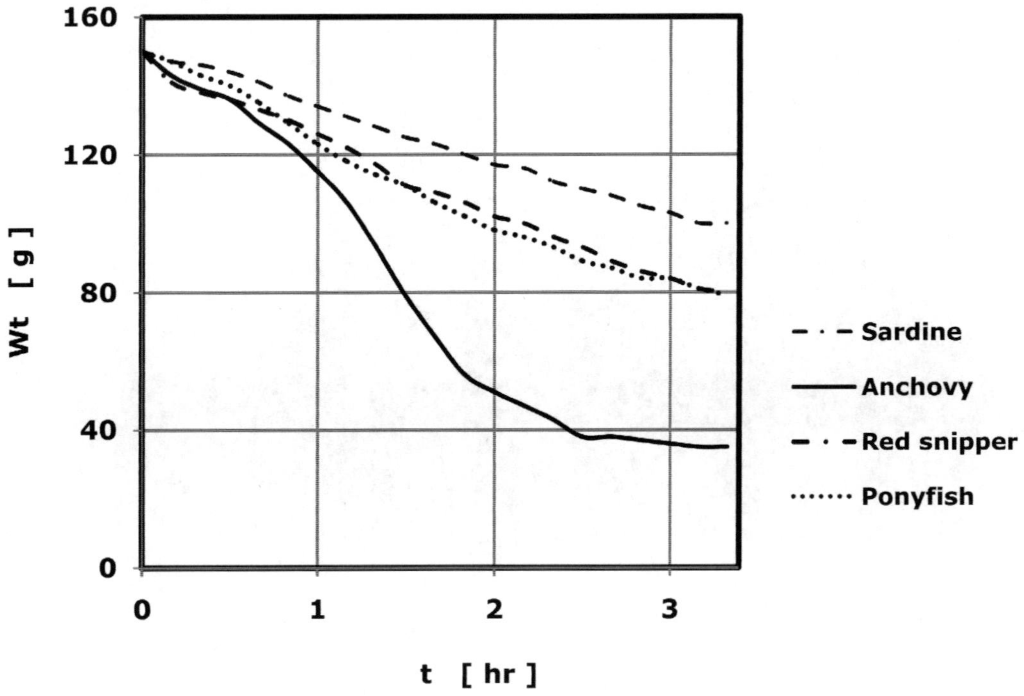

Figure 15: Drying Different Species and Sizes under Similar Conditions

The dried product was packed in laminated polypropylene bags 50 μm thick each pack weighing 250 g measuring 10 x 25 cm as shown in Figure 16. The packed products were subjected to water activity test which averaged 0.6 even when left for six weeks without any signs of mould formation. The product was also subjected to culinary tests by the researchers, UWADUSWA members, and by a group 10 volunteers picked randomly. They reported 90% favorable acceptance compared to similar products on the market for aspects like taste, appearance, and softness after cooking.

Figure 16: A Sample of Dried Anchovy Packed in a 250 g Polyethlyene Bag

CONCLUSIONS AND RECOMMENDATIONS

A tray drier technology for small-scale drying of fish up to one ton per day has been locally developed, tested and demonstrated to artisanal fish processors. Compared to existing technologies like solar drying and sun-drying, the technology is more efficient in terms of productivity and hygiene due to short drying duration and minimal handling. For example, the drier takes only 2 hours to dry 250 kg of fish compared to 5 days for medium solar tunnel drier or 2-3 days for sun-drying depending on cloud conditions. A group of processors involved in testing noted remarkable improvement in terms both quality and quantity obtained using the technology compared to sun-drying. Due to value addition they decided to sell the product only for human consumption at higher price instead of animal feed as was the case earlier.

A desired optimum temperature of 70 °C was not uniformly achieved throughout the drying chamber due to lack of mechanism to distribute the hot air. It is recommended that this bottleneck should be addressed with others which constrain optimal operation. The options suggested include:

1. Establishing suitable purging intervals using psychometric charts so as to enhance removal of moisture loaded in the hot air circulating between the heat exchanger and the drying chamber.

2. Improving the layout to create a parallel drying chamber that allows drying to take place in one chamber while unloading and loading occurs in the alternate chamber so the oven is not switched off.

At a cost of US$ 9,000 for equipment, the proposed technology is considerably expensive compared to existing technologies which range from US$ 500 for sun-drying to US$ 2,000 for a five-stage tunnel solar drier. However, as seen in the choice of technology matrix, a careful cost-benefit analysis should include socio-economic and environmental parameters. More field tests need to be carried out before widespread dissemination of the technology. However, judging from preliminary results, the technology offers higher productivity compared to subsistent methods used in East African riparian towns and the coastal belt.

ACKNOWLEDGEMENTS

The authors would like to acknowledge the financial support from the Lake Victoria Research Initiative (Vic-Res) of the Inter-University Council of East Africa (IUCEA) and the close cooperation from the Ilala Municipal Council of Dar es Salaam City for allowing installation and testing the drying facility at the Ferry Fish Market premises.

REFERENCES

Adeleye, O. A., (1992). Conservation needs of fisheries resources and reorientation for sustainable captive and culture practices. Proceedings of the 10th Annual Conference of the Fisheries Society of Nigeria, pp 230-234.

Bukola, C. A., Abiodun, A. O., and Ukpe, G. P. (2008). Mycofloral of Smoke-Dried Fishes Sold in Uyo, Eastern Nigeria, *World Journal of Agricultural Sciences*, **4** (3), 346-350.

Coulson, J. M., and Richardson J. F., (1999). Chemical Engineering, Vol. 2, Fourth Edition, Butterworth-Heinemann Publishers, Oxford.

Davies R. M., Davies, O. A., and Abowei, J. F., (2009). The Status of Fish Storage Technologies in Niger Delta, Nigeria, *American Journal of Scientific Research*, **1**, 55-63.

Eboh, L., Mepba, D. H., and Ekpo, B. M., (2006). Heavy metal contaminants and processing effects on the composition, storage stability and fatty acid profiles of five common commercially available fish species in Oron Local Government, Nigeria, *Food Chemistry*, **97**, 490-497.

Elisante, E., Chove, B. and Mchunguzi, C., (2006). Research Report submitted to Lake Victoria Research Initiative (VicRes), No. 1, 2006.

Eriahu, C. C., Kaze, S. A., and Achem, C. D.,(2006). Moisture sorption characteristics of tropical fresh water crayfish (*Procambarus clarkii*), *Journal of Food Engineering*, **75** (3), 355-363

Faid, M., Zouiten, A., Elmarrakchi, A., and Achkari-Begdouri, A., (1997). Biotransformation of fish waste into a stable feed ingredient, *Food Chemistry*, **60** (1), 13-18.

Okonta, A. A., and Ekelemu, J. K., (2005). A preliminary study of micro-organisms associated with fish spoilage in Asaba, Southern Nigeria, Proceedings of the 20th Annual Conference of the Fisheries Society of Nigeria (FISON), Port Harcourt, 14th -18th November 2005, pp 257-260.

Perry, R. H., (1985). Perry's Chemical Engineers' Handbook, 6th Edition, Mc Graw-Hill Co., New York, p 812.

Peters E. J., Esin, R. A., Immananagha, K. K., Siziya, S., and Osim, E. E., (1999). Lung function status of some Nigerian men and women chronically exposed to fish drying using burning firewood. *Central African Journal of Medicine*, **45** (5), 119-124.

BARK-BASED BIOREFINERY – FROM PILOT EXPERIMENTS TO PROCESS MODEL

Sakari Kaijaluoto, VTT Technical Research Centre of Finland, Jyväskylä, Finland.
Lotta Sorsamäki, VTT Technical Research Centre of Finland, Jyväskylä, Finland.
Olli Aaltonen, VTT Technical Research Centre of Finland, Espoo, Finland.
Tiina Nakari-Setälä, VTT Technical Research Centre of Finland, Espoo, Finland.

Introduction

There is a growing interest in liquid biofuels produced from renewable biomass resources. The bioethanol used today is mainly produced from 1st generation raw materials, such as sugar or starch crops. The abundance and relatively low cost combined with the non-food nature of agricultural, agro-industrial and forestry lignocellulosic residues make them attractive as 2nd generation feedstock for large-scale production of bio-based fuels and chemicals. Currently, bioprocessing of lignocellulosic material is focused on enzymatic hydrolysis of the cellulose fraction to glucose, followed by fermentation and post-processing to fuel-grade ethanol.

In this paper, the utilization of spruce bark as a lignocellulosic feedstock for a future integrated biorefinery was evaluated. The aim of this concept study was to develop a bark based biorefinery model exploiting results from pilot scale trials. The process concept consists of isolation of bark speciality compounds by water extraction and ethanol production by enzymatic hydrolysis and fermentation. The co-production of high value fine chemicals and ethanol may increase the economical attractiveness of the concept.

This paper focuses mainly on the modelling of water extraction of spruce bark to isolate polyphenols (tannin) and stilbene glycosides. Polyphenols and stilbene glycosides, naturally present in trees (bark, root, leaves), have been reported to possess antimicrobial, antibacterial, antifungal, antioxidant, anti-inflammatory, anticancer, and antimalarial activities, and have received much attention in nutraceutical, pharmaceutical, biochemical, food and cosmetic industry. In this paper, less weight is given to the modelling of ethanol production by enzymatic hydrolysis and fermentation.

The paper discusses the challenges in the exploitation of laboratory and pilot scale bark extraction experiments in unit operation modelling, and presents a process model of a bark-based biorefinery concept validated with results from the pilot experiments.

Methods

The effect of extraction conditions on polyphenols and stilbene glycoside yields was studied in pilot scale. The purpose of the pilot scale measurements was to gain more generic knowledge about aqueous bark extraction and to create a more predictive model for the extraction stage. The studied extraction phenomena were i) mass transfer, ii) extraction kinetics, iii) extraction yields, and iv) diffusion.

The spruce bark was collected from debarking unit of a Finnish paper mill. The extraction parameters considered were bark particle size, temperature, extraction time, and solid-liquid ratio. The choice of parameters and their values was based on preliminary lab and pilot scale experiments. Spruce bark was extracted in a batch reactor (100 L) with pure water at ambient pressure at temperatures below 100 °C. The extracted bark was pressed, and the obtained extract was separated from solids with a decanter centrifuge, and analysed to determine the stilbene glycoside and tannin concentrations. Stilbene glycoside analysis was performed with gas chromatography. Acid butanol assay was used to determine the tannin concentration. Also, monosaccharide concentrations in the filtrates were determined with methanolysis and gas chromatography.

Particle size influences the extraction rate in a number of ways. The smaller the particle size, the greater is the interfacial area between the solid and liquid, and therefore the higher is the rate of transfer of material and the smaller is the distance the solute must diffuse within the solid. On the other hand, the surface may not be so effectively used with a very fine material if circulation of the liquid is impeded [1]. In the pilot scale test series, the diffusion of the solute through the solvent in the pores of the bark to the outside of the particle was studied using two different bark particle sizes. The big particle size fraction, i.e. the coarse crushed bark, was obtained from the bark crusher found in the debarking unit. Part of this coarse crushed bark was further crushed to smaller particle size to obtain the so-called fine crushed bark fraction. Dry content of both bark fractions was 31,7 w-%.

In most cases, the solubility of the material which is being extracted will increase with temperature. Further, the diffusion coefficient will be expected to increase with rise in temperature improving the rate of extraction [1]. High-temperature solvent may promote polysaccharides on bark cell wall to distribute to solvent, and to weaken or undermine the integrity of the cell wall so that more solvent can be contacted with polyphenols. However, too high temperature may promote the oxidation of polyphenols [2]. In the pilot scale test series, the bark was first extracted with water at room temperature. The cold water extracted bark was pressed to obtain the cold water extract. The pressed bark fraction was further extracted with hot water. Extracts were analysed separately.

Extraction kinetics was studied taking samples from the extraction reactor as a function of time. The samples were analyzed to determine stilbene glycoside, tannin and monosaccharide concentrations. Total extraction time in both extraction steps was two hours.

With the solid-liquid ratio increase, the extraction rate of polyphenols also increases [2]. To study the effect of solid-liquid ratio on the extraction rate of stilbene glycosides and tannins, the extraction process was carried out using three different consistencies (4, 8 and 16 %) keeping the total batch size (55 kg) constant.

Results and discussion

The main conclusion from the test series is that bark is an extremely unhomogeneous material. Because of the considerably big bark batch size (2-9 kg dry bark) in the extractor, the bark composition in each batch differs remarkably. This hinders the observation of the effect of different process variables on the extraction yields i.e. the effect of some variable stayed unclear. Also, the storage time has its influence on the bark composition. It is known that the amount of stilbene glycosides in bark decreases as a function of storage time. In addition, human error must be considered since the extraction procedure involved steps that were performed manually. Also, the suitability of the experiment set-up was questioned. These all together complicates the interpretation of the results mathematically. However, there were couple of outcomes that improves the knowledge of the bark extraction phenomena.

The results showed that the extraction yields calculated as extracted dry matter per total dry matter were higher with coarse crushed bark than with fine crushed bark. Bark fine crushing seems not to be effective. Bark taken after the coarse crusher already found in the paper mill debarking unit is more suitable for extraction. This outcome contradicts with the theory of the diffusion length and interfacial area. The explanation for this behaviour may lie in the tendency of fine bark material to form a dense clump with poor interaction with solvent.

The effect of extraction consistency was evident; the lower the consistency the higher the extraction efficiency. However, choosing the consistency is optimization between low yields, high evaporation costs when concentrating the products and large waste water amounts. The experiments revealed that the consistency of 16 % was too high and the consistency of 4 %, on the other hand, too dilute.

The dry matter, stilbene glycoside and tannin content in the extracts were studied as a function of extraction time to determine the sufficient extraction time. The achieved flat extraction curves unfortunately did not confirm the required time. However, there was a weak peak in concentrations at time 60 minutes. It is known, that a too long extraction time may promote the decomposition of the valuable components.

One main goal of the test series was to study the mass transfer during extraction, i.e. is there a concentration profile inside and outside the bark particle or is the extract in equilibrium around the bark particle. The results were very scattered, leaving the mass transfer phenomena unclear.

Even though, only two different extraction temperatures were used (20 °C and 80 °C), it was evident, that the higher the temperature the higher the extraction efficiency. However, like a too long time, also a too hot temperature may promote the decomposition of the valuable components. The pilot scale test series was based on the assumption (derived from preliminary small scale laboratory experiments) that the major part of stilbene glycosides and only a minor part of tannins will extract at low temperatures, and the major part of tannins at high temperatures. This sequential dissolution was thought to help the post-processing steps to fractionate stilbene glycoside and tannin concentrates. This assumption led to two separate extraction steps – cold water extraction at 20 °C followed by hot water extraction at 80 °C. However, after the test series was completed, the need of the separate cold water extraction step was questioned. The stilbene glycoside yield in the cold water extract was low compared to the tannin yield, which does not justify the investment of a separate cold water extractor. An alternative process concept is to perform only hot water extraction and fractionate the hot water extract to stilbene glycoside and tannin fractions.

It seems that the dissolution of tannins and stilbene glycosides depends on their molecular weight. With increasing temperature, compounds with a higher molecular weight will dissolve. This complicates the modelling work. The component groups called "tannins" and "stilbene glycosides" must be divided to different subgroups according to their molecular weight with small molecular weight components dissolving at lower extraction temperatures, and high molecular weight components dissolving at higher temperatures.

All in all, the extraction results showed quite high variance. The contradiction between the results and common sense assumptions might be due to the inhomogeneous nature of bark, the unsuitability of the extractor used, poor repeatability of the experiments or problems in analytical methods. This adds to the uncertainty of the model.

References

[1] J. F. Richardson, J. H. Harker, J. R. Backhurst, *Coulson and Richardson's chemical engineering: Particle technology and separation processes*, 5th ed. Butterworth-Heinemann, 2002.

[2] L. Yang, J.-G. Jiang, W.-F. Li, J. Chen, D.-Y. Wang, L. Zhu, "Optimum extraction Process of polyphenols from the bark of *Phyllanthus emplica* L. based on the response surface methodology", *Journal of Separation Science*, vol. 32, no. 9, pp. 1437-1444, 2009.

(663b) Rapid Solar-Thermal Gasification of Biomass – Laboratory Curiosity to Development Facility

Alan Weimer, Chemical and Biological Engineering, University of Colorado, Boulder, CO and Christopher Perkins, Sundrop Fuels, Louisville, CO

Solar-thermal reactors using concentrated sunlight have been designed, constructed and tested on-sun for the rapid dissociation of cellulosic biomass to intermediate syngas. This presentation will focus on (1) base thermodynamic considerations for using solar energy to drive highly endothermic reactions, (2) experimental results for biomass conversion reactions, (3) materials challenges/solar reactor design, and (4) economics/process scalability. Issues to be addressed include availability of water/biomass in the desert, the capital cost of heliostats and diurnal operation dealing with limited sunlight hours.

The rapid solar-thermal conversion of cellulosic biomass has five principal advantages compared to conventional direct or indirect gasification – (1) twice the fuels yield from biomass since no partial oxidation is required to drive endothermic gasification reactions, (2) the production of undesirable tars is avoided due to high temperature operation at 1200oC or higher, (3) 30+% increase in the calorific value of the resulting syngas compared to the starting biomass, (4) a higher quality (less CO2, methane, and C2 hydrocarbons) syngas than conventional processes; and (5) the lowest carbon footprint of any biomass conversion process currently under development. The process provides for an extremely efficient utilization of valuable biomass (~ 2X that of conventional processes) with fuel energy content/ kg biomass consumed among the highest practicable.

Research in the area of solar-thermal chemical processing has been ongoing at the University of Colorado for 14 years. The main thrust of research has focused on the splitting of water using concentrated sunlight and the more recent solar-thermal biomass conversion. A remaining technical risk is the scalability of the solar receiver. The process has progressed from a laboratory curiosity in 2005 to a 1 MWthermal Receiver Development facility (RDF) currently being operated in Broomfield, CO. Copernican Energy was a spinoff out of the University of Colorado in 2006. The company was acquired by Sundrop Fuels in 2008. Commercial development efforts are underway.

See more of this Session: Pilot Plants for Green and Sustainable Processes - II
See more of this Group/Topical: Process Development Division

(663c) A Collaborative Pilot Demonstration of DOW AMBERLYST™ BD20 Catalyst in Free Fatty Acid Esterification for Biodiesel Manufacture

Aaron Sarafinas, Core R&D / Engineering Sciences, The Dow Chemical Company, Spring House, PA, Christine D. Miller, Engineering Solutions - Bristol, The Dow Chemical Company, Croydon, PA, Robert T. Hanlon, The Dow Chemical Company and Alfred K. Schultz, Dow Water & Process Solutions - R&D, The Dow Chemical Company, Spring House, PA

The use of biodiesel as a renewable energy resource that could be an economical alternative to fossil fuels is receiving great interest. Biodiesel is produced by the transesterification reaction of triglycerides with methanol to produce fatty acid methyl esters (FAME). These triglyceride oils are costly, and producers are continually searching for less expensive, crude/recycled feedstocks. Unfortunately, lower cost feedstocks for biodiesel manufacture often have unacceptably high levels of free fatty acids (FFAs). Presence of these FFAs lead to poorer process economics and increased problems in separation during the transesterification reaction. Dow has developed AMBERLYST™ BD20 catalyst specifically designed for the effective and efficient conversion of FFAs to FAME, enabling biodiesel manufacturers greater flexibility with feedstocks for transesterification, with the added benefit of meeting increasingly challenging biodiesel quality standards.

Researchers at the Dow Spring House PA facility collaborated with the Pennsylvania Department of Environmental Protection (PADEP), local PA biodiesel producers, and Bayer Technology Services (Pittsburgh PA) for a pilot trial which succeeded in demonstrating the viability of our heterogeneous polymeric catalysis technology for FFA esterification. In this paper, we will present the performance of AMBERLYST™ BD20 catalyst for the esterification of a variety of feedstocks containing a wide range of FFA concentrations. We will also discuss the long-term operability of the unit, design constraints, and recommendations for design improvements. The pilot unit is being placed back in service for interim production at a local PA biodiesel manufacturer, and is available for testing of other biodiesel feedstocks with AMBERLYST™ BD20 catalyst.

See more of this Session: Pilot Plants for Green and Sustainable Processes - II
See more of this Group/Topical: Process Development Division

(663d) Evaluation of Scaling Factors for Industrial Bioprocesses: Application to the Production of Ethanol From Glucose

Adriana Ruiz, Dept. Chemical & oil engineering, National University of Colombia, Medellín, Colombia and Hernán Alvarez, National University of Colombia, Medellín, Colombia

Before construction, biochemical plants require previous studies and demonstrations at scales smaller than industrial. Therefore, a good scaling-up process is mandatory. Traditionally, scaling-up from chemical and biochemistry process are carried out following highly heuristic procedures. Dimensionless factors are frequently used but, these procedures are associated to transport phenomenon of mass, energy and/or movement in the process. These procedures frequently have minor success in the definitive scaling-up process. The purpose of this research was to obtain a mathematical methodology, based on a Semi-physical Phenomenological Model, for the calculation of scaling up factors in biochemical processes. The methodology maintains the operating point and regime of the process constant after scaling up and is validated using the fermentation process for ethanol fuel production from glucose as a test case. In this research we obtained a highly analytical methodology based in the matrix Hankel for the determination of the hierarchy of dynamics from process to scale up bioprocesses and applied this methodology at semi-industrial scale in the production of ethanol from glucose. We determined the conditions at laboratory scale that are essential for scaling up. These conditions were used to identify with the methodology it was possible to determine that changes in the operation regime occur when the system effective volume are higher. The critical volume for maintaining the process regime is 3 liter. For higher volumes the order of importance of the effect of the different dynamics in the process is: product, biomass, substrate and temperature. The results obtained of the designs parameters are error minor to 10% according to the values report in the scientific literature for laboratory and report to industrial scale.

See more of this Session: Pilot Plants for Green and Sustainable Processes - II
See more of this Group/Topical: Process Development Division

Dual Fluidized Bed Reforming Pilot Test Results & Potential Commercial Implications

Karl Mayer, Vienna University of Technology, Austria
George Apanel, Rentech Inc., USA
Tobias Pröll, Vienna University of Technology, Austria
Mark Ibsen, Rentech Inc., USA
Hermann Hofbauer, Vienna University of Technology, Austria

Abstract

A dual circulating fluidized bed pilot plant was operated in chemical looping reforming conditions. Three types of feedstock were used. A biomass gasification product gas like mixture of methane, carbon dioxide, carbon monoxide and hydrogen (requested syngas), a gas mixture with almost the same elemental balance like the requested syngas (surrogate syngas) and a methane carbon dioxide mixture. The second reactor was operated as regenerator (air-reactor). The required heat input occurred in the air-reactor with additional fuel fed into it. The reforming catalyst was a nickel based material provided by Rentech Inc.. Reforming of the mentioned gas compositions and dry reforming have been investigated with focus on particle performance towards reforming, particle abrasion and long term stability.
Constant operation for 36 hours was possible without any decrease in catalyst activity. Particle loss through abrasion of particles was low. The overall performance of the catalyst was good, low carbon formation has been indicated in the dry reforming scenario. Based on these findings some advantages of the fluidized bed system can be indicated.

- Minimization of the reactor volume is possible.
- Less catalyst is required per unit fuel feed.
- Steam is not required or only in little amounts.
- Temperature can be maximized, since the reactors can be refractory lined.

Introduction

Reforming of hydrocarbons is widely practiced today to provide synthesis gas for production of basic chemical compounds.Today, synthesis gas is basically produced via two types of catalytic processes. First, methane steam reforming, where methane and steam are contacted to a fixed bed of catalyst which is externally heated, usually tubular reformers [1]. Second, autothermal reforming, where hydrocarbons (e.g. refinery by-products) are contacted to the catalyst and oxygen is added to release the heat for heating up the feedstock and for the reforming reactions.
Nowadays concerns about climate change and green house gas emissions are increasing. In the course of the aim to reduce CO_2 emissions, Chemical Looping Combustion (CLC) has been investigated in the recent years [2]. In the CLC process a metal oxide is used to transport oxygen from one reaction zone to another with focus on the full oxidation of a hydrocarbon based fuel. Reducing the amount of provided oxygen below stoichiometric conditions leads to the production of synthesis gas. The process is then called Chemical Looping Reforming (CLR), Figure 1 shows the principle concept of CLR. The main advantages of this concept are that neither an air separation unit nor external heating of reformer tubes is required. The technology has been first proposed in several patent applications by Standard Oil Development Company in the early 1950s (e.g. [3]). It has then been re-discovered as a variant of CLC as mentioned above and recently investigated in theory and laboratory scale at Chalmers University of Technology [4] and Vienna University of Technology [5]. In CLR, the catalyst is provided as loose particles

and is circulated between the reforming reactor, called fuel-reactor in chemical looping vocabulary, and the socalled air-reactor. The particles fulfill three main tasks:

- Reforming catalyst
- Heat transport
- Selective oxygen transport

Apart from thermodynamic and catalytic properties, the oxygen carrier has to show sufficient mechanical strength for the use in fluidized bed systems. At Vienna University of Technology, a pilot plant consisting of two interconnected circulating fluidized bed (CFB) reactors has been designed, built and operated. The unit is dimensioned for CLC of natural gas at a fuel power of 120 kW [6].
The Rentech approach shows some differences compared to CLR described above. Necessary heat input is provided via additional fuel to the air-reactor and not from oxidation of the bed-material. Oxygen transport is not desired. The feedstock to the fuel-reactor is not methane, it has already a syngas like composition and is comparable with a product gas of biomass gasification.
The aim of this work was to investigate the syngas production from a gaseous feedstock comparable with the raw product gas from biomass gasification (compare [7]) without higher hydrocarbons and no sulphur.

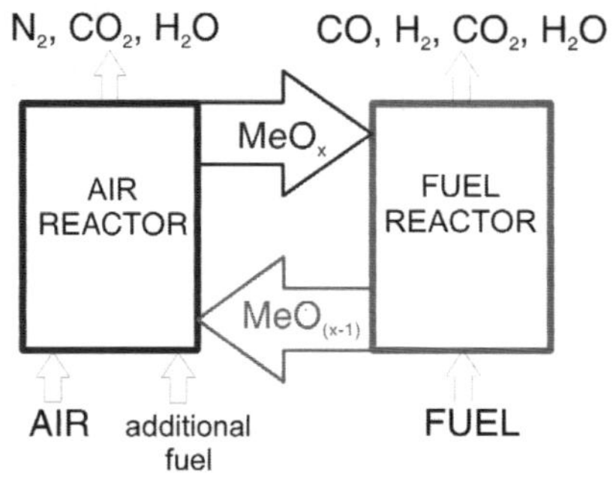

Figure 1: Chemical looping reforming principle [8]

Theory

The process approach in this work shows some specialities compared to standard CLR. The fuel-reactor feed gas shows, apart from some methane already a syngas-like composition. A decrease of the syngas energy content should be avoided, so oxygen transport is reduced as much as possible. The required heat input is provided by additional fuels to the air-reactor. This setup allows to influence the oxygen transport with the amount of additional fuel in the air-reactor. Varying the fuel feed in the air-reactor also allows to regulate temperatures of the system. Understoichiometric conditions have a cooling effect and increasing the amount of air raises the temperature. These changes, compared to CLR, reflect the approach of Rentech, as mentioned before. The goal of this concept is not the optimization of the reforming process of refinery gases, but on the upgrade of gasification product gases (e.g. from biomass). In a long term experiment of almost 33 hours of hot operation, a nickel based catalyst was tested by means of mechanical stability, catalytic performance and the influence of carbon deposition on the particles.

Main chemical reactions involved

In the fuel-reactor of CLR systems different reactions take place one after the other or in parallel. The most relevant under reforming conditions are:

- Partial oxidation of the fuel in contact with the oxygen carrier,
- steam reforming with steam present in the fuel-reactor
- CO shift reaction

- oxidation of H_2 and CO in contact with the oxygen carrier
- direct combustion of hydrocarbons in contact with the oxygen carrier and
- dry reforming when the Rentech approach is used.

In the case of methane fuelling to the fuel-reactor the main reactions are:
Direct partial oxidation with metal oxide (MeO):

$$CH_4 + MeO \leftrightharpoons CO + 2H_2 + Me \tag{1}$$

Heterogeneously catalyzed steam reforming, CO shift reaction and dry reforming reaction:

$$CH_4 + H_2O \leftrightharpoons CO + 3H_2 \tag{2}$$
$$CO + H_2O \leftrightharpoons CO_2 + H_2 \tag{3}$$
$$CH_4 + CO_2 \leftrightharpoons 2CO + 2H_2 \tag{4}$$

Internal combustion:

$$CO + MeO \leftrightharpoons CO_2 + Me \tag{5}$$
$$H_2 + MeO \leftrightharpoons H_2O + Me \tag{6}$$

In the case of CLR, too little oxygen is available for complete combustion. The overall reaction can be illustrated as a combination of Reactions 1, 5 and 6 with equilibrium of Reaction 3 prevailing. The overall reaction in the fuel-reactor is more and more endothermic the more CO and H_2 appears in the fuel-reactor exhaust gas.

Experimental

The Chemical Looping pilot plant

The rig was designed as a dual circulating fluidized bed (DCFB) system [6]. Figure 2 shows the setup of the plant. The air-reactor and the fuel-reactor, are interconnected with two fluidized loop seals. These seals are moderately fluidized with steam or nitrogen to prevent mixing of the gas streams. The air-reactor is fluidized with air and has the possibility to add a gaseous fuel for additional heat input. The fuel-reactor is fluidized with gaseous fuel, it is possible to add CH_4, H_2, CO, CO_2 and as inert component N_2. All mixtures of these gases are possible. The flow regime in the air-reactor is fast fluidization and in the fuel-reactor turbulent fluidization. Gas and solids are separated in cyclone separators in the downstream of each reactor. The gas streams are cooled separately. Fuel-reactor exhaust gas concentrations are measured with a Rosemount NGA 2000 (CO: $0-100\%$, CO_2: $0-100\%$, O_2: $0-25\%$, H_2: $0-100\%$, CH_4: $0-100\%$) and additionally with a online gas chromatograph Syntech Spectras GC955, which allows the determination of N_2 in the fuel-reactor exhaust gas stream. A Rosemount NGA 2000 (CO: $0-100\%$, CO_2: $0-100\%$, O_2: $0-25\%$) is used to analyse the gas composition of the air-reactor exhaust gas stream. Both gas streams are mixed in the post combustion chamber where all burnable gases have time to react. The downstream of the post combustion chamber cools the gas stream which is cleaned in a bag filter afterwards and sent to the chimney.

The air-reactor obtains a variable cooling system, consisting of three double jackets attached to the reactor. This setup allows variable heat extraction for varying operating conditions. Operation points from chemical looping combustion to chemical looping reforming (global stoichiometric air/fuel ratio 0.4 - 1.5) are possible. A detailed description of the reactor system, the cooling system and the other auxiliary units can be found elsewhere [6].

Experiments

A batch of 45kg of a Nickel based bed material was used in this campaign. The active solids inventory was only 17kg in each reactor. The difference between the material provided and the active solids inventory is made up by the solids present in the loop seals, downcomers and cyclones. The size distribution of the fresh particles is shown in Figure 3. In the syngas upgrade case, two gas compositions were used, a so-called preferred syngas and a surrogate syngas. The compositions and feed rates operated are shown

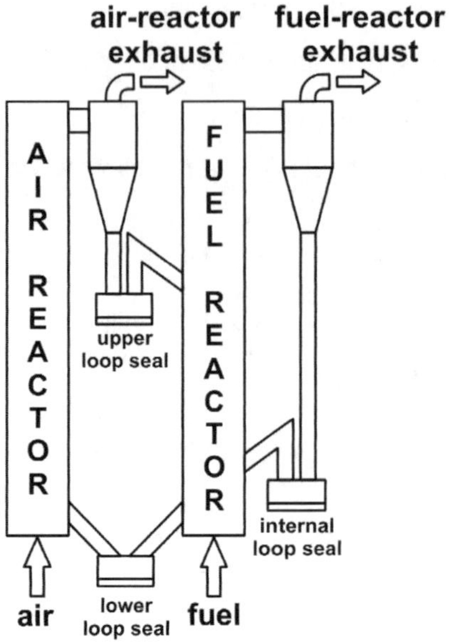

Figure 2: Scheme of a DCFB system

in Table 1. During dry reforming and syngas upgrade the loop seals were only fluidized with Nitrogen. In all operating cases the air-reactor was fluidized with air. Required heat input was provided by additional fuel (propane) fed to the air-reactor. Since oxygen transport to the fuel-reactor was not desired, the amount of excess oxygen in the air-reactor has been kept low. Operating temperature in the fuel-reactor was 1123 K at all operating points. The air-reactor temperature was 1173 K. Data are collected on-line and evaluated using a data reconciliation method for fulfillment of mass and energy balances [9].
During the experiment solids were sampled from both loop-seals to determine the degree of oxidation of

Table 1: Gas feed to fuel-reactor compositions and typical feed rates operated

Species	Requested syngas		Surrogate syngas			Dry reforming	
	[vol%]	[Nm^3/h]	[vol%]	[Nm^3/h]	[Nm^3/h]	[vol%]	[Nm^3/h]
CH_4	16	4.0	17	4.3	2.6	50	7.5
H_2	10	2.5	35	8.8	5.3	0	0.0
CO	30	7.5	8	2.0	1.2	0	0.0
CO_2	15	3.8	40	10	6.0	50	7.5
H_2O	29	7.3	0	0.0	0.0	0	0.0
Sum	100	25.1	100	25.1	15.1	100	15

the oxygen carrier X_s, which is defined as:

$$X_s = 1 - \frac{m_{ox} - m}{m_{ox} \cdot R_0} \tag{7}$$

$$R_0 = \frac{m_{ox} - m_{red}}{m_{ox}} \tag{8}$$

Where R_0 is the oxygen transport capacity which is defined as the relative mass decrease when bringing the particles to a most reduced form. Knowing X_s of solids in the upper- and the lower-loop-seal allows to calculate the global solids circulation rate and the amount of oxygen transported from the air-reactor to the fuel-reactor.
Carbon deposition is a major reason for catalyst deactivation in fixed bed reformers. The pilot plant setup makes the determination of the amount of carbon deposited on the solids leaving the fuel-reactor possible. Carbon on the solids surface reacts in the air-reactor with the oxygen present to gaseous CO_2 and CO. However, no hydrocarbon based fuel must be fed to the air-reactor during this determination,

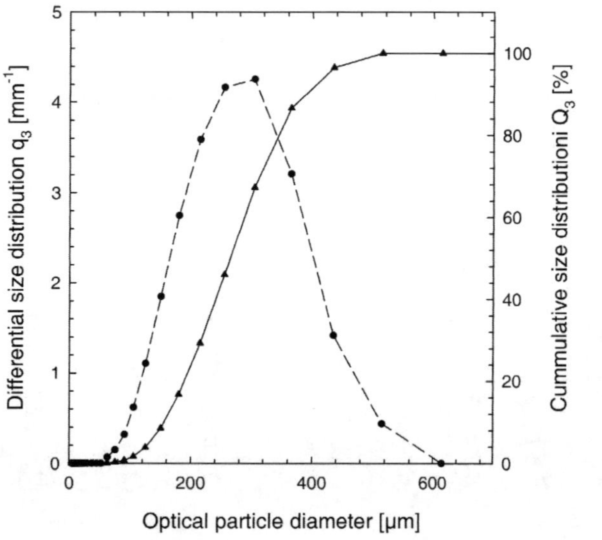

Figure 3: Size distribution of fresh solids

since combustion products will cover the signal. During the measurement propane, the primary fuel in the air-reactor, was substituted by hydrogen to provide the necessary energy to the air-reactor.

Operating results

Constant operation was possible for several hours. Figure 4 shows the temperatures of both reactors and the fuel-reactor exhaust gas composition during 7 hours of constant operation. The fuel-reactor feedstock was surrogate syngas (compare Table 1). Gas compositions go along with theoretical equilibrium gas concentrations under these conditions. The requested syngas and the surrogate syngas showed almost the same product gas composition. It seems that gas composition is not that important as the elemental composition of the feed. CO shift reaction (Reaction 3) is obviously well catalysed by the material.

As mentioned before, oxygen transport was kept low to generate an energy rich syngas. The effect of

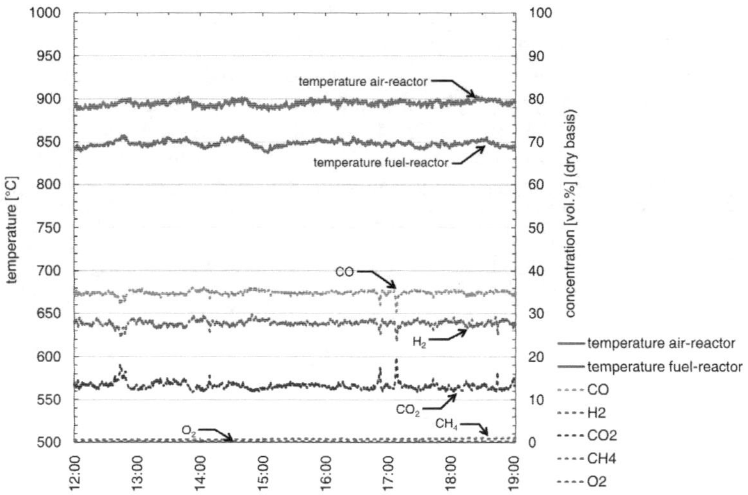

Figure 4: Gas composition of fuel-reactor exhaust gas and bed temperatures over time

oxygen transport on the gas composition of the syngas is noticeable. Methane conversion is low at all points, CO_2 is decreasing with decreasing oxygen transport and H_2 increases.

Figure 5 shows gas compositions of the fuel feed, the measured gas composition and the gas composition provided from the reconciled mass and energy balance solution. Nitrogen is from the loop seal fluidization.

The results of the determination of carbon formation on the solids surface are shown in Table 2. The

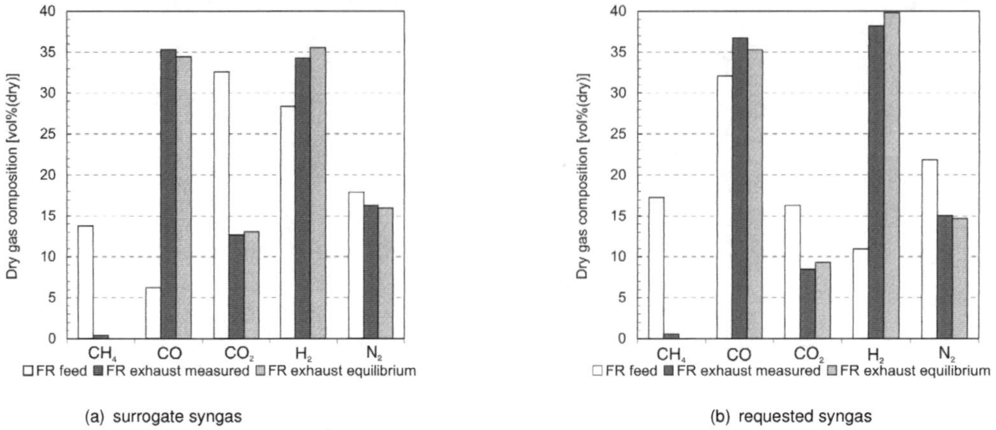

Figure 5: Comparison of dry fuel-reactor exhaust gas composition for surrogate syngas feed (a) and requested syngas feed (b)

amount of carbon formation is relatively low. After leaving the conditions of dry reforming the catalyst regenerates very fast and shows no decrease in performance.

The solids used in this campaign show high mechanical strength and good resistance towards attrition. Attrition, however, is always governed by both the particle properties and the system in which the solids are operated. Dust measurements were performed in the exhaust gas lines of both reactors using isokinetic sampling probes. The amount of sampled dust was enough to determine the dust loading but determination of size distribution was not possible. Also a bed mass balance was done comparing the amount of particles and dust in the filter-box after start-up procedure and before shut-down. The total loss of particles was 4.6% of total inventory during 36 hours of operational time. Particle size analysis showed that only about 50% of the particles in the filter-box had a reduced mean diameter, compared with the fresh material. According to this, half of the particle loss is caused from the performance of the cyclone separator.

Table 2: Results of carbon formation determination

fuel to fuel-reactor	CO in air-reactor exhaust [vol%]	CO_2 in air-reactor exhaust [vol%]	O_2 in air-reactor exhaust [vol%]	Carbon loading on solids [mg/Nm^3]
dry reforming mixture	0.02	0.13	0.03	10

Discussion

Based on the experimental findings the arrangement proposed by Rentech allows stable operation, also under long term conditions. There is no need to use a gas feed composition identically to the proposed syngas, also a gas mixture with almost the same elemental composition can be used. Equilibrium is reached under both conditions.

Carbon loss via carbon deposition on solids surface is low and occurs only under rigid conditions in the fuel-reactor. Carbon formation does not lead to an irreversible catalyst deactivation in the dual fluidized bed system. Activity is restored after leaving these extreme conditions where carbon formation occurs. Moreover, in large scale concepts the loop-seals will be fluidized with steam. That will additionally work to limit carbon formation.

With focus on the commercial scale-up of the system the fluidization regime of the fuel-reactor can be optimized in a subsequent design. Better gas-solids contact would lead to a higher amount of syngas treated in the fuel-reactor per reactor volume and solids inventory. With careful interpretation of the results of this experiment and consideration of the lack of optimization of the pilot plant it is apparent that the catalyst productivity expressed in terms of kg catalyst/MW of fuel power would be commercially attractive on the basis of a commercial feasibility study recently completed.
Physical losses due to pilot plant cyclone design inefficiencies suggests an average particle lifetime of one month in the pilot plant. On this basis, the total inventory would need to be replaced continuously within one month. However, this result may only reflect design limitations and operational circumstances in the pilot plant. Actual catalyst particle attrition losses appear to be lower. Residual olivine contamination also appears to have contributed to particle fines attrition collected in the filter box, well beyond the proportion of the residual olivine content of the circulating catalyst. During commercial scale-up of the system an optimization of the particle separation units should lead to less particle loss and a longer particle lifetime in the system.

Conclusions

Based on the fist results of the experimental campaign using the nickel based material the following conclusions can be drawn:

- The dual fluidized bed system allows stable operation also under longtime conditions no decrease in reactivity can be found.
- Surrogate and preferred syngas are both in equilibrium. CO-shift reaction prevails in the fuel-reactor. This means that, apart from a slight different energy requirement, it does not make a difference if requested or surrogate syngas is run.
- Using only the gas mixtures shown in Table 1 coking is no problem. The effect of higher hydrocarbons and the effect of sulfur on the process has to be investigated.
- The material is mechanically stable and definitely suitable for operation in circulating fluidized bed systems. The particles are more resistant than olivine which is a relatively hard material used in dual fluidized bed biomass gasification.

References

[1] Ib Dybkjaer. Tubular reforming and autothermal reforming of natural gas – an overview of available processes. *Fuel Processing Technology*, 42(2-3):85 – 107, 1995. Trends in Natural Gas Utilisation.

[2] Mohammad M. Hossain and Hugo I. de Lasa. Chemical-looping combustion (clc) for inherent co2 separations–a review. *Chemical Engineering Science*, 63(18):4433 – 4451, 2008.

[3] Welty Jr A.B. Apparatus for conversion of hydrocarbons, united states patent no.2550741, 1951.

[4] A. Lyngfelt, M. Johansson, and T. Mattisson. Chemical looping combustion Ű status ofdevelopment. In J. Werther, W. Nowak, K.E. Wirth, and E.U. Hartge, editors, *Circulating fluidized bed technology IX*, TuTech Hamburg Germany 2008, 2008.

[5] Johannes Bolhàr-Nordenkampf, Tobias Pröll, Philipp Kolbitsch, and Hermann Hofbauer. Performance of a nio-based oxygen carrier for chemical looping combustion and reforming in a 120 kw unit. *Energy Procedia*, 1(1):19 – 25, 2009. Greenhouse Gas Control Technologies 9, Proceedings of the 9th International Conference on Greenhouse Gas Control Technologies (GHGT-9), 16-20 November 2008, Washington DC, USA.

[6] P. Kolbitsch, T. Pröll, J. Bolhàr-Nordenkampf, and H. Hofbauer. Design of a chemical looping combustor using a dual circulating fluidized bed (dcfb) reactor system. *Chemical Engineering & Technology*, 32(3):398–403, 2009.

[7] Tobias Pröll and Hermann Hofbauer. H2 rich syngas by selective co2 removal from biomass gasification in a dual fluidized bed system – process modelling approach. *Fuel Processing Technology*, 89(11):1207 – 1217, 2008.

[8] Tobias Pröll, Johannes Bolhàr-Nordenkampf, Philipp Kolbitsch, and Hermann Hofbauer. Syngas and a separate nitrogen/argon stream via chemical looping reforming - a 140 kw pilot plant study. *Fuel*, 89(6):1249 – 1256, 2010.

[9] J. Bolhàr-Nordenkampf, T. Pröll, P. Kolbitsch, and H. Hofbauer. Comprehensive modeling tool for chemical looping based processes. *Chemical Engineering & Technology*, 32(3):410–417, 2009.

(707a) How Much Does My Pilot Plant Cost?

Adam Whalley, Sales & Marketing, Zeton Inc, Burlington, ON, Canada and David Edwards, Zeton Inc, Burlington, ON, Canada

Completion of a pilot plant project is a critical step in the process of scaling up and commercializing new technologies, trouble shooting problems in existing operations, and testing newer more efficient configurations. Most senior managers understand the importance of the pilot and demonstration steps, so the necessity for this essential research and development work is rarely challenged. However, after the need to build a pilot plant has been identified, project managers and engineers are given the task of developing a budget for the project before important details have been understood. This is an important step in the overall progression of the project, since a hastily designed budget will lead to financial strain that sets the stage for poor decisions in the future. Managers can become overwhelmed with many too many factors to consider in sufficient depth within the available time assigned. Zeton will address this issue by discussing the key factors that should be taken into account while setting an early budget with limited time and information.

See more of this Session: Pilot Plant Economics
See more of this Group/Topical: Process Development Division

A Fast Method for Developing Accurate Scoping and Screening Cost Estimates for Pilot Plants

Richard Palluzi
ExxonMobil Research and Engineering Company
1545 Route 22 East
Annandale, NJ 08801
908-730-2323
richard.p.palluzi@exxonmobil.com

Background

- Scoping or screening estimates for new pilot plants are often required
- Basic issues:
  - Little time
  - Little or no design information available
  - No established and proven methodology
- Subtle issues
  - Insufficient confidence in the result
  - Tendency for the result to match the perception
  - Little or no validation
  - Major decisions on potentially weak to bad data

Basic Conundrum

- Too fast an estimate often produces gross errors
- Waiting for the necessary information and spending the time and effort to produce a realistic estimate is often too time consuming or unnecessary early in a project

What Do You Need?

- An established methodology
- Which produces reasonably accurate results
- Has reasonably high confidence
- Does not require a large amount of detail
- Is fast and easy to use

Our Approach

- Assess the new unit for three basic criteria
  - Size
  - Complexity
  - Constructability
- Develop a scoping estimate based on a database of previous unit costs
- Add appropriate specialty costs
- Add appropriate contingency

Criteria: Size

- Estimate how many standard size modules or sections the proposed new unit will require
- Somewhat flexible
  - Typically skids or pallets 6 ft by 3 ft by 6 ft
  - Can range to 10 ft by 10 ft by 10 ft
  - Can be organizational preference
- Larger modules can be reasonably estimated as multiples of smaller ones

Criteria: Complexity

- Compare the envisioned unit to a series of standards based on:
  - Amount and type of instrumentation
  - Amount and type of vessels, reactors and process equipment
  - Amount and type of rotating equipment
  - Amount and type of valves and piping/tubing

Criteria: Constructability

- Materials of construction
- Type of construction
  - Piping or tubing
  - Compression fitting, threaded, welded, etc.
- Construction density (space available)
- Area electrical classification
- Other issues

Database

- Developed from actual costs for numerous units built since 1980
- Breaks units into standard size modules
- Assesses complexity of each module
- Assesses constructability of each module
- Used to develop average cost for each category

Issues

- Needs a lot of data
  - Cover entire range of classes
  - Confidence in the accuracy
  - Multiple costs/class
- Needs a good assessment
  - Some defined criteria
  - Uniformity
- Requires some analysis

Practical Problems

- Cost data is not readily available
  - Need to make conscious effort to obtain and gather
  - May take some time
- Cost data is inconsistent or suspect
  - Requires work to resolve
  - Some data may prove useless
  - Expect to spend significant effort resolving/normalizing
- Data is not terribly specific
  - Total costs with little detail
  - Expect to spend significant effort analyzing
  - May need to apply some normalizing factors

Practical Problems (cont'd)

- Assessment requires high uniformity
  - Use one person or group to assess all units in database
  - Try to define usable criteria
  - Use examples more than definitions
- Some people are better at it than others
  - Training helps
  - Experience helps
  - But ...

Specific Area	Low	Average	High
Quantity of standard size modules			

Module size is not critical and can match organizational preference/history/practice

	Complexity		
Specific Area	**Low**	**Average**	**High**
Instrumentation	•Less than normal •Standard equipment	•About normal (6-8 transducers, 10-12 TC) •Standard equipment	•More than normal •Non-standard equipment •GP equipment in a classified area
Control System	•Stand alone controllers •Standard small computer	•Standard computer	•Larger system •Non-standard system
Control Loops	•Less than normal •Standard loops	•About normal (8-10 loops) •Standard loops	•More than normal •Non-standard loops •Advanced controls
Area Electrical Classification	•General purpose •Some Class I Division 2	•Class I Division 2	•Class I Division 1 •Class II or III •Zone 0, 1 or 2
Reactors	•Single •Simple •Package autoclave or equivalent	•Single •More complexity •Larger autoclave	•Multiple •High complexity •Adiabatic •Special or novel
Reactor Temperature Control	•Standard heater	•Standard heater •Sandbath	•Non-standard heater •High complexity •Adiabatic •Special or novel
Process Vessels	•Less than normal	•Normal (4-6)	•Greater than normal •Specialty design •Novel design •Higher complexity
Rotating Equipment	•Less than normal •Standard equipment	•Normal (1-2) •Standard equipment	•More than normal •Non-standard equipment •Solids feeding
Heat Tracing (or cooling)	•None to very few	•Limited (1-4)	•Many (>4)

Complexity			
Specific Area	**Low**	**Average**	**High**
Specialty designs	•None	•Limited (1-2)	•Many
On-Line Analytical Equipment	•None or at most a single standard GC	•Normal (1-2) •Standard equipment	•More than normal •Non-standard equipment •Novel equipment •GP equipment in a classified area
Safety Interlock System	•Less than normal	•Normal (10-20) •Limited sequenced actions	•More than normal •Significant sequenced actions •High hazard shutdowns
Pressure[1]	•<500 psig	•Up to 3,000 psig	•>3,000 psig
Temperature [2]	•<250 °F	•<450 °F	•>450 °F

[1] Somewhat organization specific but clearly a break point above 5,000 psig

[2] Somewhat organization specific but clearly a break point above 450 °F

Constructability			
Specific Area	**Low**	**Average**	**High**
Piping Size	•1/4" or smaller	•1" or smaller	•>1"
Piping Type	•Tubing	•Tubing •Some pipe	•Pipe
Fittings	•Compression	•Compression •Some threaded	•Welded •All threaded •Cone and collar •Specialty fitting
Construction			
Access [3]	•Access all sides; or •Able to be modularized	•Access 3-4 sides; or •Able to be modularized	•Access only 1 side; and •Not able to be modularized
Space available	•More than required	•Satisfactory	•Less than required
Density [4]	•Less than normal •Significant open space •Most components within reach •No significant in-depth placement	•Normal •Only moderate open space •Majority of components within reach but may require ladders/stools •Moderate in-depth placement	•Higher than normal •Little to no open space •Significant components require means to access •Significant in-depth placement •Some potential problems
Height [5]	•6-10 ft	•8-20 ft	•>20 ft
Utility Feeds	•Close •Requires only ladders to access •Some exiting supports	•Within 25 ft •Requires only ladders to access •Some exiting supports	•>25 ft •Requires scaffolding or lift to access •No existing supports
Power	•Available nearby •240 VAC or less •No transformer	•Available locally •480 VAC or less •Requires single transformer 45 KVA or less	•Available within building •> 480 VAC •Requires multiple transformer of transformer > 45 KVA

[3] Modularized means able to be essentially built in a shop and then moved in place and connected
[4] In depth placement means equipment not accessible without reaching through piping/components
[5] Existing platforms reset grade

Specialty Costs

- Specialty analytical
- Packaged systems
- Special structures
- Facilities modifications
  - Based on a database of cost for similar modifications
  - Based on national estimating services such as R S Means

Contingency

- Contingency is an allowance for historically predictable but currently undefined and unknown extra costs
- It is a measure of the uncertainty in an estimate and, to some degree, the entropy of the universe
- Typical contingencies for scoping and screening estimates range from 20% to 100%

Contingency needs to be enough to cover the uncertainty but too high a contingency may ruin confidence in the estimate

Why?

- Too often contingency is based on guesswork instead of analysis or actual performance

Ideal Case					
	Scoping Estimate (k$)	Contingency	Final Costs (k$)	% Over/ (Under)	Deciding that contingencies should be raised by 5% is reasonable as the projects cover a reasonably wide range and the data is relatively consistent
Project X	$ 250	30%	$ 265	106%	
Project Y	$ 500	25%	$ 509	102%	
Project Z	$ 350	40%	$ 375	107%	
			Average:	105%	

"Less than Ideal" Case					
	Scoping Estimate (k$)	Contingency	Final Costs (k$)	% Over/ (Under)	Project A appears to have been estimated poorly while the other projects appear to have been over estimated. Decreasing contingency and a study of why Project A's estimate was so far off may be suggested
Project A	$ 100	30%	$ 250	250%	
Project B	$ 500	35%	$ 412	82%	
Project C	$ 750	40%	$ 643	86%	
			Average:	139%	

How Does One Estimate the Proper Contingency?

- Analyze existing performance
- Try to identify major variables
  - Novelty (process, equipment, chemistry, science, etc.)
  - Familiarity (process, operations, equipment, location)
  - Experience of the estimator(s)
- Try to identify areas of major over/under run and correct the next estimate instead of trying to guess the “correct” contingency

Recent Example

- Available data
  - Preliminary overall process flow steps
  - Preliminary design basis for ~15% of the unit
  - Approximate module sizes/scope available
  - Some global concerns
- Assessed unit as consisting of 35 modules

Process Step	Modules	Complexity Factor	Constructability Factor
1	2	Average	Average
2	2	Average	Average
3	2	Average	Average
4	2	High	Average
5	2	High	High
6	2	Average	High
7	4	Average	Average
8	10	High	8 High 1 Average
9	4	High	High
10	2	Low	Average
11	2	Low	Average
12	1	Low	Low
	35		

Example Only
The data is simulated for illustration

		Complexity		
		Low	Average	High
Constructability	Low	$ 35	$ 65	$ 100
	Average	$ 100	$ 155	$ 225
	High	$ 130	$ 250	$ 300

		Complexity		
		Low	Average	High
Constructability	Low	1		
	Average	4	10	4
	High		2	14

		Complexity		
		Low	Average	High
Constructability	Low	$ 35	$ -	$ -
	Average	$ 400	$ 1,550	$ 900
	High	$ -	$ 500	$ 4,200

Subtotal		$ 8,000
Contingency	30%	$ 2,400
Scoping Estimate		$ 10,400
Final Estimate		$ 10,800
		3.8%

Lesson Learned

- The better the data the better the results
- Primary assessment is critical
- Trained and experienced pilot plant estimators are critical
  - The assessor needs to be the estimator!
  - The estimator needs to fully understand the system
- Need to identify specialty concerns which are not addressed in the database
  - Structures
  - New control systems
  - Unique or novel conditions
- Cannot predict scope of basis changes

Summary

- ExxonMobil's experience with this type of system has been excellent
  - Typically ±10%
  - Gets better with time
- The database needs routine updating and analysis to remain useful
- This methodology can be used by any company in any industry

(707c) An Approach for Estimating the Value of Computational Tools in Drug Product Development

Mary T. am Ende[1], Bruno C. Hancock[1] and Kristen J. Huta[2], (1)Pharmaceutical Development, Pfizer Global Research & Development, Groton, CT, (2)Business Operations, Pfizer Global Research & Development, Groton, CT

Computational approaches to drug product development are beginning to become feasible for many types of dosage forms, and there is an opportunity to exploit these emerging tools in the pharmaceutical industry. However, as with any investment, it is important to understand the business case and potential financial benefits. In this work, an attempt has been made to objectively assess the value of emerging computational approaches to the development of solid oral dosage forms. This has involved collaboration between scientists working in this pharmaceutical development and business analysts with experience in estimating the value of productivity improvements, resource savings, and deferred expenditures. As a result of this work, the value of each of several investment areas has been estimated, so that the relative magnitude of the financial benefits can be understood and future investments can be planned appropriately. The development of a procedure for this type of financial analysis can serve as a model and benchmark for any future analyses of the benefits of computational tools in the pharmaceutical industry.

See more of this Session: Pilot Plant Economics
See more of this Group/Topical: Process Development Division

Best Practices for Reducing the Cycle Time and Cost of Piloting

David A. Berg, DuPont Center for Process Innovation, E.I. du Pont Canada Company, Kingston, ON
Michael T. Cleary, DuPont Applied Bio-Sciences, E.I. du Pont de Nemours and Company, Wilmington, DE

Introduction

The piloting of new chemical and polymer processes is a key enabling process in the development and commercialization of new technology. Most typically, piloting is done either to develop and demonstrate new process concepts, produce product samples to support customer testing and market development, or to develop basic data to support the design of a commercial manufacturing process. These different objectives can result in the need to do piloting work at multiple scales and possibly using multiple approaches (e.g. using both an integrated pilot plant and vendor testing of critical equipment at larger scale). Figure 1 describes the approach that DuPont has used to develop and scale-up the commercial process for manufacture of DuPont Cerenol® Polyols[1]. This framework illustrates the use of piloting at multiple scales to address different needs, and is a model that we have leveraged across many development projects.

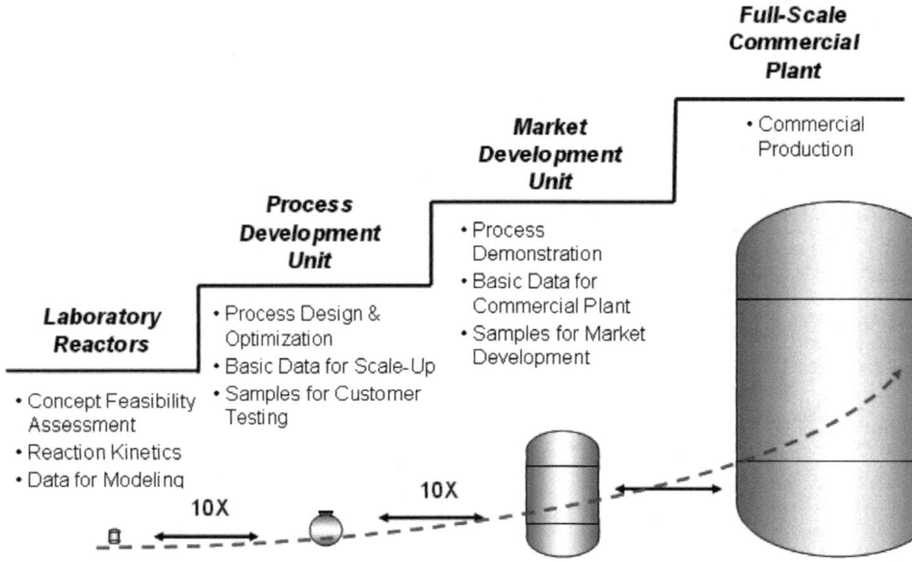

Figure 1: Approach to Development and Commercialization of Cerenol® Polyols[1]

This paper will provide an overview of several best practices that have been developed within DuPont over the years for reducing the cycle time and cost of piloting new technology. These best practices cover the span of project activities from technology selection through to pilot plant operation, focusing on the differential needs and considerations of pilot plants versus commercial plants. They are directed primarily towards process development units with an installed cost of up to three million dollars, as these facilities typically have a shorter useful lifetime and require greater flexibility in use, versus larger scale market development units that are often an order of magnitude higher cost and typically share more similarities with commercial plants.

Discussion of Best Practices

Alignment with Business Needs

All successful piloting projects start with having a clear understanding and alignment with the business needs. When these needs are not clearly developed, or there is misalignment, the consequences can range from significant delays to project failure. A key best practice for DuPont is to develop and document clear objectives at the start of all piloting projects that reflect market, technology, and commercialization needs.

- For market needs it is important to have strong voice-of-customer (VOC) input regarding product specifications, performance requirements, and an understanding of quantities of product needed for testing and qualification. The latter can often influence the required scale and capacity of the pilot plant.
- For technology needs it is helpful to develop performance scorecards or dashboards that highlight the key performance metrics, the commercial targets for each, and the progress towards meeting these targets. The metrics are typically the key process, product, and cost-of-manufacture variables that need to be achieved to have a viable commercial technology. Within DuPont we have found that this is highly effective in ensuring work is focused on the top priorities.
- For commercialization needs, it is critical to map out the approach to process scale-up and commercialization. Understanding whether the process is to be scaled up directly to commercial scale, or to an intermediate market development unit, is a necessary input to defining the objectives and scope of the piloting project.

Deciding What and When to Pilot

The development and commercialization of new technology requires the coordination and alignment of activities associated with new product development, technology delivery, and product supply. This is illustrated in Figure 2. While piloting of new technology is done within the context of technology delivery, this work requires input from, and output to parallel activities associated with new product development and product supply planning. Within DuPont we use a stage-gate development process to ensure alignment of these activities. This stage-gate process maps out the expected tasks for each parallel work stream, identifies key deliverables that need to be achieved prior to advancing to the next stage, and uses a business led steering team to make key decisions. Other key best practices for piloting around alignment of work streams include:

- Securing early input from engineering design experts regarding flow sheet options and commercial plant economics. This needs to be done prior to technology selection and developing the design basis for the pilot plant.
- Avoiding overlap between detailed engineering of the commercial plant and piloting that is being done to address key technical issues. Overlap, which is sometimes attempted with the objective of accelerating commercialization, can translate into significant rework, project delays, and cost overruns.

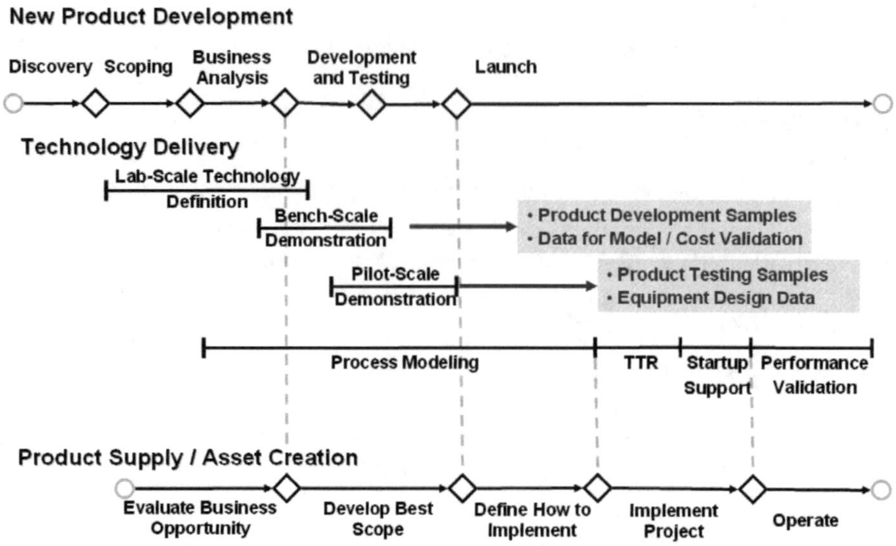

Figure 2: Alignment of Activities to Support Commercialization of New Process Technology

The decision of what elements of a new chemical process to pilot is a complex question that requires balancing of risk versus opportunity. Best practices that have been adopted within DuPont include:

- Reviewing piloting plans early and often with a peer group that is external to the project team. This peer group needs to have broad knowledge and experience, and be able to provide unbiased input. Within DuPont this is typically done through a structured risk assessment process. It is critical that the project team and business leadership listen to and use this input, and develop appropriate contingency plans before they are needed.
- Using a full integrated pilot plant for processes that involve recycle streams. These projects involve elevated risks due to buildup of impurities, and require full integrated pilot plants that include all key steps and recycle loops. One example within DuPont's lesson book of the potential consequences of skipping this step is a project in the late 1970s to develop a breakthrough process for production of methanol[2]. The commercial process experienced a myriad of technical problems, requiring over 14 months to start-up and numerous process changes.
- Using a full integrated pilot plant for processes that involve either handling or processing of solids. The challenges associated with solids handling can be significant, but are often overlooked. Piloting is required to assess issues associated with scale-up, fouling, equipment wear, and up-time.
- Building as small a pilot plant as possible to meet all of the project objectives. Smaller designs can be installed more quickly, have lower investment and operating costs, and can be easier to modify after start-up.

Use of External Partners

With the growing recognition of the potential benefits that can flow from Open Innovation, many companies are becoming increasingly motivated around using external partners for piloting of new technology. This can include several different types of partner relationships, ranging from contract research and toll processing, to joint development of technology. Each type of partner relationship involves unique considerations. In this paper our comments are directed towards the use of toll processors. While the use of toll processors is typically motivated by a desire to avoid the time and expense associated with building a new pilot plant through accessing existing equipment in place at a toll processor, a second motivating factor can be to facilitate the access to specialized know-how and skills that may be lacking within the project team. Within DuPont we are still gaining experience around how best to use toll processors most effectively. Practices that we have found to be helpful include:

- Having a structured process for identifying potential toll processors, screening promising candidates, and selecting the best toll processor to work with.
- Developing detailed technology packages that describe equipment specifications, process operating conditions, analytical methods, and specific chemical handling and safety considerations.
- Having feet-on-the-ground at the toll processor, both to provide guidance around testing and to support the identification and transfer of new learnings back to the project team.

When the decision is made to proceed with building a new pilot plant, consideration should also be given to contracting out the engineering design and fabrication to a company that specializes in pilot plants. Within DuPont we have had good success with this approach. Benefits we have realized include reducing project uncertainty through securing a fixed price quotation, accessing highly skilled resources that have experience in working on multiple pilot plant projects of similar design, and fast-tracking the project schedule. Best practices that should be considered with this approach include:

- Selecting contractors that have prior experience in designing and fabricating multiple pilot plants for the same industry. In this way you can benefit from their existing expertise versus supporting the development of this.
- Completing the pilot plant basic data up-front prior to engaging a company to start design work. Basic data is not something that can be contracted out, and engaging a contractor too early can inflate the project costs.
- Establishing an in-house multi-disciplinary project team in order to work with the contractor. The project team has accountability for the overall project, including any OSBL facilities that are outside the scope of what is contracted out. This team plays a critical role in working with the contractor on the ISBL equipment, including development of basic data, review and sign off on drawings, process hazard reviews, selection of piping codes for design, vendor selection, and writing of operating procedures.

Staffing for Success

Project staffing is a critical issue for both pilot and commercial projects, and many of the best practices for staffing of commercial plant projects apply equally to pilot plant projects. These practices include establishing a multi-disciplinary project team that includes participation from technology, business, and all relevant engineering disciplines, base loading the team with experienced resources, and establishing a project steering team that includes strong business representation and has the authority to make key decisions. Additional best practices that apply specifically to piloting include:

- Selecting a project lead that has prior experience with both Engineering (design and construction of pilot plants) and Research and Development (operation of pilot plants). These are very much experienced based competencies, in which people learn through doing, and needs to be a key consideration in developing staffing and succession plans. Within DuPont, it is sometimes difficult to find people with experience in both areas, as Engineering and Research and Development are operated as separate functions.

- Having continuity of key leadership roles from laboratory technology development, through piloting, and then to commercialization. Managing these activities with separate teams, with hand-off between each, can result in significant delays and elevated risk of failure.
- Assigning scientists to work as shift leads in starting up new pilot plants. We have found this to be a good development opportunity for newly hired scientists and can help in significantly reducing the time to troubleshoot process issues and establish stable pilot plant operation.
- Getting people from the commercial plant involved in the pilot plant, and using people from the pilot plant to support the design and start-up of the commercial plant. This can significantly accelerate start-up, and applies not just to scientists and engineers, but also to technologists and operators.

Design for Flexibility and Change

One of the most significant differences between pilot plants vs. commercial plants is the need for flexibility and change. With pilot plants, there is the expectation that future design and operation changes will need to be implemented in order to deal with unanticipated problems that are identified during pilot plant start-up, and/or changes and improvements to the technology that are made subsequent to freezing the pilot plant design basis. There is also the expectation that any specialized equipment (e.g. reactors) that is purchased will likely be re-used in future pilot plants. There are a number of best practices that can be employed during the design phase to accommodate future flexibility:

- Building to a large footprint where there is room to accommodate new equipment.
- Using skid-mounted modular equipment that can be quickly moved in and out.
- Building supporting infrastructure such as utilities, equipment for handling of raw materials and products, PLC control systems, and emission control systems that are highly flexible and can be reconfigured for use with future pilot plants.
- Considering the future-use of specialized equipment such as reactors in specifying the design temperature, pressure, and materials of construction.

Use Value Improving Practices Appropriately

There are a number of value improving practices that have been adopted industry wide as best practices for the design and construction of chemical processes. While these practices can be effective when applied to larger scale commercial plants, they are not always appropriate for use with pilot plants and can actually increase the project cost and cycle time. Selecting which practices to use must be done in consideration of the differential needs of pilot plant projects, which are typically schedule versus cost driven, involve significant design uncertainty, and have a relatively short useful life. Having an experienced project leader is critical in deciding which practices to use. Examples of value improving practices that need to be questioned include:

- Using 3D CAD for process design. For most pilot plants it is typically more cost effective to field fit tubing versus using 3D CAD for process layout and piping.
- Building to a small footprint. As discussed above, pilot plants, by their nature, frequently undergo both small and large design modifications following start-up. Building to a small footprint can increase the cost and time required to make these changes.
- Sending out all work for competitive bids. While this is a best practice within many companies and can be entirely appropriate for larger commercial plants, it can lead to delays for pilot plants that involve specialized analytical equipment, pumps, or reactors that have complex specifications and long lead times for delivery. Often the resulting impact of these delays on the overall project is disproportionate to the cost savings that are derived. A good practice is to engage the Sourcing function early to stream-line the purchase of long-lead equipment.

While the above practices are best avoided in many cases, there are several value improving practices that are particularly relevant for pilot plant projects:

- Conducting process simplification reviews. This practice is best conducted during technology selection for the planned commercial process, with the objective of combining or making unnecessary one or more processing steps. It needs to be completed prior to designing the pilot plant.
- Using modular skid-mounted equipment. This practice has several advantages versus building-in-place: modules can be fabricated and tested off-site, and then quickly installed once site preparations are complete; modules can be easily removed and re-used in different pilot plants following the completion of testing. Figure 3 includes a photograph of a modular pilot plant that was operated in DuPont's R&D center in Kingston, ON for the development of Versipol® LAO technology[3].

Figure 3: Modular Pilot Plant for Development of Versipol® LAO Technology[3]

Effective Use of Process Modeling

We have found process modeling to be invaluable for accelerating process development work and reducing the amount of piloting that is required to support scale-up to commercial operation. Figure 4 contains an illustration of how modeling was used to support the development of Versipol® LAO Technology. This approach is consistent with that used to develop most new processes within DuPont.

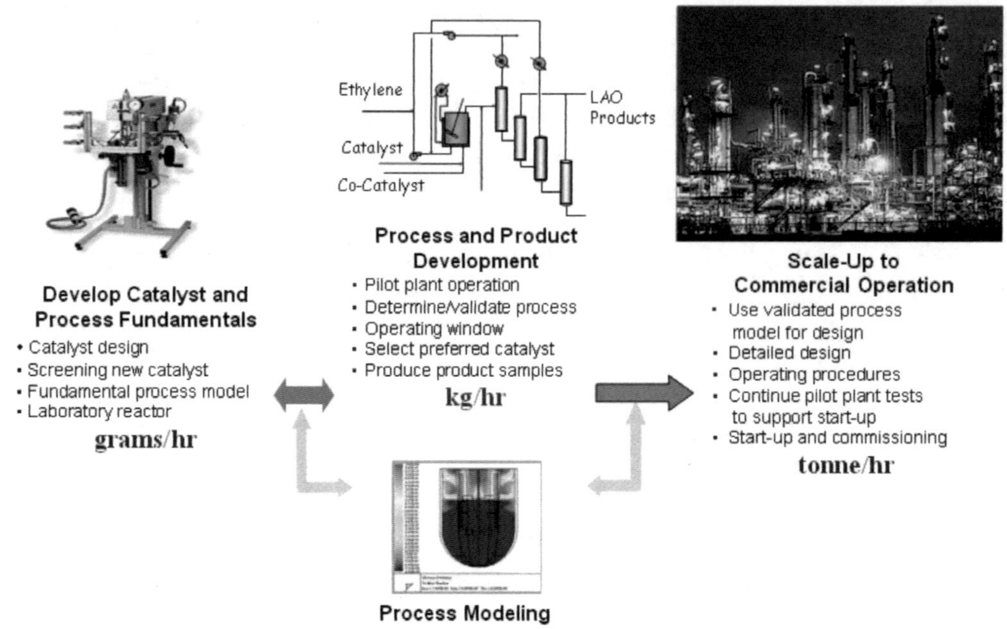

Figure 4: Role of Process Modeling in Support Process Development and Scale-Up of Versipol® LAO Technology[2]

For the Versipol® LAO project we initially developed process simulation models based upon reaction kinetics developed using semi-batch laboratory reactors. These models were then further developed to simulate the planned continuous process, and used to (a) evaluate different commercial process flow sheet options, (b) design the pilot plant shown in Figure 3, and (c) support the design and interpretation of pilot plant experiments. In parallel with this work, we also conducted extensive cold flow and CFD modeling work to study hydrodynamics and mixing within both the pilot plant and commercial reactors. This latter work proved to be invaluable in addressing polymer deposition issues in the pilot plant, and significantly accelerating the development work.

Summary

Process development and piloting are very much experience based competencies, in which we learn from both our past successes and failures, and draw upon these lessons for future projects. Within DuPont we have a rich selection of lessons to draw upon, and have developed a number of best practices for piloting that have been reviewed in this paper. When used appropriately, we have found these practices to be highly effective in reducing both the overall cost and cycle time for piloting of new technology. These practices focus on not just how to design and build pilot plants, but more importantly deciding what process elements needs to be piloted, and then how to support the effective operation of the pilot plant and technology transfer to commercialization.

References

1. Xie, Tuyu (2009), "Development of Bio-Based Polymers: Fundamentals, Process Scale-Up, and Technology Challenges for Cerenol®", *Polymer Reaction Engineering VII,* Niagara Falls, Ontario
2. Norling, Parry M. (1998), "Lessons in New Product Development From DuPont", *Engineering Management Journal*, Vol. 10 No. 3
3. Fisher, John, David Berg, Tuyu Xie, Darwin Kiel, and Trevor Hilderman (2006), "Pilot Plant Success: Versipol® Alpha Olefin Technology Development, AIChE Spring Conference

(707e) Compact Heat Exchange Reactor for Process Efficiency Improvement

Zhijun Jia, Chart Energy & Chemicals, La Crosse, WI

Chart's compact heat exchange reactor (CHER) is an advantaged technology which combines heat exchanger and reactor together in a compact structure. The device is very suitable for highly exothermic or endothermic reactions, or even scenarios required process intensification. One of the significant benefits by using CHER in chemical processes is to improve the process efficiency. Chart's CHER devices have been used in various applications. This paper is to illustrate the efficiency improvement through those examples.

See more of this Session: Pilot Plant Economics
See more of this Group/Topical: Process Development Division

(734a) Green Chemistry with Microreactors and Novel Catalysts

Sunitha Tadepalli and Geatesh Tampy, Research and Development, International Flavors & Fragrances, Union Beach, NJ

Microreactors are significantly more energy efficient than conventional batch reactors. Furthermore, their reduced heat and mass transfer resistances may be exploited to suppress unwanted side reactions, leading to high conversion, low waste, increased safety and also reduced resource needs in subsequent separation/purification steps.

Microreactors may also be effectively coupled with novel catalysts, resulting in the use of flow reactors even for relatively low volume products. With recent advances in catalyst design, conventional toxic, corrosive and hazardous reactants and catalysts may be replaced with friendlier alternatives for many synthetic chemical processes.

Example processes illustrate how microreactors may be coupled with novel catalysts to enhance both the sustainability and economics of existing manufacturing processes for fragrance ingredients or other specialty chemicals.

See more of this Session: Microreactors and Microprocessing
See more of this Group/Topical: Process Development Division

(734b) Hydrodeoxygenation of Pyrolysis Oil in a Microreactor

Narendra Joshi, Chemical Engineering and Material Science, New Jersey Center for Microchemical Systems, Stevens Institute of Technology, Hoboken, NJ and Adeniyi Lawal, Chemical Engineering and Materials Science, New Jersey Center for MicroChemical Systems, Stevens Institute of Technology, Hoboken, NJ

Hydrodeoxygenation studies of pyrolysis oil are conventionally carried out in continuous macro reactor systems, which have mass and heat transfer limitations that lead to uneconomical operating conditions such as long residence times, extremely high pressures, and excessively high H2 flow rates. Due to its high surface-to-volume ratio, microreactor provides enhanced heat and mass transfer. Pyrolysis oil obtained from sawdust was hydro-treated in a microreactor to reduce its oxygen content. Sulfided NiMo catalyst, reduced at high temperature and pressure, was used for the hydrodeoxygenation of the pyrolysis oil. In this study, changes in the hydrogen consumption and the product composition as a function of temperature, hydrogen partial pressure, residence time, and mass transfer rate were determined quantitatively using analytical tools such as gas chromatography, mass spectroscopy, Karl-Fischer titration, and elemental analyses.

See more of this Session: Microreactors and Microprocessing
See more of this Group/Topical: Process Development Division

(734c) Polymer Membrane Separations for Microreactors

Thomas Schafer, Institute for Polymer Materials (Polymat), University of the Basque Country, San Sebastián, Spain

Membrane separation processes constitute an established variant in downstream processing of the chemical industry, particularly as gas/vapour separations, pervaporation, nanofiltration and membrane contactors. Operating conditions and separation task permitting, in large-scale membrane processes polymeric membranes are preferred over inorganic ones as the latter tend to be less mechanical stress-resistant, resulting in higher maintenance costs. In micro reactor technology, the opposite trend can hitherto be observed. Of the few works dealing with membrane separations in microreactors, most involve inorganic membranes. This may be motivated predominantly by the type of separation, e.g. hydrogen removal, but also by the fact that dimensions are dramatically reduced in comparison to large scale applications. As a consequence, mechanical stability is less of an issue. Polymer membrane separations can nevertheless be an interesting option for micro reactor separations. With an ample range of chemical and physico-chemical properties available for tailoring these membranes toward a particular application, and with the possibility of a straightforward production of large batches which warrants a reproducible selectivity and permeation flux, it will be discussed in this presentation how polymeric membranes may enhance the separation efficiency in microreactors, for example through removal of gases/vapours or reaction by-products. It will also be shown that the use of polymeric membranes in microreactors is not a mere scaling down and transfer of know-how from large-scale processes. For instance, membrane processes offering a high surface-to-volume ratio for mass transfer, such as membrane contactors, might find limited applications in microreactors which boast precisely the same asset. Existing membrane processes therefore require a careful reassessment before introducing them into a microreactor system.

ACKNOWLEDGEMENTS The author would like to thank the European Research Council for funding this work through an IDEAS starting grant (MATRIX-209842).

See more of this Session: Microreactors and Microprocessing
See more of this Group/Topical: Process Development Division

(734d) Enzyme Catalyzed Polymerization Reactions in Microreactors

Santanu Kundu[1], Atul S. Bhangale[2], William E. Wallace III[1], Richard A. Gross[2] and Kathryn L. Beers[1], (1)Polymers Division, National Institute of Standards and Technology (NIST), Gaithersburg, MD, (2)Polytechnic Institute of NYU, Brooklyn, NY

Application of microreactor technologies enable improved safety, selectivity and yield in a wide range of chemical reactions in addition to new measurement methods that are often faster, cheaper and more accurate than traditional methods. In this study, we have considered a model system, Candida Antartica Lipase B (CAL B) catalyzed ring opening polymerization of ε-caprolactone to polycaprolactone. Commercially available macroporous polymethyl methacrylate (PMMA) beads immobilized with CAL B were loaded in the inexpensive metal (aluminum) microchannel reactors. Metal microreactors enabled us to perform polymerization reactions in organic media and at high temperatures. Reaction products were characterized as a function of residence time using Raman spectroscopy and gel permeation chromatography (GPC). By implementing a suitable reactor design we achieve (1) faster polymerization rate and (2) higher molecular weight of polycaprolactone compared to that obtained in batch reactors at similar experimental conditions. Experiments were also performed in continuous mode for an extended time to understand the change of enzyme activity as a function of time.

See more of this Session: Microreactors and Microprocessing
See more of this Group/Topical: Process Development Division

(734e) Membrane Microbioreactor for Enzyme-Catalyzed Degradation of Pectin

Muhd. Nazrul Hisham Zainal Alam, Manuel Pinelo, Anne S. Meyer, Gunnar E. Jonsson and **Krist V. Gernaey**, Department of Chemical and Biochemical Engineering, Technical University of Denmark (DTU), Kgs. Lyngby, Denmark

A membrane bioreactor offers several advantages for process development of enzyme-catalyzed degradation of pectin [1]. With a membrane bioreactor, enzymes can be retained within the reactor thus reducing the operating cost of the system (reuse of enzymes) [1]. Furthermore, integration of the membrane separation with the reactor allows for continuous separation of small low-molecular inhibitory products from the reaction mixture – a functionality that cannot be achieved with traditional batch reactors. This minimizes the effect of end product inhibition and consequently increases the reaction yield as well as reaction rates [1]. Nevertheless, when evaluating new enzymes for enzyme-catalyzed production processes, the costs related to obtaining a sufficient amount of enzyme forms a major constraint. This limitation becomes more pronounced in cases where a large amount of experimental data is required (e.g. screening of new enzyme processes). Microbioreactor technology offers the prospect of circumventing these limitations and the possibility to develop a miniaturized version of the membrane bioreactor. The fundamental idea underlying the concept of a microbioreactor is to fabricate a reactor with very small working volumes (typically less than 1 mL) that emulates all the basic features of a typical bench-scale membrane bioreactor whilst retaining the low cost and easy handling features of microtiter plate operations [2].

This work reports on the development of a continuous membrane microbioreactor for combined pectin degradation and separation. The work focuses on two aspects. First of all, the main technical challenges related to establishing the membrane microbioreactor are described, together with the solutions that were implemented. Secondly, experimental results obtained with the membrane microbioreactor will be presented, demonstrating the workability of the reactor.

REFERENCES [1] Pinelo, M., Jonsson,G., Meyer, A. S., Membrane technology for purification of enzymatically produced oligosaccharides: Molecular and operational features affecting performance, Separation and Purification Technology. (2009) 70. 1-11. [2] Schäpper, D., Zainal Alam, M. N. H., Szita, N., Lantz, A. E., Gernaey, K. V. Application of microbioreactors in fermentation process development: a review. Analytical and Bioanalytical Chemistry. (2009). 395. 679-695.

See more of this Session: Microreactors and Microprocessing
See more of this Group/Topical: Process Development Division

(734f) Separation Processes in Micro Reaction Technology

T. Dietrich[1], Andreas Freitag[1], Ralf Scholz[1], Stefan Link[1] and Jan-Wilhelm Thies[2], (1)mikroglas chemtech GmbH, Mainz, Germany, (2)Institute for Separation and Process Technology, Clausthal University of Technology, Clausthal-Zellerfeld, Germany

INTRODUCTION

Micro reaction technology has been investigated for nearly 20 years. Several micro mixers, micro heat exchangers, micro reactors for different purposes and applications have been developed. For a variety of reactions it could be shown that there are certain advantages of this technology, namely better control of the reaction, leading to a more efficient, economical and ecological way of producing chemical products. Looking into more detail, it turns out that most of the published examples are single step reactions. In some cases multiple reaction steps could combined, but only when the first steps showed total conversion and by-products could be avoided. Even though micro reaction technology allows a much better conversion and yield, still by-products and unused starting compounds are present in the product stream. Therefore, it is necessary to provide a continuous separation unit which allows to collect the unwanted parts of the mixture before the product stream enters the next process step. Although first developments have been published already [1] most of the necessary downstream processes are not available, yet. This presentation will show new separation modules and their first test results for the separation of gas-liquid and liquid-liquid mixtures.

SEPARATION MODULES

All developed and tested modules of this work are made out of FOTURAN glass, a special photosensitive material which allows to manufacture micro channels in glass with high precision [2]. The material was chosen because of its chemical stability and its optical transparency. With these modules it was possible to observe their behavior during the separation process.

It will be shown in the presentation that the separation quality depends on different physical parameters. Gas-liquid and liquid-liquid separations in a triangular shaped separation chamber are strongly flow rate dependent. At low flows the mixture has time to settle: the heavy phase can be extracted from the bottom, the light phase from the top. Above a certain velocity of the solution (in the range of a few ml/min, depending on the liquids) the residence time is not sufficient for a good separation. The efficiency can be increased by a surface modification according to the properties of the used liquid: e.g. hydrophilic for water or hydrophobic for organic solvents. This kind of surface modification also helps to separate liquids in a "Y"-separator. In this case, the reaction channel leads into a Y-shaped structure with two exits. The surface of one branch of the "Y" has been modified by a plasma process to make it hydrophobic, the other one is hydrophilic. In the reaction channel before the "Y" separator, the mixture forms plugs, which then leave the separator through the hydrophilic or the hydrophobic branch according to their respective properties. Another very effective separation module for gas-liquid mixtures is the "capillary separator". At the end of a micro channel reactor, a system of capillaries take out the liquid phase by capillary forces. Depending on the geometry and the number of capillaries as well as the surface properties and the flow of the components, it is possible to get a stream of gas-free liquid out of the separator. All different separation modules showed good behavior in multi-step reactions. The separation efficiency was in an acceptable range and could be increased by using more than one module in a row. The separation module could also be used for cleaning and re-use of process solvents

ACKNOWLEDGEMENTS

The authors would like to thank the German Federal Ministry of Education and Research and the Deutsche Bundesstiftung Umwelt for funding parts of this work within the projects MIRFA, HotChip, and TRANSKAT.

REFERENCES

[1] T.R.Dietrich (Ed.), Microchemical Engineering in Practice: A. Gavriilidis, J.E.A.Shaw, Separation Units, John Wiley & Sons, Hoboken NJ, 2009, pp. 131-163 [2] T.R.Dietrich, W.Ehrfeld, M.Lacher, M. Krämer, .Speit, Fabrication technologies for Microsystems utilizing photoetchable glass, Microelectr. Eng. 30 (1996) 497-504.

See more of this Session: Microreactors and Microprocessing
See more of this Group/Topical: Process Development Division

(734g) Structurally Optimized Microbioreactors: A Design Example for Immobilized Yeast Cultivations

Rita L. Fernandes[1], Daniel Schäpper[1], Fridolin Okkels[2], Anna Eliasson Lantz[3], Henrik Bruus[2] and **Krist V. Gernaey**[1], (1)Department of Chemical and Biochemical Engineering, Technical University of Denmark (DTU), Kgs. Lyngby, Denmark, (2)Department of Micro- and Nanotechnology, Technical University of Denmark (DTU), Kgs. Lyngby, Denmark, (3)Department of Systems Biology, Technical University of Denmark (DTU), Kgs. Lyngby, AL, Denmark

During the last decade there has been a clear interest in using microfluidic systems for cultivation of microorganisms. This works exploits the increased design flexibility offered by microsystems for optimizing the operation of a microbioreactor. In this optimization project, the production of a recombinant protein in a continuous culture of immobilized Saccharomyces cerevisiae was used as a case study. A topology optimization methodology [1] was applied in order to obtain the spatial distribution of immobilized yeast that maximizes protein concentration at the outlet. Although yeast is typically cultivated as a suspension, immobilized S. cerevisiae has been successfully used in continuous and stable cultivations [2]. In the chosen case study, a population of yeast grows at steady state on a carrier and is partially released into the flow phase forming a suspended population. A plasmid-encoded recombinant protein is produced when yeast consumes glucose. However, at high glucose concentrations, respiration becomes limiting and glucose is preferably fermented to ethanol. Furthermore, when glucose is depleted, ethanol is consumed and its oxidation leads to further production of the protein. These events were translated into a mathematical model based on two models available in the literature describing the growth of a brewing yeast biofilm [2] and an aerobic culture of a recombinant yeast [3], respectively and the model was incorporated into the topology optimization routine. The maximum local protein production rate was set as the optimization target, which results in a maximization of the protein concentration at the reactor outlet. In order to benchmark the optimized solutions obtained for various feed concentrations, protein concentrations at the optimized reactor outlet were compared to the corresponding concentrations obtained for reactors where the immobilized biomass was homogeneously distributed. For a glucose feed concentration of 0.1 g/L, an increase of nearly 10 fold was observed for the optimized yeast distribution.

References [1] Okkels, F. & Bruus, H. Phys. Rev. E 75, 016301 1-4 (2007) [2] Brányik, T. et al. Biotechnol. Prog. (2004) 20 1733-1740 [3] Zhang, Z., Scharer, J.M., Moo-Young, M. Bioprocess Eng. (1997) 17 235-240

See more of this Session: Microreactors and Microprocessing
See more of this Group/Topical: Process Development Division

AUTHOR INDEX

AUTHOR INDEX

AUTHOR INDEX

AUTHOR INDEX

American Institute of Chemical Engineers
3 Park Avenue
New York, New York 10016-5991

CURRAN ASSOCIATES INC
proceedings
.com